T032446

Till: A Glacial Process Sedimentology

Wiley-Blackwell Cryosphere Science Series

Permafrost, sea ice, snow, and ice masses ranging from continental ice sheets to mountain glaciers, are key components of the global environment, intersecting both physical and human systems. The study of the cryosphere is central to issues such as global climate change, regional water resources, and sea level change, and is at the forefront of research across a wide spectrum of disciplines, including glaciology, climatology, geology, environmental science, geography and planning.

The Wiley-Blackwell Cryosphere Science Series comprises volumes that are at the cutting edge of new research or provide a focused interdisciplinary review of key aspects of the science.

Series Editor

Peter G. Knight, senior lecturer in geography, Keele University

Related Titles

Recent Climate Change Impacts on Mountain Glaciers

Mauri Pelto, November 2016

978-1-119-06811-2

Remote Sensing of the Cryosphere

Marco Tedesco, December 2014

978-1-118-36885-5

Till: A Glacial Process Sedimentology

David J A Evans
Department of Geography
Durham University
United Kingdom

 The Cryosphere Science Series

Series Editor: Dr Peter Knight, University of Keele

WILEY Blackwell

Registered Offices
John Wiley & Sons Ltd, The Atrium, Southern Gate, Chichester, West Sussex, PO19 8SQ, UK

Editorial Office
111 River Street, Hoboken, NJ 07030, USA
9600 Garsington Road, Oxford, OX4 2DQ, UK
The Atrium, Southern Gate, Chichester, West Sussex, PO19 8SQ, UK

For details of our global editorial offices, customer services, and more information about Wiley products visit us at www.wiley.com.

Wiley also publishes its books in a variety of electronic formats and by print-on-demand. Some content that appears in standard print versions of this book may not be available in other formats.

Library of Congress Cataloging-in-Publication Data applied for

ISBN - 9781118652596

Cover Design: Wiley
Cover Image: Courtesy of David J A Evans

Set in 10/12pt, WarnockPro by SPi Global, Chennai, India

10 9 8 7 6 5 4 3 2 1

Contents

Acknowledgements

In any area of scientific research, you are always aware that you are standing on the shoulders of giants, but it is a daunting task to undertake the challenge of writing a research monograph when many of the most influential of those giants are still around watching over your deliberations! I have benefitted hugely from many people in my glacial sedimentology endeavours, a large number of whom have contributed to this book in a variety of ways. Whether it has been through providing images or diagrams, reading bits of text, passing on years of experience and wisdom or just exchanging ideas over a pint, I am indebted to you all. I hope you get something out of this book, to make up for all the material you have indirectly or even inadvertently put in; you may not agree with it all, and I certainly do not have all the answers, but I am sure we can have more conversations about that in the future. You are too numerous to mention individually, but I would like to acknowledge the inputs of some stalwarts specifically.

My career-long Durham colleagues Dave Roberts and Colm Ó Cofaigh have spent longer at till sections with me than anyone else, and over the years we have excavated, chiselled, scraped, debated vigorously (never argued, of course), logged, fabricked and clast-formed our way through hundreds of square metres of dirt – where most see nothing but mud, we see a compelling story waiting to be unravelled! Also worthy of mention for time served with me at the outcrop are John Hiemstra, Brice Rea and Emrys Phillips, who have kept me straight not only on my physics but also on microscopic and geotechnical details. My old textbook-writing buddy, Doug Benn, has long since seen sense and moved away from till (he already had the answers anyway), but I remain ever grateful for those early years we spent together in Scottish quarries and on loch shores learning all about clast fabric shape triangles, A and B horizons, glacitectonites (sorry, 'Drymenite') and fine malt whiskeys – I can appreciate that glacier caving is a little more racy than till fabrics, but who's making the beans on toast? Finally, thanks to Tessa, Tara and Lotte for putting up with the inevitable 'just have a look at this cliff down here' while we were supposed to be on holiday!

1

Glacigenic Diamictons – A Rationale for Study

The Glacial Drifts … are known to us all but too well. We cannot escape them; their clays, their sands and gravels confront us at every turn, so masking the underlying rock that they are a positive curse to the 'solid' geologist.

Carruthers (1947–48, p. 43)

The process sedimentology of tills is crucial to the understanding of the glacier ice–bed interface as a complex depositional, erosional and shear boundary layer. Consequently, it also plays a central role in deciphering the genesis of enigmatic subglacial bedforms such as drumlins, flutings and ribbed terrain. Yet, unlike the study of other boundary layers such as those that operate at the bed of fluvial, aeolian and deep water systems, our knowledge of subglacial process–form relationships is relatively impoverished, largely due to the inaccessibility of glacier and ice sheet beds. Notwithstanding the important contributions now being made to this research problem by remotely sensed and localised borehole observations as well as reductionist laboratory experiments, it is critical that glacial scientists continue to refine their interpretations of ancient archives of subglacial processes, specifically those that are represented by tills and associated deposits, as these archives form the most widespread and accessible record of processes at the ice–bed interface (Figure 1.1).

Such an inductive approach to the reconstruction of former subglacial processes has some considerable shortcomings, largely because it relies on actualist principles that are in turn based on process–form relationships that we cannot as yet unequivocally validate. This often has been compounded by the glacial geomorphology literature, wherein the traditional, uncritical acceptance of thick sequences of diamicton as 'lodgement tills' has assumed a definitive knowledge of process–form relationships even though that knowledge base is far from definitive. This has been exposed more recently in the apparent incompatibility between modern process measurements (indicating thin subglacial deformation/till construction) and ancient glacigenic sequences interpreted to contain often very thick subglacial tills. Moreover, the existence of ambiguous diagnostic criteria for identifying processes of subglacial sedimentation in ancient diamictons does not inspire confidence in the glacial research community when turning to till sedimentology for some guidance!

What we can now be confident in espousing are the concepts of debris entrainment and transport pathways together with concomitant clast modification within the glacial debris cascade (Figure 1.2). Till sedimentology should reflect the nature of the debris cascade, or more specifically: (1) the entrainment and transport history; (2) the continuum of clast modification during various phases or repeat cycles of transport and deposition; (3) the debris release processes and (4) any secondary displacement processes such as deformation. Glacial systems are complex in that these three aspects of the

Till: A Glacial Process Sedimentology, First Edition. David J A Evans.
© 2018 John Wiley & Sons Ltd. Published 2018 by John Wiley & Sons Ltd.

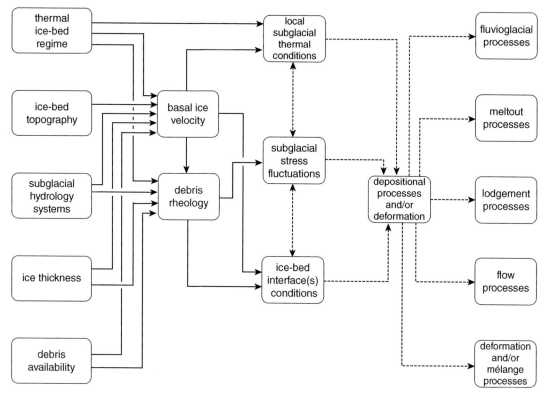

Figure 1.1 Flow diagram to illustrate the inter-relationships between the main glaciological and sedimentological processes associated with the subglacial environment (modified from Menzies and Shilts (1996) to acknowledge glacitectonic processes as deformational rather than depositional).

debris cascade are juxtaposed in a contemporaneous process–form regime. The result is a temporal and spatial mosaic of process operation that can be shut down at any stage of development once a landscape undergoes deglaciation. This temporal and spatial mosaic is a recurring theme throughout this book, because it allows us to visualise the till forming environment more appropriately as a process–form hierarchy more akin to the principles of sequence stratigraphy as they are applied to other geomorphological systems.

Given the ambiguity of the diagnostic criteria that have been proposed as interpretive aids in the study of tills, it is not surprising that controversy abounds and is manifest in some surprisingly contradictory alternative explanations of glacigenic sediment genesis. This is nowhere better illustrated than with glacigenic diamictons and the sediments that are stratigraphically and genetically associated with them (Figure 1.3), all of which lie at the heart of some significant debates in reconstructions of glacial depositional environments pertaining to the whole range of the geological timescale. The crux of these debates commonly can be distilled into disagreements over the subaqueous and/or mass flowage versus subglacial (till) origins of diamictons (e.g. Visser *et al.*, 1984; Eyles, 1987; C.H. Eyles *et al.*, 1985; Eyles *et al.*, 1987, 1988a, b, 1990; Shaw, 1988). Some high-profile examples of such dichotomies include the deposits of the Late Proterozoic Snowball Earth (e.g. Spencer, 1971; Schermerhorn, 1974; Eyles and Eyles, 1983a; Hoffman *et al.*, 1998; Hoffman and Schrag, 2000, 2002; Benn and Prave, 2006; van Loon, 2008; Carto and Eyles, 2012a, b) and the Pleistocene glaciated basins of the Great Lakes (e.g. Eyles and Eyles, 1983b, 1984a; Dreimanis, 1984), North Sea (e.g. Eyles *et al.*, 1989,

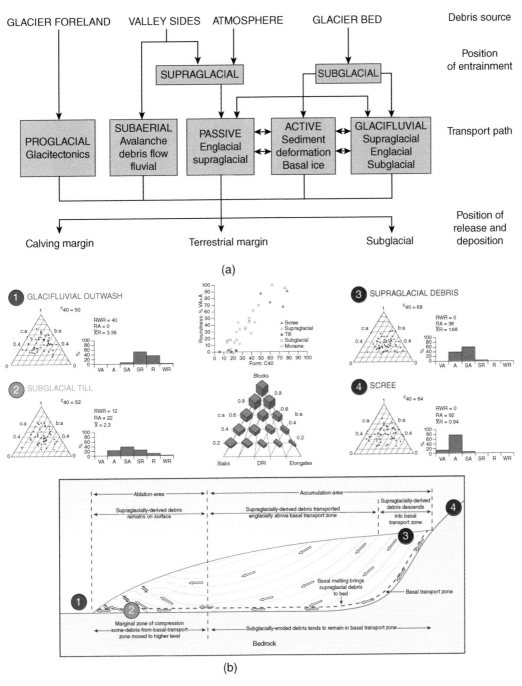

Figure 1.2 The glacial debris cascade and transport pathways: (a) the glacial debris cascade (from Benn and Evans, 2010); (b) simplified diagram to show the main debris transport pathways through a simple valley glacier, indicating that some debris may bypass the subglacial traction zone and follow a passive transport route (after Boulton, 1978). The impacts of various transport routes on clast form signatures are illustrated using: (1) glacifluvial outwash, (2) subglacial till, (3) supraglacial debris and (4) scree. Clast form data is depicted in the commonly used graphics of ternary diagrams (depicting clast shape based upon principle A – long, B – intermediate and C – short axes), histograms (depicting roundness, VA-WR or 0–5) and co-variance graph (plotting RA roundness or VA+A% against C40 form or % clasts below 0.4 c:a axial ratio). Other statistics are RWR = R+WR% and $\dot{X}R$ = average roundness.

Figure 1.3 Examples of the range of deposits generally referred to as *glacigenic diamictons* (including tills) and sediments that are stratigraphically and genetically related to them: (a) stratified diamicton, Filey Bay, eastern England; (b) pseudo-laminated diamicton with gravel clot/intraclast, Red Deer Lake, Alberta, Canada; (c) stratified diamicton and horizontally bedded interbeds, Drayton Valley, Alberta, Canada; (d) fissile, clast-rich diamicton, Glen Varragill, Isle of Skye, Scotland; (e) discontinuous boulder pavement beneath massive, matrix-rich diamicton, Whitburn, northeast England; (f) heterogeneous, tectonically laminated and shale-rich diamicton, near Kinsella, Alberta, Canada; (g) mélange of stratified sands and gravels, pseudo stratified diamictons and sand and gravel intraclasts (rafts), West Runton, East Anglia, England; and (h) heavily deformed and attenuated stratified diamicton, Sheringham, East Anglia, England.

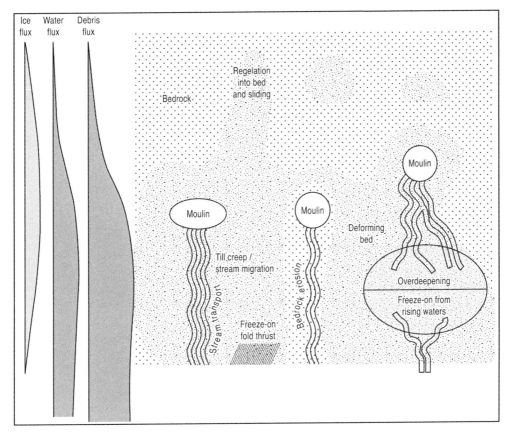

Figure 1.4 Summary schematic diagram showing the range of subglacial sediment transport mechanisms that can operate near the sub-marginal zone of an ice sheet or glacier (from Alley *et al.*, 1997). The debris flux and its typical relationships with ice and meltwater flux are also depicted.

1994; Hart and Roberts, 1994; Lunkka, 1994) and Irish Sea Basin (e.g. Eyles and Eyles, 1984b; Eyles and McCabe, 1989a, b, 1991; Wingfield, 1992; McCarroll, 2001; Ó Cofaigh and Evans, 2001a, b; Scourse and Furze, 2001; Evans and Ó Cofaigh, 2003). Additionally, parallel debates have ensued over the precise origins of subglacial tills, focussed on the relative roles of lodgement, melt-out, deformation and meltwater (c.f. Paul and Eyles, 1990; Piotrowski and Tulaczyk, 1999; Munro-Stasiuk, 2000; Boulton *et al.*, 2001; Piotrowski *et al.*, 2001, 2002, 2004; Ruszczynska-Sjenach, 2001; Evans *et al.*, 2006a).

 In summary, despite a long history of investigation and a plethora of process-based nomenclature, it is clear that glacial sedimentologists have yet to reach a consensus on the diagnostic criteria for identifying till genesis in the geological record. More than 30 years after Dreimanis and Lundqvist (1984) posed the question '*What should be called Till?*', this book attempts to ask and answer the same query. In the interim, advances in physical glaciology (see Alley *et al.*, 1997) have clarified the nature of ice–bed interactions as they pertain to the entrainment and transport history in the glacial debris cascade (Figure 1.4), highlighting the spatial patterns of regelation, meltwater drainage, bed deformation, marginal freeze-on and supercooling, and englacial folding and thrusting. Other aspects of the debris cascade, particularly debris release processes and deformation signatures, have similarly received concerted attention from glacial geomorphologists who have elucidated on the depositional and structural impacts of such aspects, but the accumulated knowledge has yet to fully permeate the realm of glacial sedimentology or at least to be systematically assimilated into clear diagnostic criteria for the reading of the glacial depositional record. This book addresses these issues through critical reviews of the till literature, laboratory- and experiment-based assessments of subglacial processes, and the theoretical constructs that have emerged from process sedimentology. These deliberations are then employed in the erection of a contemporary till nomenclature in which process–form relationships are founded on a coherent synthesis of a wide range of knowledge bases.

2

A Brief History of Till Research and Developing Nomenclature

With relief one remembers that, after all, the facts gathered with such infinite care, over so many years, are in no ways affected: their permanency is untouched, their value as high as ever. It is the interpretation which has gone astray.

Carruthers (1953, p. 36)

A benchmark publication in the development of till nomenclature was contained in the final report by the INQUA Commission on Genesis and Lithology of Glacial Quaternary Deposits, entitled '*Genetic Classification of Glacigenic Deposits*' (Goldthwait and Matsch, 1989; Figure 2.1). Most significant in this report was the paper by Aleksis Dreimanis (Figure 2.2), entitled '*Tills: Their Genetic Terminology and Classification*', a summary of the findings of the Till Work Group, which operated over the period 1974–1986. It was a synthesis of knowledge and a rationale for a unified process-based nomenclature but at the same time afforded the presentation of alternative standpoints on till classification, and hence delivered a selection of frameworks containing complex and overlapping genetic terms. More broadly, 'till' at this juncture was defined as:

> a sediment that has been transported and is subsequently deposited by or from glacier ice, with little or no sorting by water.
>
> (Dreimanis and Lundqvist, 1984, p. 9)

As a way forward, the Till Work Group, through Dreimanis (1989), arrived at a series of nomenclature diagrams (Figure 2.3), which aimed at an inclusive but at the same time simplified and unambiguous, process-based till classification scheme. More specifically, Dreimanis (1989), within the same volume, compiled a table of diagnostic characteristics for differentiating what he termed 'lodgement till', 'melt-out till' and 'gravity flowtill'. Although this book later advocates a fundamentally different set of sedimentological terms for the deposits being described by Dreimanis (1989), the contents of his summary table are nonetheless still highly relevant to the differentiation of subglacial versus mass flow origins for diamictons on the one hand and subglacial traction versus melt-out processes on the other, and hence are reproduced here in Table 2.1.

Prior to the production of the Goldthwait and Matsch (1989) volume, till nomenclature had developed out of a small number of local case studies, not all of which were based on modern process, as was reviewed by Dreimanis (1989). We shall return to the issue of process-based till nomenclature schemes throughout this book, but first it is important to provide historical context for the deliberations of the Till Work Group and beyond.

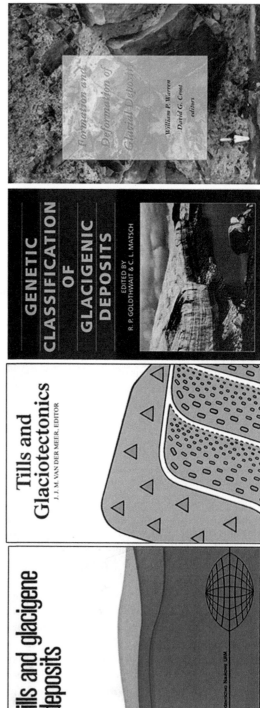

Figure 2.1 Symposium volumes compiled on the subject of tills during the 1970s to the early 1990s.

Figure 2.2 Researchers involved in the historical development of till sedimentology up to, and in some cases beyond, the late 1970s.

Table 2.1 The main criteria identified by Dreimanis (1989) as critical to the differentiation and classification of 'lodgement till', 'melt-out till' and 'gravity flow till'.

Criterion	Lodgement till	Melt-out till	Gravity flow till
Position and sequence in relation to other glacigenic sediments	Advancing glaciers – lodged over pre-advance sediments and glacitectonites, unless eroded. Retreating glaciers – lowermost unit if pre-advance deposits and advance till has been eroded. Locally underlain by meltwater channel deposits. May be overlain by any glacigenic sediments.	Usually deposited during glacial retreat over any glacially eroded substratum or over lodgement till. May be interbedded with lenses of englacial meltwater deposits. Locally underlain by syndepositional subglacial meltwater sediment and subglacial flow till.	Most commonly the uppermost glacial sediment in a non-aquatic facies association. Also associated locally with subglacial tills, where cavities were present under glacier ice or where the glacier had overridden ice-marginal flow till. May be interbedded or interdigitated with glacifluvial, glacilacustrine or glacimarine sediments.
Basal contact	Since both lodgement and melt-out tills begin their formation and deposition at the glacier sole, their basal contact with the substratum (bedrock or unconsolidated sediments) is similar in the large scale, being usually erosional and sharp. Glacial erosion marks underneath the basal contact and clast alignments immediately above the contact have the same orientation. Glacitectonic deformation structures formed by the till-depositing glacier may occur under both till types and strike transverse to the direction of local glacier stress. Basal contact represents sliding glacier base and is generally planar if substratum is over-consolidated but may be grooved. Bedrock contact is usually abraded, particularly on stoss sides of protrusions. Sliding glacier base is a shear plane, so sheared and strongly attenuated substrate material may be deposited as a thin layer along this plane, and in places is sheared up into lodgement till. Clast pavements may be present along the basal contact but may also occur higher up in lodgement till. If lodgement till becomes deformed by glacial drag shortly after its deposition, the basal contact may become involved in the deformation with tight recumbent folding, overthrusting and shearing.	If the basal contact of glacier ice was tight with the substratum during melting, the pre-depositional erosional marks are well preserved under lodgement till. However, subsole meltwater may modify the basal contact locally, and produce convex-up channels and various other meltwater scour features.	Variable; seldom planar over longer distances. Flows may fill shallow channels or depressions. Contact may be either concordant or erosional, with sole marks parallel to the local direction of sediment flow. Loading structures may be present at the basal contact of waterlain flow till and the underlying soft sediment.

Thickness	Typically one to a few metres. Relatively constant laterally over long distances.	Single units are usually a few centimetres to a few metres thick, but they may be stacked to much greater thicknesses.	Very variable. Individual flows are usually a few decimetres to metres thick, but they may locally stack up to many metres, particularly in proglacial ice-marginal moraines and some lateral moraines.
Structure, folding and faulting	Typically described as massive but on closer examination a variety of consistently orientated macro- and microstructures indicative of shear or thrusting may be found. Folds are overturned, with anticlines attenuated down-glacier. Deformation structures are particularly noticeable if underlying sediments are involved or incorporated in the till, developing smudges. Sub-horizontal jointing or fissility is common. Vertical joint systems, bisected by the stress direction, and transverse joints steeply dipping down glacier, may be formed by glacier deforming its own lodgement till. The orientation of all the deformation structures is related to the stress applied by the moving glacier, and therefore it is laterally consistent for some distance.	Either massive or with palimpsest structures partially preserved from debris stratification in basal debris-rich ice. Lenses, clasts and pods of texturally different material preserve best, for instance soft-sediment inclusions of various sizes and englacial channel fills. Loss of volume with melting leads to the draping of sorted sediments over large clasts. Most large rafts or floes of substratum are associated with melt-out tills and they may be deformed by glacial transport and by differential settlement during melting.	Structures depend upon the type of flow and other associated mass movements, the water content and the position in the flow. Either massive or displaying a variety of flow structures such as: (a) overturned folds with flat-lying isoclinal anticlines; (b) slump folds or flow lobes with their base usually sloping down flow; (c) roll-up structures; (d) stretched out silt and sand clasts; (e) intraformationally sheared lenses of sediments incorporated from substratum, with their upper downflow end attenuated if consisting of fine-grained material, or banana shaped.
Grain size composition	Usually a diamicton, containing clasts of various sizes. Grain size composition depends greatly on the lithology and composition of the substrata up-glacier and the distance and mode of transport from there. Comminution during glacial transport and lodgement has produced a multimodal particle size distribution. Most resulting subglacial tills are poorly to very poorly sorted, described also as well graded, and their skewness has a nearly symmetrical distribution, except for those tills that are rich in incorporated pre-sorted materials.		Usually a diamicton with polymodal particle size distribution. Texturally similar to the primary till to which it is related, but with a greater variability in grain size composition, due to washing out of, or enrichment in fines, or incorporation of soft-substratum sediment during the flow. Some particle size redistribution takes place during the flow. Grain size composition depends greatly upon the type of flow and the position or zone within it. Sorting, inverse or normal grading may develop in some zones of flows, and parts of clasts may sink to the base of the flow.

(continued)

Table 2.1 (Continued)

Criterion	Lodgement till	Melt-out till	Gravity flow till
	Abrasion in zone of traction during lodgement produces silt size particles. Most lodgement tills have relatively consistent grain size composition, traceable laterally for kilometres, except for the lower 0.5–1 m that strongly reflects the local material. Clusters or pavements of clasts are common.	Winnowing of silt and clay size particles in the voids during melt-out may reduce their abundance in comparison with their lodged equivalents. Some particle size variability is inherited from texturally different debris bands in ice. Extreme variations in grain size may occur over short distances in the vicinity of large rafts and other inclusions of soft sediment.	
Lithology of clasts and matrix	Lithological composition tends to be less variable than in other genetic varieties of tills; most constant is the mineralogical and geochemical composition of the till matrix. Materials of local derivation increase in abundance towards the basal contact with the substratum.	Since glacial debris of distant derivation is more common in the englacial zone than in the basal zone of a glacier and since the englacial zone is more likely to be deposited as melt-out till rather than by lodgement, materials of distant derivation may be more abundant in the melt-out than in the lodgement component of the same till unit, particularly in supraglacial melt-out till. Great compositional variability occurs in the vicinity of incorporated megaclasts, rafts or floes of sub-till material. Soft-sediment clasts, for instance consisting of sand, may be found in melt-out till, but not in typical undeformed lodgement till.	Lithological composition is generally the same as that of the source material of the flowed till – a primary till or glacial debris, plus some substratum material incorporated during the flowage. Material of distant derivation dominates in flowed tills derived from supraglacial and englacial debris, but dominance of local materials indicates derivation from basal debris. Soft-sediment clasts derived from the substratum or from sediment interbeds in multiple flows are common.
Clast shapes and their surface marks	The following criteria apply where most clasts are derived from single cycle transport: sub-angular to sub-rounded shapes dominate, depending mainly upon the distance of transport in the basal zone of traction. Bullet-shaped (flat-iron, elongate pentagonal) clasts are more common than in other tills and non-glacial deposits and their tapered ends usually point up glacier. Some elongate clasts have a keel at their base. Glacial striae are visible mainly on medium to hard, fine-grained rock surfaces. Elongate clasts are striated mainly parallel to their long axes, unless they have been lodged or transported by rolling.		If present, soft-sediment clasts are either rounded or deformed by shear or dewatering. The more resistant rock clasts are in the same shape as they were in the source material when re-sedimented by the flowage. Therefore, the relative abundance of glacially abraded, sub-angular to sub-rounded clasts versus completely angular clasts in flowed tills in mountain glaciers will indicate the approximate participation of basal debris versus supraglacial debris in the formation of the flowed till. Some rounded, water-reworked clasts without striations may derive from meltwater stream deposits.

	The bullet-shaped and facetted clasts, also crushed and sheared clasts, are more common in lodgement tills than in other tills. Lodged clasts are striated parallel to the direction of the lodging glacial movement, and they have impact marks on both the upper and lower surfaces, but in opposite orientation; on the surface, the stoss end is up glacier, but on the underside the stoss end is down glacier. Clast pavements with sets of striae parallel to the direction of the latest glacier movement over them may occur at several lodgement levels. Their top facets are either parallel with the general plane of lodgement or they dip up glacier.	If, in an area of mountain glaciation, the source of supraglacial melt-out till is englacially or even supraglacially transported and supraglacially derived debris, then the clasts are angular. Most commonly, supraglacial melt-out till in such areas also contains an admixture of glacially abraded basal debris, also englacially transported.	
Macro- and microfabrics	Strong macrofabrics with the log axis parallel to the local direction of glacier movement in diamictons reported as either lodgement or melt-out tills. Occasionally transverse maxims have developed, associated with folding and shearing. Fabric strength may also vary, depending upon till grain size, the abundance of clasts and post-depositional modification. Lodgement till fabric may be of complex origin: produced by lodgement or by deformation of the already deposited dilated till under the same glacier. If both stress directions coincide, a strong fabric will develop; if not, the lodgement fabric becomes weakened. Typically, the a-b planes dip slightly up glacier if lodgement alone is involved. Microfabric is usually as strong as the macrofabric.	Melt-out till fabric is inherited from glacier transport. Where fabric dominates it is parallel to the direction of glacier movement, unless deformation changes it to transverse fabric locally. Melt-out process can, however, weaken the fabric, particularly the microfabric. Also, the dip inclination of clasts becomes reduced by the reduction of the volume of ice during melting.	Variable, and dependent greatly upon the type of flow and the position in the flow. May range from randomly orientated to strong fabric in thin flows. Fabric maxima are either parallel or transverse to the local flow direction, unrelated to glacier movement; the a – b planes are either sub-parallel to the base of the flow or they dip up-flow. Fabric maxima may also differ laterally over short distances.
Consolidation, permeability and density	Most lodgement tills, particularly the poorly sorted, matrix-supported varieties, are over-consolidated, provided there was adequate subglacial drainage. Their bulk density, penetration resistance and seismic velocity are usually high, and permeability low, relative to other varieties of till of the region.	Supraglacially formed melt-out tills are usually less (normally weakly) consolidated than the subglacially formed, commonly over-consolidated melt-out tills, provided there was adequate drainage of meltwater. Bulk density and penetration resistance may be lower and more variable than in related lodgement till. Permeability is also more variable.	Primarily normally consolidated and relatively permeable. If clayey, may become over-consolidated due to post-depositional desiccation. Density lower than in primary tills.

(continued)

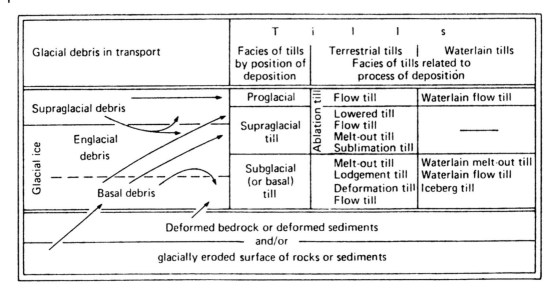

Glacial debris in transport	T i l l s		
	Facies of tills by position of deposition	Terrestrial tills | Waterlain tills Facies of tills related to process of deposition	
Supraglacial debris →	Proglacial	Flow till	Waterlain flow till
	Supraglacial till	Lowered till Flow till Melt-out till Sublimation till	—
Englacial debris	Subglacial (or basal) till	Melt-out till Lodgement till Deformation till Flow till	Waterlain melt-out till Waterlain flow till Iceberg till
Basal debris			
Deformed bedrock or deformed sediments — and/or — glacially eroded surface of rocks or sediments			

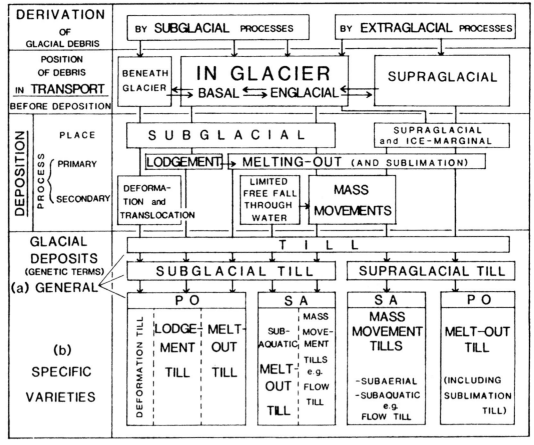

Figure 2.3

RELEASE OF GLACIAL DEBRIS AND ITS DEPOSITION OR REDEPOSITION			DEPOSITIONAL GENETIC VARIETIES OF TILL		
I. ENVIRONMENT	II. POSITION	III. PROCESS	IV. BY ENVIRONMENT	V. BY POSITION	VI. BY PROCESS
GLACIO-TERRESTRIAL	ICE-MARGINAL: -FRONTAL -LATERAL	A. PRIMARY MELTING OUT SUBLIMATION LODGEMENT SQUEEZE FLOW SUBSOLE DRAG	TERRESTIAL NONAQUATIC TILL	ICE-MARGINAL TILL	A. PRIMARY TILL MELT-OUT TILL -SUBLIMATION TILL
	SUPRAGLACIAL	B. SECONDARY: -GRAVITY FLOW		SUPRAGLACIAL TILL	LODGEMENT TILL DEFORMATION TILL OR
GLACIOAQUATIC		SLUMPING	SUBAQUATIC OR WATERLAIN TILL	SUBGLACIAL TILL	GLACITECTONITE SQUEEZE FLOWTILL
	SUBGLACIAL	SLIDING AND ROLLING			B. SECONDARY TILL
	SUBSTRATUM	FREE FALL			FLOWTILL -GRAVITY FLOWTILL

Figure 2.3 A variety of till and till process classification schemes compiled by the Till Work Group of the INQUA Commission on Genesis and Lithology of Glacial Quaternary Deposits (from Dreimanis, 1989). The upper diagram is the groups' genetic classification of tills in 1979 (after Dreimanis, 1969, 1976) which attempted to summarise both the position and migration of debris during glacial transport on the left and the final deposits, firstly in relation to position of deposition (middle column) and secondly in terms of facies nomenclature on the right. Note that deformed materials (not including 'deformation till') occur along the base of a diagram that mimics the vertical stratigraphic succession related to any one glacial phase (Hambrey and Harland, 1981). The middle diagram is the summary in 1982 of the groups' deliberations on process–form relationships, with factors that influence till production in the top half and the till genetic classifications at the base (Dreimanis, 1982). The lower diagram is the 'depositional genetic classification of till' compiled by Dreimanis (1989), with debris release and deposition on the left and till type on the right (no horizontal correlation is implied).

The term 'till' was first used by the Scots to refer to rough and agriculturally impoverished ground conditions or stoney clay and as a consequence was then adopted by Archibald Geikie (1863) as a geological term to refer to glacial deposits, specifically those that appeared as 'stiff clay full of stones varying in size up to boulders produced by abrasion carried on by the ice sheet as it moved over the land' (Geikie, 1863, p. 185). Since that time, the term 'till' has always been associated with glacial debris, and hence the frequently used variant 'glacial till' is a redundancy. However, 'till' was soon replaced by the term 'boulder clay', mostly by British geological mappers, a map unit classification that has remarkably endured on many British geology maps despite its grain-size implications being applicable predominantly only to lowland settings. Classification schemes for glacial and glacifluvial deposits were originally proposed by Chamberlin (1894a), who subdivided what was called 'glacial drift' into stratified and unstratified categories and also, together with Upham (1892), first used the term 'lodge'. The genetic qualifiers of 'lodgement' and 'ablation' were proposed for till by some early workers, such as Upham (1891a, b, 1894b), Chamberlin (1894a, b), Salisbury (1902), Tarr (1909) and Shaw (1912), after the sedimentological observations of Torell (1877). The characteristics of till were remarkably well described by James Geikie (1894), when he documented features such as 'broken' (glacitectonised) and plucked bedrock, bedrock rafts in contorted drift (previously highlighted by the British regional geologists C. Reid, S.V. Wood, J.L. Rome and F.W. Harmer), lee-side bedding, 'stone lines' (clast pavements), crude stratification, crag-and-tails and clast wear features like striae, facets and uneven edge rounding. Striated clast pavements, in places accompanied by slickensided or striated clay matrixes, were first documented by Stoddard (1859). Although till fabric analysis was not developed until the mid-twentieth century, Hind (1859) and then Miller (1884) appear to be the first to have identified non-random clast distribution in tills. Miller identified fissility in what he called 'fluxion structure', a signature of shearing in fine matrixes. Observations on the modification of clasts during glacial transport were first elucidated by T.C. Chamberlin (in Upham, 1894a) where he states the following:

> the material on the surface and slopes of a considerable number of glaciers … [are] … invariably of sharp, angular, unworn forms. … The englacial material that comes to the surface on the terminal slope of the Rhone glacier. … I found to be altogether angular and entirely without any evidence that it had been at the bottom of the glacier. … The basal material of the same glacier was, however, well rounded, and the moraines just below contained large quantities of this rounded material.
>
> Chamberlin (in Upham, 1894a, p. 85).

Although the term 'boulder clay' was widely applied to tills in the late nineteenth and early twentieth centuries based upon their massive appearance and fine-grained matrix, the apparently stratified nature of the deposits was a subject of some significant investigations, particularly by George Lamplugh (1881a, b, c, 1882, 1884a, b, 1890, 1919) in Britain but also by Crosby (1890, 1896) in the eastern USA. Lamplugh's detailed sketches of the large coastal exposures on the Holderness coast of eastern England (Figure 2.4) indicated that the depositional processes involved in the production of 'boulder clay' in the lowland glaciation record were strongly influenced by subaqueous or glacifluvial mechanisms.

The inappropriateness of the term 'boulder clay' was inherent within the definitions of its early proponents as perceived by Flint (1957) when he stated that it:

> is not a good designation for the range of deposits we know as till. It is not good because some till contains no boulders, some contains little or no clay, and some … contains neither boulders nor clay, but only silt, sand and pebbles.
>
> Flint (1957, p. 109)

At the same time, Charlesworth (1957) acknowledged that 'boulder' represented all size grades larger than pebbles (>15 cm diameter) and that the matrix, rather than exclusively always clay, varied according to the bedrock source:

> on sandstones … it is liable to be loose and sandy; on granites, gneisses and quartzose schists … it is stoney, coarse and gravelly and often hardly distinguishable from decayed rock in situ. In these cases, the term 'clay' is less appropriate than in areas of limestone, clay or shale.
>
> Charlesworth (1957, p. 377)

Hence, 'typical boulder clays' were regarded as prevalent in areas of relatively softer or finer-grained bedrock, whether that was in slate upland settings like the northern English Lake District or on coastal lowlands such as the post Cretaceous bedrock terrain around the western margins of the North Sea (cf. Lamplugh, 1881a, b, c, 1882, 1884a, b, 1890, 1919). Boulder clay matrix characteristics also displayed regional patterns, which were broadly interpreted (cf. Charlesworth, 1957) as the result of the generation of finer matrix by progressive wear over distance (i.e. from ice sheet dispersal centre to margin). The matrix generated, even over relatively short distances, was termed 'rock-flour', immediately evident in the milky, turbid nature of streams draining glaciers on hard beds. The relationship between crushing and abrasion processes and fine-grained matrix production in tills was later elucidated through the concept of 'terminal grade' (see Chapter 5). It is worth noting in this respect

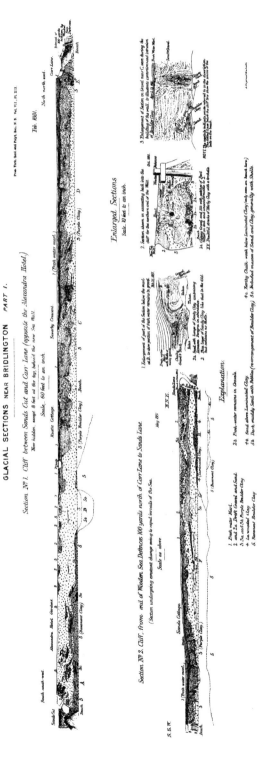

Figure 2.4 Examples of section sketches through the coastal exposures through the East Yorkshire coast tills at Bridlington, England by Lamplugh (1881a). The extent of stratified sediments and their deformation are well illustrated and were influential in Lamplugh's (1911) use of surging Svalbard glacier snouts and their proglacial deformation of foreland deposits as a modern analogue in his interpretations of till genesis.

that Lamplugh (1911) associated the clay-rich matrix of the English east coast 'boulder clays' with cannibalisation of offshore muds by the advancing North Sea glacier, akin to the processes around surging snouts he had observed on Svalbard (cf. Garwood and Gregory, 1898). The importance of matrix generation locally in this way was acknowledged by early workers (e.g. Salisbury, 1900) and was quantified by Flint (1947), who concluded that 75–90% of boulder clays had not travelled further than 80 km and long-distance erratics must have travelled mostly supraglacially or englacially and/or been reworked from earlier deposits.

By the time Dreimanis (1989) had reported on the findings of the Till Work Group, it was clear to all glacial researchers that 'till' was an extremely diverse deposit, and due to its similarities with other poorly sorted materials, it had great potential to be widely mis-identified or mis-classified. Problems continue to arise wherever the term 'till' is applied to a wide range of material types merely because they are associated with glaciation. This stratigraphic approach is particularly suited to terrain mapping where poorly sorted surficial materials are classified as till units, such as 'till plains', 'ground moraine', 'till and moraine', and so on, a procedure that can be justified if the most recent process to act upon a till is not used to genetically classify it; for example, mass movement deposits developed in till can be classified as 'flow till'. Even sedimentologists, for example, Harland *et al.* (1966) used the terms 'till' and 'tillite' for all diamictic (poorly sorted) sediment containing glacially transported material. Such an approach may avoid semantic arguments but hamper attempts to delve into the process sedimentology of glacier beds. More appropriate is the employment of a non-genetic, descriptive classification prior to genetic labelling, as is standard procedure in other realms of sedimentology. At landform and landscape scales, especially in relation to mapping, the ancient term 'drift' still has applicability despite its linkages with diluvial theory; it merely communicates that the ground surface is covered by debris of likely glacial provenance (e.g. drift mound, drift ridge, drift belt, drift limit) and continues to be employed as a non-genetic descriptor, especially in the British Isles, although alternative non-genetic terms such as 'discrete debris accumulation' (Harrison *et al.*, 2008; Whalley, 2009) are becoming popular. In sedimentology, a similar procedure has long been employed by using the descriptive term 'diamicton' or 'diamict' ('diamictite' for lithified materials) for poorly sorted sediment with a wide range of grain sizes (Flint *et al.*, 1960; Harland *et al.*, 1966; Flint, 1971; Eyles *et al.*, 1983a; Evans and Benn, 2004). Other terms proposed included 'glacial conglomerates' under the group of 'cataclastic rudites' (Pettijohn, 1949), 'conglomeratic mudstones' (Miller, 1953; Crowell, 1957; Wayne, 1963), 'paraconglomerates' (Pettijohn, 1957) and 'mixtites' (Schermerhorn, 1966; Martin *et al.*, 1985; Spencer, 1985). For engineers, the term 'diamicton' when unqualified has restricted utility because it does not convey the grain size characteristics of what is a hugely variable material, which ranges from clast-supported and gravelly to matrix-supported and clay-rich deposits. Hence, Eyles *et al.* (1983a) initiated the procedure of facies codes that communicated diamicton characteristics using qualifiers (e.g. matrix-supported or clast-supported, massive or stratified diamictons or laminated diamictons, etc.).

Particularly problematic in the analysis of glacial depositional process–form regimes has always been the ubiquitous appearance of stratified material or stratified diamictons (e.g. Lamplugh, 1879), a subject that was creatively addressed by Goodchild (1875) and Carruthers (1939, 1947, 1953). Whereas thin bands of stratified sediment were readily acknowledged as the product of thin films of water created at the ice–till interface (Charlesworth, 1957), the more substantial stratified inter- and intrabeds that were associated with many outcrops of 'boulder clay' appeared to require far more subaqueous sedimentation than was compatible with the 'lodgement' process *per se*. Even the early three-fold classification scheme of Chamberlin (1883) recognised subglacial till, upper (englacial or supraglacial) till and subaqueous till. The thickness of many such till sequences, and also thick massive tills, is not only a subject of significant debate developed throughout this book but also a

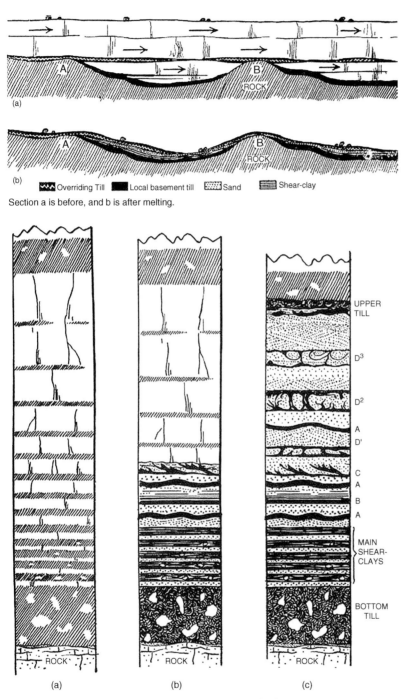

Section a is before, and b is after melting.

Progressive stages in the undermelting of englacial detritus. (a) Glacier section with "bottom", "banded" and, at the top, " overriding" dirts; (b) undermelt in progress, and (c) nearly completed.

Figure 2.5 Diagrams produced by Carruthers (1947–1948, 1953) to explain his undermelt theory.

problem that emerged early on in the study of glacigenic deposits. Early studies by Penck (1882), Heim (1885), Drygalski (1897, 1898) and Wahnschaffe (1901) questioned whether or not thick 'boulder clays' could be transported as subglacial materials, and Crosby (1900) proposed only thin layers of till deposition based upon the volume of sediment emerging from beneath contemporary glacier snouts. Shaler (1870) hypothesised that the maximum depth of till that could be transported beneath glacier ice was 30 m, but the importance of englacial debris (e.g. Chamberlin, 1895; Crosby, 1896) in the formation of thick 'till' sequences was soon identified as a potential source of debris released during final glacier melt (Upham, 1891a, b, 1892; Hershey, 1897). Thus was born the problematic and controversial 'melt-out' concept.

The grandfather of the melt-out concept is widely acknowledged to be J.G. Goodchild (1875), as he was the first to propose that at least some 'boulder clays' were derived from the melting of debris-rich ice similar to that being observed at that time in Svalbard and Greenland. The principle that a subglacial till could be overlain by an englacially derived till was popular amongst those (e.g. Torell, 1877, Hitchcock, 1879; Upham, 1891a, b, 1895, Russell, 1895; Salisbury, 1896, 1902, Tornquist, 1910; Shaw, 1912) who had observed sequences of lower, dense and compact tills ('typical boulder clay') overlain by coarser and loosely packed 'upper tills', the two tills often separated by stratified deposits. From such vertical sequences came the genetic terms 'lodgement till' and 'ablation till'. What some regarded as a more extreme variant of the melt-out concept was the 'undermelt theory' of George Carruthers (1939, 1947–1948, 1953), designed to explain the more problematic clay-rich tills ('typical boulder clays') and their intra- and interbeds of stratified sediments, typified by the Holderness tills of the Eastern England coast. Although his model implied, we now understand implausibly, that even the most delicate sedimentary bedforms and laminations ('shear clays') could be perfectly preserved after englacial melt-out (Figure 2.5), Carruthers was advocating nothing more than a more passive variant of the melt-out process that was championed by earlier geologists with their 'ablation till'. What Carruthers had succeeded in articulating were the sedimentological attributes necessitated by the melt-out theory if it was to be used to explain thick, partially stratified sequences of clay-rich tills. That the model was a step too far (i.e. it was an outrageous geological hypothesis that could be falsified; Davis, 1926) was demonstrated by the fact that his last offering on the theory in 1953 had to be self-published as a pamphlet (cf. Wordie, 1950; Anderson, 1967; Bennett and Doyle, 1994). Nevertheless, the concept of passive melt-out was to return for a fresh airing in the 1970s and continues to be debated by glacial sedimentologists (see below). Additionally, the final stage in the undermelt process (Figure 2.5) depicted a style of stratigraphy that was already widely observed in the ancient glacial record (lower and upper tills separated by stratified sediments) and which was to be addressed through the application of modern Arctic analogues to ancient till stratigraphies in the late 1960s to 1970s.

After the early recognition by Torell (1877), Chamberlin (1883, 1894a, b), Upham (1891a, b, 1895), Salisbury (1902), Tarr (1909) and Shaw (1912), amongst others, that glaciers appeared to produce both basal 'lodgement' and supraglacial 'ablation' tills, it was accepted that a specific set of processes operated during the subaerial release of debris from glacier ice, leading to flowage and resedimentation (Sharp, 1949; Flint, 1957; Harrison, 1957). The introduction of the term 'flow till' by Hartshorn (1958) to classify this type of sediment was a significant benchmark in glacial sedimentology, providing a genetic label for those glacial diamictons not created by lodgement or melt-out. A significant step was then taken in the late 1960s and early 1970s when Geoffrey Boulton reported on his systematic observations on sedimentary processes operating on some Svalbard glacier snouts. His widely used conceptual model (Figure 2.6a, b; Boulton, 1972a) conveyed a process sedimentology that acknowledged the overwhelming importance in polythermal glaciers of supraglacially reworked

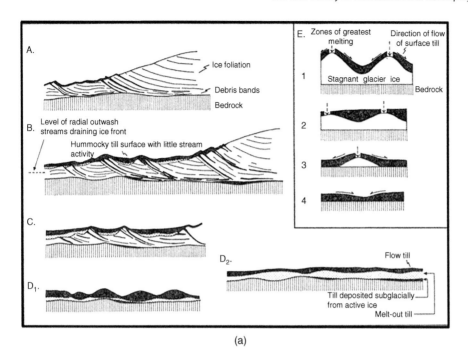

(a)

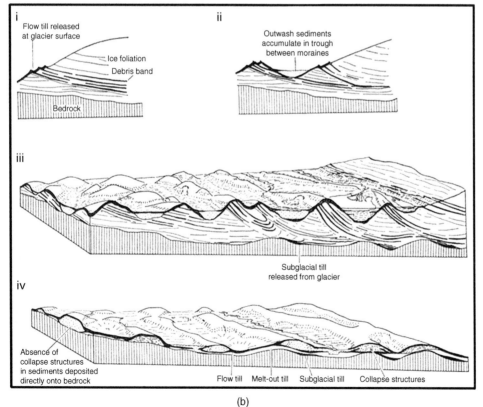

(b)

Figure 2.6 (*Continued*)

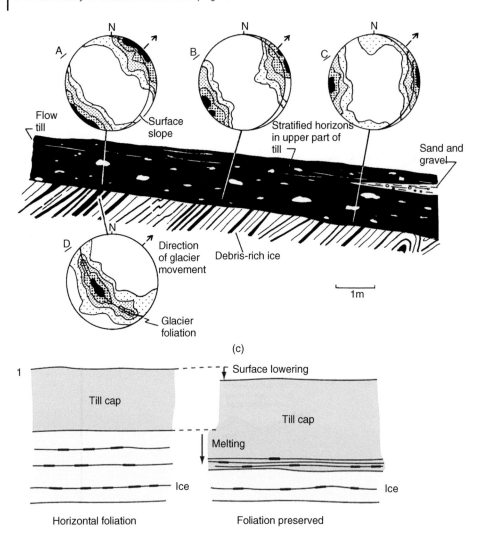

(c)

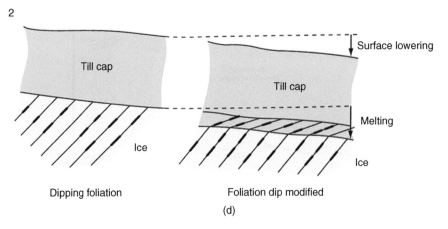

(d)

Figure 2.6 (*Continued*)

debris or 'flow tills' (Figure 2.6c; Boulton, 1967, 1968) as well as elucidating on the melt-out process in debris-rich basal ice sequences, re-affirming the broad concepts of englacial till production of Goodchild (1875) and Carruthers (1939, 1947–1948, 1953) and introducing for the first time the term 'melt-out till' (Figure 2.6d; Boulton, 1970a). The subglacial or lodgement component of the till stratigraphy in these glaciers was subordinate (Boulton, 1970a, 1971), but nevertheless combined with Boulton's melt-out and flow tills to form a tripartite sequence of till emplacement relating to one phase of glaciation, a stratigraphic model that he applied to a thick glacigenic sediment sequence at Glanllynnau, North Wales (Figure 2.7; Boulton, 1977; cf. McCarroll and Harris, 1992), which strongly resembled the final stage of the undermelt process proposed by Carruthers (1953; Figure 2.5).

A publication benchmark in till sedimentology was the release in 1971 of R.P. Goldthwait's edited volume entitled *Till: A Symposium* (Figure 2.1), wherein Boulton (1971) delivered one of his two seminal pieces on the tripartite till sequence and a range of other contributions on till sedimentology focussed on aspects such as macrofabrics, grain size and minerology of basal and 'ablation' tills; the concept of melt-out till had not been fully embraced by the glacial community at this stage even though Boulton's paper was already substantiating it. Even the knowledge base on subglacial till production processes was regarded as being in its infancy by Goldthwait (1971b) in his introduction to the volume. He summarised the state of the art on subglacial deposition as follows: (1) particles collect one by one due to frictional interference with bumps on the bed; (2) sheets of till or ice–debris mixes can be emplaced and sheared over when basal ice flow velocity drops to zero; (3) continuous basal melt brings material to the bed by the release of particles from debris-rich ice, as demonstrated in the seminal offering by Nobles and Weertman (1971) in the same volume, who developed ideas by Robin (1955), Gow *et al.* (1968) and Weertman (1964) on the melting beneath ice sheets due to thermal gradients beneath thick ice and frictional melt beneath sliding ice. On the basis of this, Goldthwait (1971b) proposed that Flint's (1957) term 'lodgement till', derived from Chamberlin's (1894a, b) term 'lodge', was most appropriate for the material produced by the combination of processes at the ice–bed interface. Goldthwait also offered three more state-of-the-art pointers that signposted future developments in subglacial deformation: (1) macrofabrics may be re-orientated within soft tills (e.g. MacClintock and Dreimanis, 1964; Evenson, 1971); (2) ice flowing over saturated tills might produce wave-like structures from which till could be injected into folds of moving ice (cf. recent proposals that till is subject to instabilities that create subglacial bedforms like drumlins and ribbed terrain; Dunlop *et al.*, 2008; Fowler, 2000, 2009, 2010; Sergienko and Hindmarsh, 2013; Stokes *et al.*, 2013); (3) till flows into corrugations in the ice base in sub-marginal locations (e.g. Ramsden and Westgate, 1971). Finally, Goldthwait acknowledged the importance of englacial debris in the delivery of material to thick till sheets by the final melt-out of debris-rich sub-polar ice.

Figure 2.6 Results of field observations on till production on Svalbard glacier snouts by Geoffrey Boulton: (a) diagrammatic sequence of depositional events related to the downwasting of debris-charged ice. A-D$_1$ depicts the development of hummocky terrain due to the continuous topographic inversions created by flowage of debris once melted out from discrete debris-rich ice folia. A-D$_2$ depicts the alternative scenario of till plain production due to more fluid 'flow till'. E shows the process of topographic inversion and till flowage due to uneven surface melting (from Boulton, 1972a); (b) the classic supraglacial process–form (landsystem) model of Boulton (1972a), showing the spatial relationships between subglacial, melt-out and flow tills and associated glacifluvial sediments due to the downwasting of a polythermal, debris-charged glacier snout; (c) field sketch and macrofabric data of 'flow till' observed to be accumulating on the surface of a debris-charged polythermal glacier on Svalbard by Boulton (1971); (d) simplified diagram to show the development of melt-out till as observed on Svalbard polythermal glaciers (from Boulton, 1971).

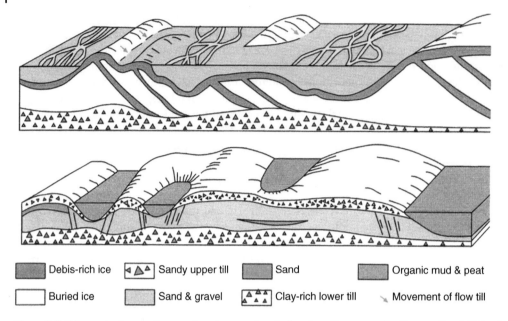

▓ Debis-rich ice	◄▲▲ Sandy upper till	▓ Sand	▓ Organic mud & peat
□ Buried ice	▓ Sand & gravel	▲▲▲ Clay-rich lower till	↘ Movement of flow till

Figure 2.7 Schematic diagram interpreting the complex glacigenic sediments at Glanllynnau, North Wales (lower panel), guided by the process–form relationships observed in Svalbard (upper panel) and conveying the principle of tripartite (till-stratified sediments-till) sequences relating to one glacial advance (from Boulton, 1977).

The next symposium volume dedicated entirely to till was Legget's (1976) edited book entitled *Glacial Till: An Inter-disciplinary Study* (Figure 2.1). In terms of till sedimentology, this volume contained a review of till origins and properties by Dreimanis (1976) and a brief application of the tri-partite till classification scheme to geotechnical properties by Boulton (1976); otherwise the chapters were aimed at engineering practicalities. Significant in the Dreimanis paper was the modification of his till classification scheme (Figure 2.3; upper panel) in which Boulton's (1970a, b) subglacial melt-out tills were firmly established, but the term 'melt-out' was applied to two variants, 'ablation melt-out till' and 'basal melt-out till', the former previously having been termed '(superglacial) ablation till'. Beyond Legget's (1976) collection, a number of case studies firmly established the 1970s as a decade of progress in till sedimentology in two main realms. First, Boulton's (1967, 1968, 1970a, 1972a, b) concept of 'melt-out till' was rapidly verified and consolidated by studies on modern Alaskan glaciers (Mickelson, 1971, 1973; Lawson, 1979a, b) and assessments of complex diamicton and stratified sedi-ment sequences in ancient deposits (Shaw, 1972, 1979). Stemming from this, the benchmark study of melt-out till in modern and ancient settings was that of Shaw (1982), followed up with detailed stud-ies by Haldorsen and Shaw (1982) and Shaw (1983). Second, the process–form regime involved in basal or 'lodgement' till production was investigated directly at the glacier bed in pioneering studies by Geoffrey Boulton, following on from his more holistic studies of till types on Svalbard (Boulton, 1974, 1975, 1979, 1982; Boulton *et al.*, 1974, Boulton and Dent, 1974; Boulton and Jones, 1979). From this work came Boulton's (1974) 'critical lodgement index' whereby increasing effective pressures bring about increasing frictional resistance and concomitant grain-by-grain lodgement at the sliding ice–bed interface. Notions amongst till researchers that the subglacial bed could deform and could also initiate lodgement due to clast ploughing (e.g. Boulton, 1975, 1976, 1982) were converted into firm understandings during the late 1970s with daring experiments on soft glacier beds.

Although the subglacial deformation paradigm (Boulton, 1986) is widely acknowledged as having been initiated in the late 1970s to 1980s (the paradigm status is specifically related to contributions of deformation to glacier flow), the sedimentological signatures of till deformation were proposed much earlier in the late nineteenth and early twentieth centuries by some very perceptive observers. For example, McGee (1894) speculated on the possibility of 'differentially moving ground moraine'. Shortly afterwards, Geinitz (1903) and then Hollingworth (1931) alluded to viscous drag in sub-glacial materials by proposing a vertical deformation profile that increased in magnitude from the base of the deforming layer to a zone of maximum displacement near the top, after which the displacement again dropped off towards the ice–till boundary; this pattern of vertical displacement is now widely recognised in subglacial tills as we shall discuss at various places throughout this book. Other features indicative of subglacial deformation were Geinitz's (1903) and Hollingworth's (1931) 'lee tails' and 'pre-crags' (pressure boudins), Alden's (1905) cleavage slip planes or fissility, and Reid's (1885) crushed clasts. The juxtaposition of lodgement and deformation was also proposed by Virkkala (1952), and the till classification scheme of Elson (1961) clearly acknowledged the subglacial crushing and deformation processes in his terms 'comminution till' and 'deformation till', the former relating to densely crushed and ground bedrock and the latter to the partially homogenised upper layers of glacitectonically disturbed pre-existing materials. In an attempt to explain the genesis of drumlins, Smalley (1966) and, classically, Smalley and Unwin (1968) explored the discipline of soil mechanics to make till sedimentologists aware of the importance of dilatancy (expansion and contraction in response to porewater pressure changes) in the deformation of granular materials in addition to constant-volume deformation. The implications of this behaviour were clearly demonstrated in the celebrated subglacial experiments at Breiðamerkurjökull, Iceland, as reported by Boulton (1979) and Boulton and Jones (1979) and then later by Boulton and Hindmarsh (1987). At the same time, Engelhardt *et al.* (1978) identified a deforming substrate beneath Blue Glacier in Washington, USA. The Breiðamerkurjökull experiment identified a two-tiered structure in the subglacial deforming diamicton, comprising a low-strength, high-porosity upper layer (A horizon) and a stronger, higher-density lower layer (B horizon), thought to represent ductile and brittle deformation, respectively. Despite the importance of Boulton's (1970a, b, 1979) observations on subglacial deforming till, his was the only paper in the *Journal of Glaciology*'s (1979) glacier-bed processes special issue that was on the topic of deformable beds; Iverson (2010) has more recently reflected on this apparent early reticence by glaciologists to recognise till deformation, suggesting that it was an intellectual bias whereby

> experts on the flow and thermodynamics of ice were perhaps predisposed to not muddy the sliding problem with dirt.
>
> Iverson (2010, p. 1104).

Although glacial sedimentologists did not fully understand till deformation until the 1970s, glacitectonic deformation of materials was widely recognised (e.g. Torell, 1872, 1873; Johnstrup, 1874; Merrill, 1886; Sardeson, 1906; Fuller, 1914; Slater, 1927a–e; Kozarski, 1959; Elson, 1961; Moran, 1971; Rotnicki, 1976; Banham, 1977; Berthelsen, 1978), and till classification schemes at that time (e.g. Dreimanis, 1976) included 'deformation till', defined as:

> characterized by an abundance of glacio-dynamic structures such as folds, overthrusts, shear planes, injections, breccias, and mylonites formed by differential movement or compressive stresses during … lodgement processes.
>
> (Dreimanis, 1976, p. 37).

The definition arrived at by the Till Work Group was summarised by Elson (1989; cf. 1961) as follows:

> Deformation till comprises weak rock or unconsolidated sediment that has been detached from its source, the primary sedimentary structures distorted or destroyed, and some foreign material admixed.
>
> (Elson, 1989, p. 85).

This clearly referred to pre-existing materials deformed by overriding ice, although Elson (1961) originally used the term to refer to a continuum that included the deformed pre-existing materials as well as the more homogenised shear zone developed within them, which he later (Elson, 1989) compared to the deforming bed of Boulton and Jones (1979). Indeed, it was not until the recognition of subglacial deforming till layers in the late 1970s that the term 'deformation' till was applied more to tills rather than glacitectonised substrates. This application of the term 'deformation till' to subglacial deforming diamicton necessitated a development and expansion of the sedimentological nomenclature, an outcome that was extensively and exhaustively debated in the Goldthwait and Matsch (1989) volume, specifically by Dreimanis (1989), Elson (1989), Pedersen (1989) and Stephan (1989), and which strongly features Banham's (1977) term 'glacitectonite'. This was defined by Pedersen (1989) as:

> a brecciated sediment or a cataclastic sedimentary rock formed by glaciotectonic deformation.
>
> (Pedersen, 1989, p. 89).

It became clear at this time that a complex nomenclature designed to capture every nuance of a deforming glacier bed was in reality an attempt to draw sharp dividing lines within a sediment continuum and hence was increasingly of limited utility and certainly difficult to exercise in sedimentological practice. Pedersen (1989) advocated the adoption of Banham's (1977) term 'glacitectonite' instead of 'deformation till' and highlighted that Banham's scheme contained an internal nomenclature that recognised a continuum of deformation intensity (Figure 2.8); 'exodiamict glacitectonite' was sheared material that retained some primary parent structure and 'endiamict glacitectonite' was material sheared to the point where all primary structure was destroyed. Deformation till was from this time to become the homogenised or cannibalised upper contact of glacitectonites or, in other words, the endiamict glacitectonite.

Despite the fact that glacially deformed material has long been recognised, and more recently 'glacitectonite', or at least 'deformation till', is an established genetic glacial sedimentological term, there are still significant shortfalls in establishing diagnostic criteria for the description and interpretation of complex diamictons. As discussed above, terms such as 'glacial conglomerate'/'cataclastic rudite', 'conglomeratic mudstone', 'paraconglomerate' and 'mixtite' have all been proposed but tend to underplay some of the most significant common attributes of diamictons such as discontinuous stratification, pseudo-lamination, inter- and intra-bedding, soft-sediment rafts and a range of deformation structures from low-strain soft-sediment deformation to high-strain shear structures and fissility. Consequently, as we will see in the following chapters, glacial diamictons at their most heterogeneous display an often bewildering array of structures and sedimentary attributes (Figure 1.3), an appearance that some glacial researchers (e.g. Aber, 1982) have described as a 'mélange'. Although this term has genetic connotations in metamorphic rocks (Hsu, 1974), it can be employed as a non-genetic descriptive label and hence sedimentologically is a sound foundation for objective field investigations (see Chapter 4).

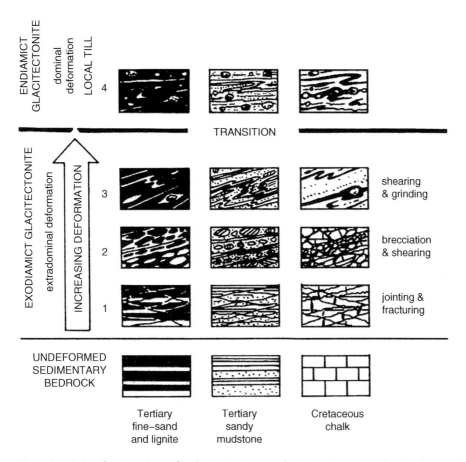

Figure 2.8 A classification scheme for glacitectonite compiled by Pedersen (1989) using the previous proposals of Banham (1977) and Berthelsen (1978).

Stratified, predominantly diamictic deposits have also long been recognised by glacial researchers (e.g. Geikie, 1894; Tarr, 1897; Kindle, 1924; Miller, 1953; Armstrong and Brown, 1954) and were classified by Harland *et al.* (1966) for those geologists working with the hard rock record as 'pseudo tillites', in contrast to the directly glacially related 'ortho tillites'. Harland *et al.* (1966) also coined the term 'para till' to cover ice-rafted debris. The lack of a suitable term for what appeared to be the submarine equivalents of terrestrial tills prompted Miller (1953) to use the term 'Yakatagite' for pre-Quaternary till-like stratified materials on Middleton Island, Alaska. Dreimanis (1969) later introduced the term 'waterlaid till' ('waterlain till' *sensu* Francis, 1975) to be used for 'a crudely stratified variety of till deposited in water' (Dreimanis, 1976, p. 39). A wide range of terms were then introduced for such deposits including 'subaquatic/subaqueous till', 'aqua till', 'underwater till', 'lacustro-till', 'marine till', 'iceberg (dump) till' and 'subaquatic ablation till' (see Dreimanis, 1976 and references therein) but were treated as unsatisfactory at the time by, for example, Flint (1971) and Boulton (1976). A process-based nomenclature has since been developed for such stratified sediments deposited clearly in deep water (cf. Evenson *et al.*, 1977; Dreimanis, 1979; Gravenor *et al.*, 1984; Powell, 1984), which gradually has moved away from using the term 'till' (e.g. 'dropstone

Figure 2.9 Photograph of the typical features of the Sveg till (J. Lundqvist).

diamicton', 'undermelt diamicton'), but the origins of many exposures through partially stratified and pseudo-stratified diamictons pose significant difficulties for glacial sedimentologists, especially at the complex interface between subglacial and subaqueous depositional environments. A classic example is that of the Catfish Creek Drift Formation in Ontario, Canada, a complex stratigraphic sequence of interbedded diamictons and stratified sediments, including the 'waterlaid Catfish Creek till' of Dreimanis (1976), which have been variously interpreted as subaqueous mass flow deposits, ice shelf undermelt and subaqueous flow deposits and alternating subglacial tills and meltwater cavity infills (cf. Evenson *et al.*, 1977; Gibbard, 1980; Dreimanis, 1982; Dreimanis *et al.*, 1987; Hicock, 1992, 1993; Boyce and Eyles, 2000; Dreimanis and Gibbard, 2005). Similarly, regionally significant 'stratified' till types like the 'Kalix' and 'Sveg' tills of Scandinavia (Figure 2.9) have been variously interpreted as subglacially deformed and/or waterlain (cf. Beskow, 1935; Hoppe, 1959; Lundqvist, 1969a, b; Virkkala, 1969; Shaw, 1979). In such settings, and more particularly in deep water marine settings, thick sequences of extensive massive and stratified diamictons were first highlighted by Craddock *et al.* (1964) as potentially constituting a particular problem for glacial sedimentology, because their origin as subglacial till/glacitectonite versus subaqueous rain-out is often notoriously difficult to demonstrate, as we shall see throughout this book.

This chapter has concentrated briefly on the history of till sedimentology, predominantly up until and including the deliberations of the Till Work Group, as published in Dreimanis (1989), as well as the general findings of the benchmark subglacial deformation experiments of the late 1970s. The deliberations of the INQUA 'Till Work Group' on genetic classifications for till were summarised by Dreimanis (1989; Table 2.1), identifying terrestrial and aquatic environments of deposition and the three forms of process-related till labelling (lodgement, melt-out and flow) championed by Boulton's

Svalbard observations. It is important to note that not all glacial sedimentologists were entirely convinced by the breadth of sediment types being called 'till', as demonstrated by the seminal work of Dan Lawson (1979a, b, 1981a, b, 1982) on sediments evolving around the margin of the Matanuska Glacier, Alaska. After observing the development of glacigenic diamictons in supraglacial and englacial settings, Lawson (1979a) regarded till as:

> a sediment deposited directly from glacier ice that has not undergone subsequent disaggregation and resedimentation.

> (Lawson, 1979a, p. 28)

Nevertheless, Table 2.1 has been employed to derive complex till types whose names reflected environment, position and process of deposition, transport process and derivation (supraglacial or subglacial). For example, Dreimanis (1989) highlighted 'glacioterrestrial subglacial melt-out till, of basal transport and subglacial derivation', from which the term 'subglacial melt-out till' would presumably suffice, because even Dreimanis acknowledged that the names 'are long and they appear cumbersome'. In their assessment of Dreimanis's (1989) genetic classification scheme for tills, Benn and Evans (2010) concluded that:

> In reality, field and laboratory techniques are not actually capable of refined assessments of the exact genesis of tills, making complex classification schemes such as that proposed by the Till Work Group difficult to apply in practice; more specifically, such schemes give a false sense that the glacial research community has accomplished a foolproof forensic procedure for the reconstruction of ancient process–form relationships.

> (Benn and Evans, 2010, p. 369)

Far simpler to use visually as well as communicatively was Dreimanis's (1989) tetrahedron or end member pyramid (Figure 2.10), which conveyed the three Svalbard till types of lodgement, melt-out and flow till in tandem with the more recently proposed deformation till, implying at the same time that melt-out invariably led on to lodgement, deformation or mass flows.

The preceding review forms a context for the significant details of modern till sedimentology, which is now covered in the remainder of this book. Although we will from here on concentrate on modern, or at least the most recent, studies of till sedimentology, it is appropriate also to digest in greater detail some older literature that remains pertinent today but which has been reviewed very briefly in this chapter.

Figure 2.10 Till type tetrahedron (from Dreimanis, 1989).

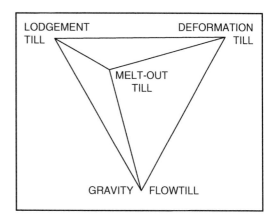

3

Till – When is it an Inappropriate Term?

I am not optimistic that all the different tills in the present classification can always be recognized in the field, even by meticulous descriptions and laboratory analysis.

Elson (1989, p. 85)

Before progressing, we need to digest the implications and subsequent developments that stemmed from the proposition by Lawson (1979a, 1981a, b) that a till is a sediment that has been deposited directly from or by glacier ice and has not been subject to subsequent disaggregation and resedimentation. Lawson (1989) summarised the implications of his Matanuska Glacier work in Goldthwait and Matsch's (1989) volume by highlighting the logic of sub-dividing glacigenic deposits into primary and secondary categories; primary deposits include those laid down uniquely by glacial agencies, whereas secondary deposits include those which have undergone reworking by non-glacial processes. Till in its various guises of lodgement, deformation and melt-out fits the definition of 'primary' but all other forms of glacigenic diamicton, because they are remobilised by a combination of gravitational mass flow and fluvial processes are 'secondary'. However, the dividing line between primary and secondary diamictic deposits is notoriously blurred, as is inherent within the deliberations of the Till Work Group, who arrived at its definition of till as 'a sediment that has been transported and is subsequently deposited by or from glacier ice, with little or no sorting by water' (Dreimanis and Lundqvist, 1984; Dreimanis, 1989), thereby side-stepping the 'flow till' problem despite Lawson's (1989) strong case to deal with it at that stage. Nevertheless, the unease with which the term 'till' was used to classify supraglacially deposited diamictons had developed quickly after Boulton's Svalbard-based verifications of Hartshorn's (1958) 'flow till'. This was communicated not only in Lawson's (1979a, b; 1981a, b) process-based classification scheme for mass flows (Figure 3.1a) but also in the development of alternative terms aimed at acknowledging a non-primary origin for glacigenic diamictons. For example, Eyles (1979) proposed the term 'supraglacial morainic till complex' (Figure 3.1b), and more recently efforts have been made to remove the term 'till' completely to arrive at 'supraglacial mass flow diamicton' or 'glacigenic mass (or debris) flow diamicton', the latter being a form of 'sediment flow diamicton' (Lawson, 1989). In subglacial settings, the flow of diamictic and associated materials into cavities at the ice–bed interface produced a crudely stratified deposit that was named 'lee-side till' by Hillefors (1973) and Haldorsen (1982) but regarded as 'flow till' by Boulton (1971) and later incorporated by him into the lodgement process–form regime (Boulton, 1982). Hence, the term 'flow till' has gradually been retired; although glacial researchers knew what the term 'flow till' tried to communicate (*sensu* Boulton, 1967, 1968, 1970a, b, 1972a, b), it was nevertheless fundamentally flawed simply because it implied the characteristics of both primary and secondary deposits. More significantly,

Till: A Glacial Process Sedimentology, First Edition. David J A Evans.
© 2018 John Wiley & Sons Ltd. Published 2018 by John Wiley & Sons Ltd.

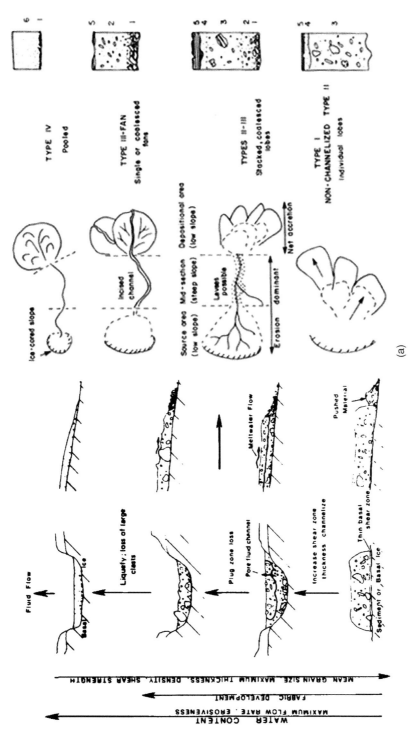

Figure 3.1 The sedimentary characteristics and process based classification schemes for supraglacial diamictons: (a) Lawson (1979a, b) classification scheme, showing four sediment flow types identified at the Matanuska Glacier, Aslaska.

(a)

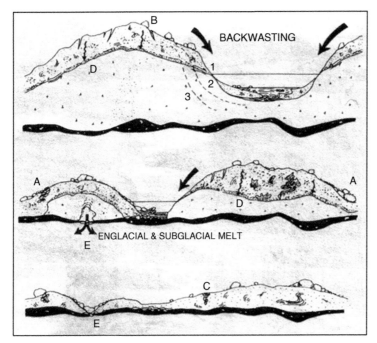

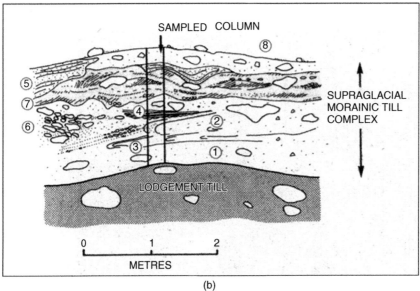

(b)

Figure 3.1 (b) the processes (upper panel) and resulting deposits (lower panel) of the 'supraglacial morainic till complex' of Eyles (1979).

glacially derived, flowed diamictons look like many other non-glacial gravitational mass-flow diamictons and hence the application of the word 'till' has a very high likelihood of being inappropriate in most stratigraphic settings and becomes less secure the longer the period of time since deglaciation. In other words, there is a very high likelihood that diamictons in a glaciated catchment, which are not interpreted as subglacial, automatically get classified as 'flow till' when their true origin is often from paraglacial gravitational mass wasting and hence they are 'sediment flow diamictons' (e.g. Eyles and Kocsis, 1988; Ballantyne and Benn, 1994, 1996; Harrison and Winchester, 1997; Curry and Ballantyne, 1999; Menzies and Zaniewski, 2003; Figure 3.2). Even in supraglacial settings, Lawson (1989) makes the strong case that 'glacigenic mass flow diamictons' are emplaced by repeated gravitational mass wasting and hence display the sedimentological signatures of the various mechanisms of secondary depositional processes (Figure 3.1a).

The inappropriateness of the term 'till' for deposits laid down subaqueously has also been recognised for some time (cf. Evenson *et al.*, 1977; Dreimanis, 1979; Gravenor *et al.*, 1984; Powell, 1984), resulting in a process-based nomenclature that recognises the disaggregation and/or remobilisation of glacigenic material once it is released into ice-contact lake and marine environments. The stratified nature of such materials, as well as their internal structures related to iceberg activity, have given rise to: (1) suspension settling and iceberg-related terms such as 'dropstone diamicton', 'undermelt diamicton', 'iceberg contact deposits' and 'ice-keel turbate'; and (b) mass-flow-related terms such as 'subaqueous fall deposits' or 'grain flows' and 'olistostromes', 'subaqueous slumps' or 'slides' (including soft-sediment deformation structures), 'subaqueous debris flows' (cohesive and cohesionless) and 'turbidites'.

Stemming from these proposed modifications to the nomenclature of both terrestrial and subaqueous glacigenic deposits, Evans and Benn (2010), following guidelines recommended by Evans *et al.* (2006b), compiled a classification scheme for primary glacigenic deposits that comprise only three end members: 'glacitectonite', 'subglacial traction till' and 'subglacial melt-out till'. This is compatible with a number of previous recommendations (e.g. Anderson *et al.*, 1980, 1986; Kemmis, 1981; Bergersen and Garnes, 1983; Dreimanis, 1983; Lundqvist, 1983; Stephan and Ehlers, 1983; Ringberg *et al.*, 1984; van der Meer *et al.*, 1985; Rappol, 1985; Hansel and Johnson, 1987) that 'subglacial till' was an appropriate term to use in referring to those materials laid down by the processes of lodgement, melt-out, deformation and undermelt, because the sedimentological properties of such materials were not sufficiently unequivocally diagnostic to discern a specific process–form regime. Assessments of the conceptually pivotal Catfish Creek Drift Formation (specifically the 'till' component) in Ontario, Canada, have reflected this conundrum, specifically in the conclusion by May *et al.* (1980) and Dreimanis *et al.* (1987) that the imprint of processes like lodgement, melt-out and flow could be observed but not used to sub-classify the till. The following chapters are organised around, and expand upon, the case for the threefold classification scheme for till as proposed by Evans *et al.* (2006b), and hence the terms 'glacitectonite', 'subglacial traction till' and 'melt-out till' are employed as genetic terms beyond the descriptive nomenclature outlined in Chapter 4. We will nevertheless develop the case for 'subglacial traction till' as a genetic term in Chapters 9–11 before using it definitively in Chapter 17.

Figure 3.2 Modern-day examples of remobilisation and resedimentation of glacigenic deposits, including tills and their sedimentological products: (a) retrogressive flow sides in supraglacial debris on the ice-cored Little Ice Age lateral moraines of Horbyebreen, Svalbard; b) Late Wisconsinan debris-rich buried glacier ice creating debris flows in englacial and supraglacial materials, northern Banks Island, Arctic Canada; (c) crudely stratified diamicton created by paraglacial debris flows in former glacigenic deposits, Leirdalen, Norway (photo by A.M. Curry); (d) debris flows emanating from the distal face of a push moraine during its construction by the advancing snout of Fláajökull, Iceland; (e) debris flows on the proximal slope of the Little Ice Age lateral moraines of Kvíárjökull, Iceland.

4

Glacigenic Diamictons: A Strategy for Field Description and Analysis

*In many places where it occupies a succession of lofty cliffs, it puts on a rude appearance of strat-
ification, or at least may be sub-divided into separate masses which possess distinct characters.*

Sedgewick (1825, p. 25)

Before dealing in detail with the processes and forms of till sedimentology, it is pertinent to reflect
on the contents of Chapter 2 and establish a workable nomenclature for simply and systematically
describing glacigenic diamictons in the field and laboratory. The motivation for including such a
schema at this juncture is to avoid the recurring issue in the vast literature on glacial sediments
and stratigraphy of dwelling on genetically inherent terminology at the expense of practical descriptive terms. For example, the comprehensive reviews of, and proposals for, till nomenclature, such as
Dreimanis (1976, 1980, 1989) and Dreimanis and Schluchter (1985), predominantly dwell on genetic
labelling, thereby offering only restricted guidance to those intending to initiate and conduct analyses
of glacigenic stratigraphies in an entirely objective (i.e. descriptive) way. Sedimentological proce-
dures designed to aid in the study of glacigenic deposits more generally are reviewed elsewhere
(Eyles *et al.*, 1983a; Evans and Benn, 2004; Hubbard and Glasser, 2005), so provided here are the
proposed descriptive classification menus and techniques aimed at procuring objective assessments
of glacigenic diamictons and their associated deposits.

4.1   Diamicton

The descriptive term 'diamicton' or 'diamict' ('diamictite' for lithified materials) has been widely
employed amongst glacial sedimentologists to describe in short, poorly sorted sediment with a wide
range of grain sizes (Flint *et al.*, 1960; Harland *et al.*, 1966; Flint, 1971; Eyles *et al.*, 1983a; Evans and
Benn, 2004). The utility of the term is less valued by engineers, because it does not convey the grain
size characteristics, and as a result the scheme derived by Moncrieff (1989) is sometimes preferred
(Figure 4.1). The descriptive scheme of Eyles *et al.* (1983a; modified by Evans and Benn, 2004),
based upon facies codes comprising grain-size definitions (i.e. D for diamicton as opposed to B for
boulders, G for gravel, GR for granules, S for sand and F for fines) and matrix/structure qualifiers
(matrix-supported or clast-supported; massive, stratified or laminated), is presented in Figure 4.1
together with the grain-size-specific refinements from the Moncrieff scheme (see Figure 1.3 for some

Till: A Glacial Process Sedimentology, First Edition. David J A Evans.
© 2018 John Wiley & Sons Ltd. Published 2018 by John Wiley & Sons Ltd.

Descriptive facies codes for diamicton, melange and predominantly non-fluvial coarse-clastic deposits

Diamictons (very poorly sorted admixture with wide range of grain sizes)

Dmm	matrix-supported, massive
Dcm	clast-supported, massive
Dcs	clast-supported, stratified
Dms	matrix-supported, stratified
Dml	matrix-supported, laminated
Dmf	matrix-supported, fissile
D - - (p)	diamictons containing clast pavements
D - - (cf)	diamictons arranged in clinoforms

Boulders

Bms	matrix-supported, massive
Bmg	matrix-supported, graded
Bcm	clast-supported, massive
Bcg	clast-supported, graded
Bcf	boulder units arranged in clinoforms
BL	boulder lag or pavement

Gravels

Gms	matrix-supported, massive
Gmg	matrix-supported, graded

Melanges

MtI	type I
MtII	type II
MtIII	type III
MtIV	type IV

Interpretive facies codes (glacigenic settings)

Dcs(c)	current reworked, clast-supported, stratified diamicton
Dcs(r)	re-sedimented, clast-supported, stratified diamicton
Dcs(s)	sheared, clast-supported, stratified diamicton
Dms(c)	current reworked, matrix-supported, stratified diamicton
Dms(r)	re-sedimented, matrix-supported, stratified diamicton
Dms(s)	sheared, matrix-supported, stratified diamicton
Dml(c)	current reworked, matrix-supported, laminated diamicton
Dml(r)	re-sedimented, matrix-supported, laminated diamicton
Dml(s)	sheared, matrix-supported, laminated diamicton
Dmf(s)	sheared, matrix-supported, fissile diamicton
Fm(d)	ice-rafted muds with dropstones (dropstone diamicton)
B/Gtmf	bouldery/gravelly, terrestrial mass flow deposit
B/Gsmf	bouldery/gravelly, subaqueous mass flow deposit
Mgt(a)	Type A glaciteconite
Mgt(b)	Type B glacitectonite
Mtmf	terrestrial mass flow melange (slump or slide)
Mtsmf	subaqueous mass flow melange(olistostome, slump or slide)

Figure 4.1 Classification schemes and facies coding systems for diamictons and related materials. Upper panel shows a facies coding scheme modified from Eyles *et al.* (1983a) and Evans and Benn (2004). Lower panel shows the Moncrieff (1989) scheme for classifying poorly sorted materials emphasising specific grain-size characteristics.

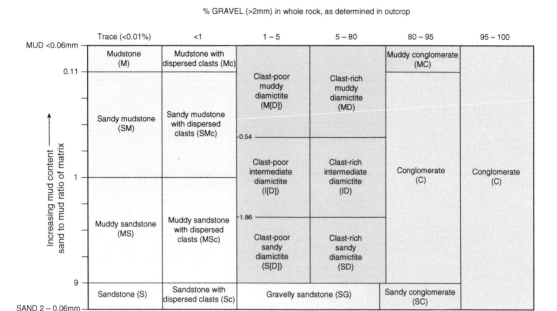

Increasing gravel content ⟶

% GRAVEL (>2mm) in whole rock, as determined in outcrop

Figure 4.1 (*Continued*)

typical field appearances). These facies codes can be employed in tandem with the various signature types for mélanges (Section 4.2) and internal structures (Section 4.3) in field-based descriptive assessments of diamictons regardless of their depositional origins. Some genetic implications are inferred by the structure qualifiers in Figure 4.1, wherein evidence of current reworking (localised stratified sediments), re-sedimentation (grading or crude sorting) and shearing (based on matrix fissility and slickensides) are noted, the extent of each providing immediate impressions of primary versus secondary origins. The terms 'homogeneous' and 'heterogeneous' are often used as prefix qualifiers for diamictons, referring to their overall appearance as either predominantly invariable or variable, respectively. This classification refers to a spectrum, whereby highly heterogeneous diamictons display the greatest range in textural variability. Where that high degree of textural variability comprises discrete packages of sub-units (intraclasts and/or discontinuous intrabeds), the term 'melange' is more appropriate. The occurrence of intraclasts and intrabeds can be particularly diagnostic of specific till forming processes and also common in other types of glacigenic and non-glacigenic diamictons, hence their characteristics and relationships with the host diamicton are important forms of descriptive information (Figure 4.2).

4.2   (Glacigenic) Melange

The application of the term 'melange' (Greenly, 1919) to glacial deposits generally (Menzies and Shilts, 1996), because of its linkage to metamorphic petrology (e.g. Hsu, 1974), was strongly criticised by Pedersen (1989), but nevertheless it adequately describes the complex internal structures often

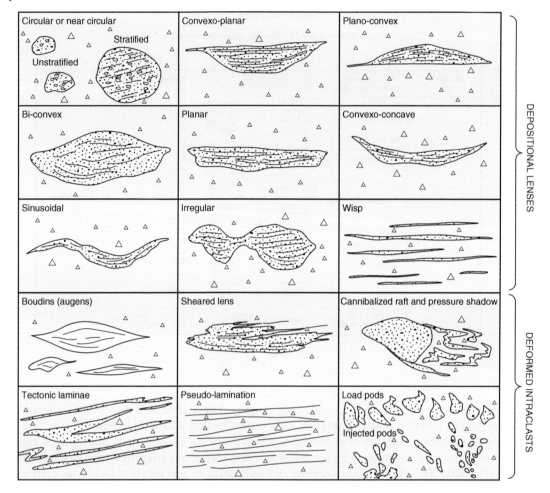

Figure 4.2 Various types of intraclasts and intrabeds or lenses.

observed in heterogeneous glacigenic diamictons, because they have invariably been subject to deformation in the same style as that recognised in mudstones at convergent plate margins. Clearly, the use of the pre-script 'glacigenic' in a purist sense implies a genesis. Hence, the term 'glacigenic melange' might be avoided unless research is being undertaken in an entirely glacially influenced depo-centre or basin. Nevertheless, the term 'melange' accurately describes materials that Cowan (1985) broadly defines as fragments enveloped within a fine-grained matrix; typically of obscure stratigraphy, stratal and/or chaotic 'block-in-matrix' fabric.

Application of the term 'glacigenic melange' can then be justified as a first stage of interpretive progress if the process of deposition/deformation was glacially induced. The more advanced stage of interpretation involves the identification of a specific process (e.g. glacitectonite; glacigenic mass flow diamicton; subaqueous mass flow diamicton, etc.), if possible. The use of the term 'melange' for subglacial deforming layers in their widest sense has also been proposed by Menzies and Shilts (1996). Four types of melange were recognised by Cowan (1985), the definitions of which are entirely practical in terms of pursuing a descriptive approach to glacigenic diamictons (Figure 4.3).

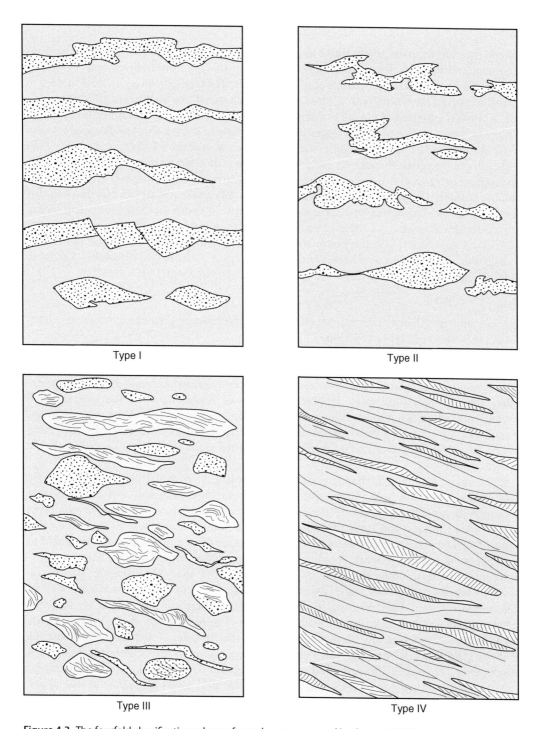

Figure 4.3 The fourfold classification scheme for melanges proposed by Cowan (1985).

4.3   Physics of Material Behaviour

As glacigenic diamictons, particularly those in subglacial settings, are typically subject to various spatial and temporal scales of deformation, it is important at this juncture to investigate the physical behaviour and associated nomenclature of such materials. The term 'rheology' is used to refer to the way in which strain rate varies with applied stress for a material like diamicton, after Eugene C. Bingham's (e.g. 1933) initial use of Heraclitus' expression *panta rei*, meaning 'everything flows'. More broadly, *rheology* refers to the study of changes in form and flow of matter, through the appreciation of elasticity, viscosity and plasticity. The basal shear stresses created at the ice–bed interface result in a strain signature in glacigenic diamictons and melanges that is dictated by material rheology, which can be understood in terms of two basic properties: the yield strength and the stress–strain relationships at higher stresses. The deformation structures (Section 4.4) and macrofabrics (Section 4.5) that are visible in glacigenic diamictons and melanges are a relative measure of their permanent strain, which was initiated whenever basal shear stresses exceeded their yield strengths. The yield strength is the value of the applied stress at the point when permanent deformation takes place and is controlled by a combination of material cohesion and internal friction, as defined by the Mohr–Coulomb equation (after Charles Augustin de Coulomb 1736–1806 and Otto Mohr 1835–1918):

$$\tau_{\text{yield}} = c + N \tan \Phi$$

where τ_{yield} is yield strength of a material; c is cohesion; N is the effective pressure or effective normal stress and $\tan \Phi$ is the coefficient of friction.

Once stress is at the yield strength of the material, permanent deformation takes place either as 'brittle failure' (breakage along a fracture) or 'ductile deformation' (flow or creep). With respect to the latter process-response, subglacial tills have traditionally been regarded as being prone to 'dilation' (e.g. Smalley, 1966; Smalley and Unwin, 1968; see Chapter 2 of this book), which is an increase in a material's void ratio when being sheared (Figure 4.4). This increase in volume of the material arises from the tendency for grains to climb over one another during shear. As strain rate falls, the grains collapse in on the void spaces to produce a denser material once again, but it is unlikely to regain the pre-deformation state.

Stress–strain relationships at higher stresses vary according to material properties and some debate has ensued since the benchmark subglacial experiments by Boulton and co-workers at Breiðamerkurjökull, concerning the mode of failure/deformation and the concomitant development of a flow law for till (Boulton and Paul, 1976; Boulton, 1979; Boulton and Jones, 1979; Boulton and Hindmarsh, 1987). Details of this debate and its implications for till sedimentology are presented in following chapters, but at this juncture it is important to define the critical process-based nomenclature relevant to understanding the style of deformation above the yield strength. The debate has focussed on whether subglacial materials, especially tills, adhere to a plastic or viscous style of failure/deformation (Figure 4.5a). *Viscosity* is defined as the constant of proportionality between stress and strain rate and is a measure of the ability of a fluid to resist force. A *viscous rheology* is defined as a linear increase in strain rate with shear stress; this includes Newtonian, linear-viscous materials if the yield stress is zero. Variable deformation behaviour in response to changing flow conditions, for example, in mass flows, is often termed a Bingham viscoplastic response, whereby an initial plastic response changes to viscous behaviour once the yield stress is reached. A *non-linear viscous rheology* is defined as a non-linear increase in strain rate with shear stress, wherein the viscosity is not constant but a function of the strain rate. In contrast, a material with a perfect plastic rheology will only deform at the yield stress and at whatever rate is required to prevent the shear stress from exceeding the yield stress; note that, unfortunately, the term 'plastic' is used in a

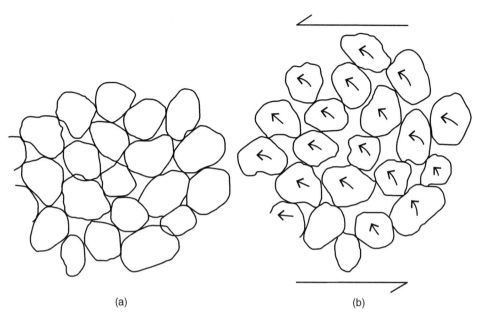

(a)   (b)

Figure 4.4 Schematic diagram showing the process of dilation during shear, whereby grains climb over one another and increase the sediment volume (from Benn and Evans, 1998).

somewhat confusing way in geotechnical nomenclature in that it refers both to a material with a yield strength and to 'plastic flow' or viscous flow of a plastic above its yield strength. In terms of the style of deformation signatures visible in glacigenic diamictons and melanges, a viscous rheology gives rise to sediment flow or creep (ductile deformation), whereas a plastic rheology results in irreversible/irrecoverable sliding of particles or sediment packages past one another (brittle failure). Viscoplastic rheologies involve spatial and temporal variability in strain responses, resulting in overprinted signatures and/or brittle–ductile styles of deformation; in subglacial environments where porewater pressures vary and dilatancy is possible, viscoplastic behaviour should perhaps not be unexpected.

Another important concept that relates to the behaviour of subglacial materials is that of 'consolidation'. This refers to the state of a material when it has been compressed to the point where all of its void spaces are closed and its particles densely packed so that there is a permanent loss of volume, as would be the case under a permanent overburden. Once a material is fully adjusted in this way to the overburden, it is referred to as 'normally consolidated'. If the overburden is for some reason then reduced and the material does not increase its void space to create a volume that is adjusted to the new, reduced overburden pressure, then it is called 'over-consolidated'. As we will see in following chapters, subglacial tills undergo porewater pressure changes and concomitant deformation that effectively dilates the material; additionally, the zone of dilation may migrate vertically over time. Hence, normally consolidated till can undergo additional consolidation when it is subjected to shear stress, leading to 'strain hardening', and conversely over-consolidated till may increase its voids ratio during dilation.

Till, when it is sheared to a sufficient strain, attains a 'critical state', where its porosity and shear resistance (ultimate strength) remains steady with further strain (see Iverson, 2010 for a full explanation; Figure 4.5b). Experiments have repeatedly proved that till ultimate strength is highly insensitive to strain rate but linearly dependent on effective pressure, supporting a Coulomb-plastic rheology

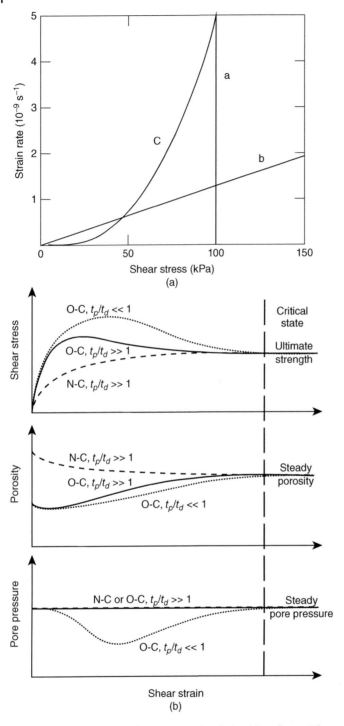

Figure 4.5 Graphs to show the stress–strain relationships of materials common to glacial environments: a) examples of typical stress–strain relationships. Line 'a' is a perfect plastic which remains rigid until the shear stress reaches the yield stress. This example shows that at 100 kPa the material deforms at whatever rate is required to prevent the shear stress exceeding the yield stress. Line 'b' is a Newtonian, linear-viscous material, for which strain rate is linearly proportional to shear stress. Line 'c' is a non-linearly viscous material, like ice, for which the strain rate is non-linearly proportional to shear stress (from Benn and Evans, 2010, after Paterson, 1994); b) graphs depicting the concept of 'critical state' and attainment of ultimate strength (from Iverson, 2010).

(Iverson *et al.*, 1998; Clarke, 2005). If a till is weakened by high porewater pressures, the glacier driving stress may exceed the till ultimate strength, potentially triggering viscous failure, but it appears that viscous deformation is resisted at large scales, for example, beneath ice streams, because flow does not become unstable. This resistance must come from 'sticky spots' on the bed (Alley, 1993; MacAyeal *et al.*, 1995; Stokes *et al.*, 2007), which operate to reduce any instability that might accrue through viscous deformation, if it operates at all. The episodic nature of subglacial bed deformation (see Section 6.2) due to changing effective pressures also initiates changes in till critical state porosity and strength, reflected in the material dilatancy. Iverson (2010) provides an explanation of what he terms 'pseudo-viscous' till behaviour that arises likely from changes in its critical state. He highlights that an increase followed by a decrease in effective pressure on a till at critical state porosity will cause consolidation and then swelling, although the swelling will only be a fraction of the consolidation due the irreversible nature of grain realignment during consolidation (Clarke, 1987, 2005). The till is then less porous than in the critical state and therefore could be subject to strengthening during the early stages of later deforming phases, thereby introducing the possibility of time-averaged pseudo-viscous behaviour. We also need to be aware of pseudo-plastic behaviour, whereby a material can display decreasing viscosity with increasing shear, also known as 'shear thinning'.

The operation of dilation in a shearing medium like till, and the associated changes in porosity states, demands that we understand the process of liquefaction. Engineers define *liquefaction* as:

> the transformation of a granular soil (unconsolidated sediment) from a solid state to a liquefied state as a consequence of increased porewater pressure and reduced effective stress.
> (Committee on Soil Dynamics, 1978; Youd, 2003)

The knowledge base on the liquefaction of unconsolidated sediments, particularly in response to earthquake shocks, is well established, as it constitutes a major geological hazard (Youd, 1978, 2003; Holzer *et al.*, 1989; Miwa *et al.*, 2006). Indeed, beyond the subglacial environment, a range of deposits, potentially including non-glacial diamictons, formed by palaeoseismic liquefaction events (seismites) are widely reported from the geological record (Obermeier, 1998; Menzies and Taylor, 2003; Green *et al.*, 2005; Obermeier *et al.*, 2005). Subglacial materials are prone to liquefaction because (1) they are typically composed of unconsolidated, granular sediments; (2) they possess a high water content and are at, or near, saturation; (3) their water table is high or perched, because it is constrained within the soft-bed materials by the underlying less permeable bedrock and the overlying ice (Phillips *et al.* in press). Hence, any trigger for energy release at the ice–bed interface, for example, during stick–slip motion (see Chapter 6), is likely to result in soft-bed liquefaction on a localised scale due to the spatial and temporal variation in sediment grain size, composition, porosity, permeability and water content. Temporally, such liquefaction would also be restricted due to the confining pressure exerted by the overlying ice, as evident in the stick–slip model for glacier motion over a soft bed. As explained above, the increase in porewater content and/or pressure in tills leads to a reduction in grain contacts and an increase in the distance between the component grains, referred to by glacial geologists as dilation. In a saturated till, Evans *et al.* (2006b) and Phillips *et al.* (in press) suggest that an increase in porewater pressure will ultimately lead to localised liquefaction. In terms of till deposits, this process, although short-lived and localised, can overprint pre-existing structures such as bedding, lamination and folding and thereby also lead to homogenisation of subglacially deforming materials.

4.4   Typical Structures

The style of melange identified in Figure 4.3 is just one method of characterising the appearance of glacigenic diamictons and a range of other more specific structures at both macro- and microscales can be identified and then used to procure genetic interpretations and ultimately apply process-based classifications to such deposits. Where they are massive in character and therefore lack strain markers like bedding, diamictons will rarely display clear deformation signatures at macroscale, but more stratified diamictons and melanges can contain well-developed signatures. The style of deformation can be interpreted using the structures displayed within glacigenic deposits or sequences of deposits (McCarroll and Rijsdijk, 2003) as summarised in Figure 4.6. This classification scheme provides a shorthand coding for structures associated with pure shear (P – overburden driven), simple shear (S – overburden and basal shear stress driven, typical of subglacial environments), compressional deformation (C – common in proglacial settings) and vertical deformation (V – gravity-driven processes), all of which can be contrasted with undeformed material (U). Tills and subglacially deformed materials, because they are produced in the subglacial shear zone, display a range of simple shear indicators at a range of scales (Figure 4.7; van der Wateren *et al.*, 2000). The larger-scale or stratigraphic influence of simple shear on subglacial materials is depicted in Figure 4.6 as a vertical continuum from non-deformed to folded or faulted sediments at the base (dependent on either ductile or brittle material behaviour) grading upwards to attenuated (boudinage) materials or melange and then to homogenised diamicton (till), a style of deformation signature recognised by Banham (1977) and Pedersen (1989; Figure 2.8). The directional elements created by simple shear in subglacial environments can be measured using the variety of structures compiled in Berthelsen's (1978) kinetostratigraphic protocol (Figure 4.8). The signatures of shearing, soft-sediment deformation and porewater movement identified at macroscale in Figure 4.6 can all be detected also at microscale through the analysis of thin sections (van der Meer, 1993; Menzies, 2000; Carr, 2004), a particularly important technique in diamictons that appear predominantly macroscopically massive (Figure 4.9).

4.5   Clast Macrofabrics and Microfabrics

Non-random clast distributions in tills were first identified by Hind (1859) and Miller (1884) and the assessment of strain signatures in glacial deposits using clast macrofabrics is an area of glacial sedimentology with a long pedigree (cf. Holmes, 1941; Harrison, 1957; Andrews and Shimizu, 1966; Andrews and Smithson, 1966; Andrews, 1971; May *et al.*, 1980), although the fabric of a sediment is a reflection of all manifestations of internal strain (Berthelsen, 1978; Benn, 2004a). This strain signature is just one of a number of characteristics that are used in combination to infer the genesis of glacigenic diamictons and to assess the evidence for subglacial deformation (Figure 4.10), but we need to review the analysis of strain signature using clast fabrics at this juncture in order to provide context for the following sections on subglacial sedimentary processes.

Particles in deforming media like subglacial tills and melanges, clearly preferentially orientate themselves so that their long (A) axes align parallel with the principal stress direction and A/B planes dip towards the stress direction or are imbricated (Figure 4.11), although many studies highlight dips that may vary according to the protuberances in the substrate such as bedrock bumps (e.g. Catto, 1990, 1998) or overridden sediment piles (e.g. Evans and Twigg, 2002), as has been explained by two alternative theoretical models (Figure 4.12). First, Jeffery (1922) proposed that particles roll continuously in response to velocity gradients in a shearing viscous medium, a process subsequently

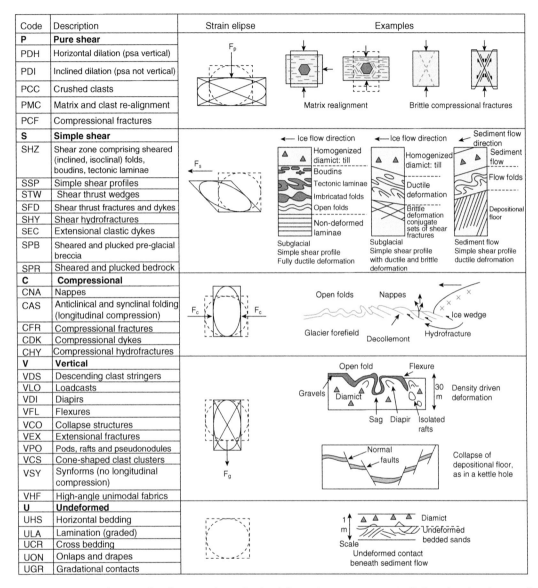

Figure 4.6 Comprehensive classification scheme for the deformation structures typically found in glacial materials (from McCarroll and Rijsdijk, 2003), showing the different forces on strain ellipses (*Fp*, *Fs*, *Fc* and *Fg*) as pure shear, simple shear, compressional and gravitational.

termed 'Jeffery-type rotation'. Second, March (1932) proposed that particles act as passive strain markers and rotate towards parallelism with the principal axis of extensional strain, a process now called 'March-type rotation'. The latter is consistent with a plastic rheology (see Section 4.3), where strain is accommodated by slippage between grains (Ildefonse and Mancktelow, 1993; Hooyer and Iverson, 2000). In terms of sediment properties, Jeffery-type rotation should produce weak fabrics, as many particle axes would align transverse or oblique to the flow direction. In contrast, March-type

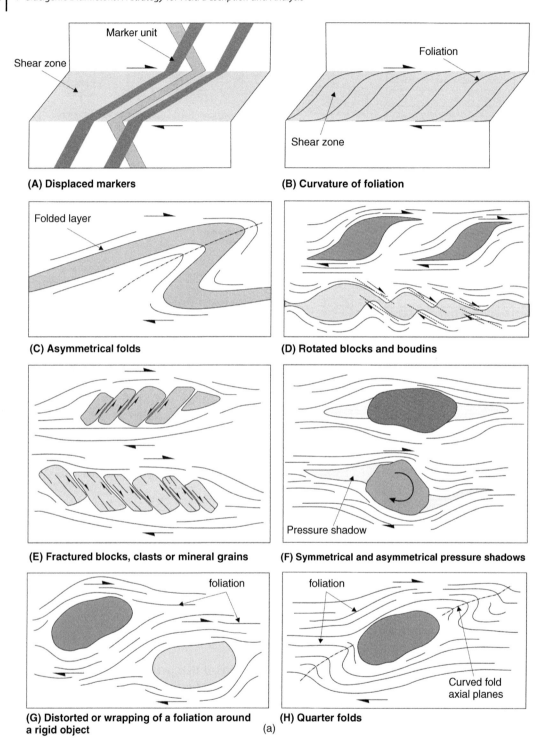

(a)

Figure 4.7 Styles of deformation fabrics and structures observed in glacially deformed materials (from van der Wateren *et al.*, 2000).

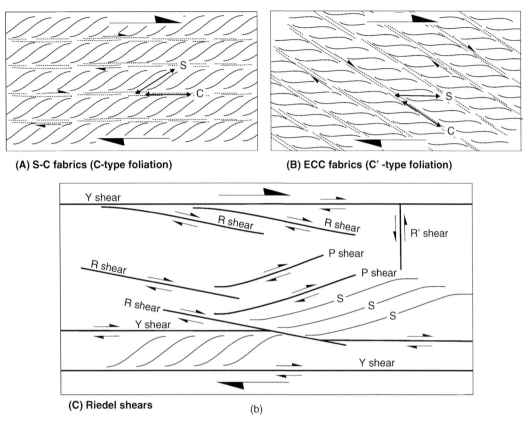

(A) S-C fabrics (C-type foliation)

(B) ECC fabrics (C' -type foliation)

(C) Riedel shears

(b)

Figure 4.7 (*Continued*)

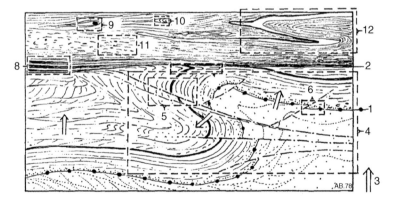

Figure 4.8 Glacitectonic signatures compiled by Berthelsen (1978): (1) boundary between an upper kineto-stratigraphic unit with domainal deformation and subjacent strata with extra-domainal deformation; (2) base of subglacial till/deforming layer; (3) way-up in glacifluvial sediments; (4) zone of structural investigation into glacitectonite; (5) overthrusts; (6) conjugate thrusts; (7) sub-sole drag; (8) lineations/tectonic laminae; (9) torpedo (boudin) structure; (10) intrafolial folds; (11) macrofabric; (12) macroscale glaciodynamic structures, visible only in pseudo-laminated till.

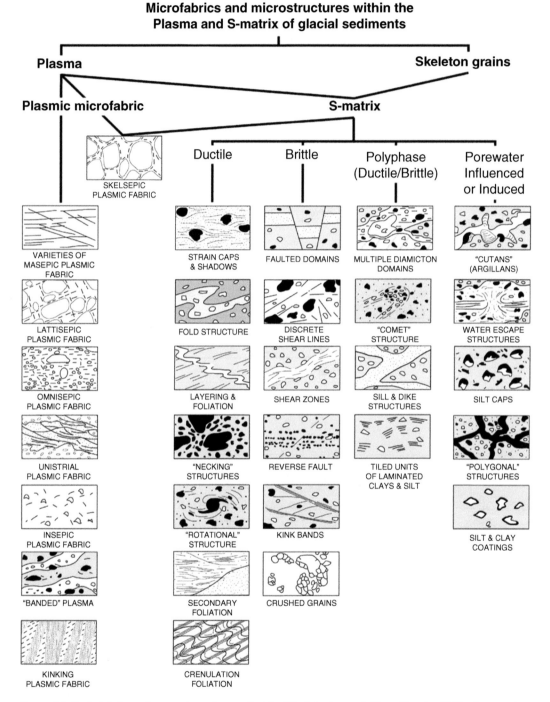

Figure 4.9 Typical deformation structures that can be identified at microscale (from Menzies, 2000, after van der Meer, 1993).

Figure 4.10 Typical directional elements that can be identified in glacial deposits (from Berthelsen, 1978): (1) striae on bedrock; (2) striae on clast pavement; (3) glacitectonic folds and faults; (4) clast macrofabric displaying ice flow-parallel (a) and transverse (b) and mixed (c) orientations; (5) lineations or tectonic laminae in sheared materials.

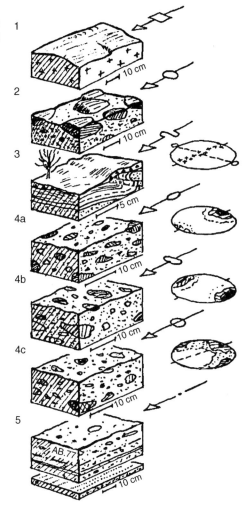

rotation produces strong or cluster fabrics, with a single mode parallel to the direction of shear. At the ice–bed interface, the assumption is that clast orientations will reflect the development of the strain ellipsoid within a typical simple shear zone (Figure 4.13). At outcrop, macrofabrics are routinely assessed by employing the measurement of the orientations of clast long (A) axes, the alignment of A/B planes, polished facets and stoss-lee forms (e.g. Hicock, 1991; Benn, 1994a, 1995; Li *et al.*, 2006; Piotrowski *et al.*, 2006; Evans *et al.*, 2007), and field techniques are detailed in Benn (2004b). Data are primarily displayed in rose diagrams or contoured stereonets (Figure 4.11), which provide visual impressions and primary statistics of macrofabric orientation and strength. It has become routine to further quantify fabric 'shape' using the relative magnitudes of the three derived eigenvalues S_1, S_2 and S_3 in ternary plots (cf. Mark, 1973, 1974; Woodcock, 1977; Woodcock and Naylor, 1983; Dowdeswell *et al.*, 1985; Dowdeswell and Sharp, 1986; Benn, 1994a, 2004b; Benn and Ringrose, 2001; Figure 4.14), thereby recognising: (i) isotropic fabrics, with no significant preferred orientation in any direction ($S_1 {\sim} S_2 {\sim} S_3$); (ii) girdle fabrics, with most clast orientations confined to a

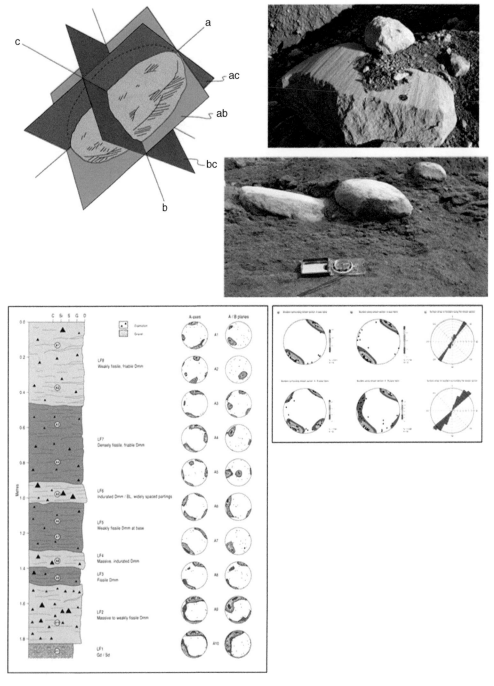

Figure 4.11 The principle of preferred clast macrofabric orientations in subglacial tills, whereby either A axes or A/B planes (defined in top-left sketch) orientate themselves parallel with the principal stress direction. Photographs show typically lodged boulders with striated upper facets in a multiple till sequence such as that depicted in the vertical profile log (from Evans *et al.* 2016). The macrofabrics measured using the orientations of the clast A axes and A/B planes of 50 clasts in each till are represented in contoured stereonets alongside the profile log. The macrofabrics and surface striae of the lodged boulders only are depicted in contoured stereonets and rose diagrams on the right, a procedure that isolates the lodged component from the deformation component of the tills.

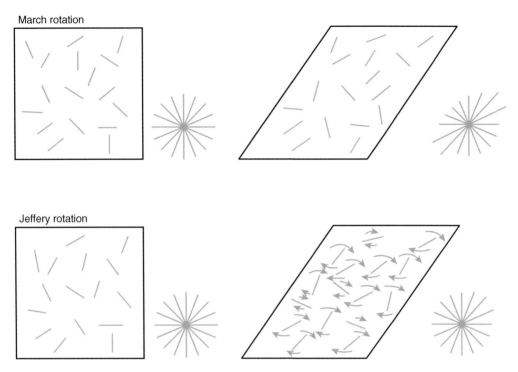

Figure 4.12 Schematic diagrams to compare Jeffery and March type models of particle orientation in a deforming medium. March type rotation involves particles rotating passively so that the fabric ellipsoid reflects the deformation ellipsoid. Jeffery type rotation involves particles continuously rolling so that the fabric ellipsoid may be more or less elongate than the deformation ellipsoid (from Benn and Evans, 1996).

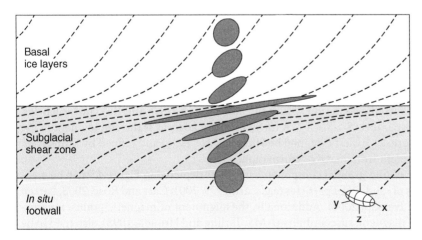

Figure 4.13 Conceptual diagram of the subglacial shear zone with typical strain ellipses and strain trajectories (from van der Wateren *et al.*, 2000).

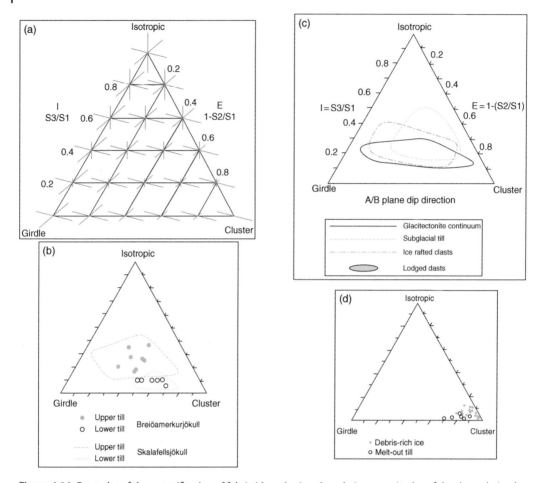

Figure 4.14 Examples of the quantification of fabric 'shape' using the relative magnitudes of the three derived eigenvalues S_1, S_2 and S_3 in ternary plots: (a) key to fabric shape triangles, in which sample populations are plotted according to their isotropy and elongation (from Benn, 1994b); (b) clast A-axis samples from two Iceland glaciers (after Benn and Evans, 1996); (c) clast A/B plane samples from glacitectonites, Icelandic subglacial tills, ice-rafted sediments and subglacially lodged clasts (from Evans *et al.*, 2007); (d) clast A axis samples from melt-out till and debris-rich ice (data from Lawson, 1979b).

plane but no significant preferred orientation within the plane ($S_1 \sim S_2 > S_3$); and (iii) cluster fabrics, with most clast orientations parallel ($S_1 > S_2 \sim S_3$). A further level of analysis is the employment of the modality/isotropy plot (Figure 4.15), first designed by Hicock *et al.* (1996) and refined by Evans *et al.* (2007) to isolate clast lodgement from matrix deformation.

At microscales, the fabric of fine-grained components in tills as depicted in Figure 4.9 can be measured using the orientations of sand-sized particles (e.g. Carr, 1999, 2001; Carr and Rose, 2003; Larsen *et al.*, 2006; Thomason and Iverson, 2006). Additionally, the alignment of magnetic grains, measurable using the anisotropy of magnetic susceptibility (AMS; Tarling and Hrouda, 1993), also tend to be aligned parallel to the direction of shear and commonly show close, but not always consistent, agreement with clast microfabric and macrofabric orientations (e.g. Boulton, 1976; Hooyer *et al.*, 2008; Shumway and Iverson, 2009; Gentoso *et al.*, 2012).

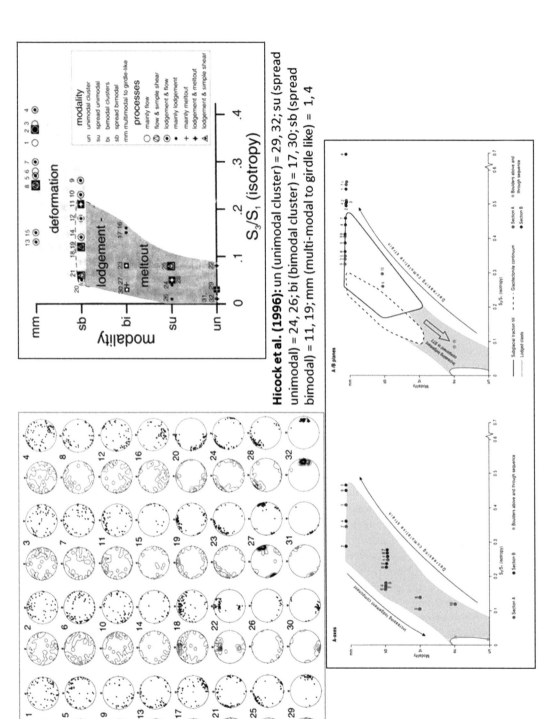

Figure 4.15 Quantification of fabric 'shape' using the modality/isotropy plot. Upper diagrams show stereonets and plot as first employed by Hicock et al. (1996). Lower diagram shows refined plot as proposed by Evans et al. (2007) in order to isolate clast lodgement from matrix deformation. Data plotted on these examples are A axes (left) and A/B planes (right) from Icelandic multiple till sequence of Evans et al. (2016), with comparative data for A/B planes from till, lodged clasts and glacitectonite from Evans et al. (2007).

5

Subglacial Sedimentary Processes: Origins of Till Matrix and Terminal Grade

Normal till is made up predominantly of materials which have not been transported many miles, though some of the minor constituents have often come greater distances. Roughly speaking, the more distant the contributing rock formation, the less its contribution to the till at any given point.

Salisbury (1900, p. 426).

The movement of particles through various pathways in the glacial debris cascade (Figure 1.2a) brings about recognisable clast form changes (Figure 1.2b), the most significant of which in terms of a unique glacial signature is the abrasion and breakage of particles during active transport in the traction zone (Boulton *et al.*, 1974). This produces a diagnostic grain-size distribution (Figure 5.1) which, despite exhibiting a wide range of particle sizes from clay or silt up to cobbles and boulders, are typically bimodal or polymodal. The earliest attempts to characterise till grain size or texture, for example, Krumbein (1933), Goldthwaite (1948) and Deane (1950), acknowledged the role of the glacial crushing process as well as the incorporation of pre-existing sediments in the production of till grain-size distributions, but it was not until the more systematic statistical analyses of, for example, Legget (1942), Dreimanis and Reavely (1953), Shepps (1953, 1958), Elson (1961) and Frye *et al.* (1969) that a consistent pattern in till grain size was recognised (cf. Yi Chaolu, 1997). We now understand that the broadly diagnostic till grain-size distribution reflects the progressive reduction or 'comminution' of particles as they are fractured during shear (e.g. Hooke and Iverson, 1995). This comminution grain-size signature has been interpreted by Boulton (1978) and Haldorsen (1981) as the product of two processes in combination: (i) 'crushing', which is the breakage of two interlocking grains and (ii) 'abrasion', which is the production of fine-grained fragments as two grains move past each other. Laboratory experiments by Haldorsen (1981) indicated that crushing produces particle sizes of $0.016-2$ mm (6 to -1 Φ or medium silt to very coarse sand) and abrasion produces fragments of $0.002-0.063$ mm ($9-4$ Φ or silt). These particle size modes are visible in both modern-day glacial debris loads (Lawson, 1979a; Figure 5.1a) and tills at outcrop (Haldorsen, 1981; Figure 5.1b). However, natural tills also display an important third mode (initially recognised by Beaumont, 1971) coarser than 4 mm (-2 Φ or pebbles), which represents the 'residual clast mode' or the rock fragments which have not been reduced to the optimum comminution product but eventually would be, given further abrasion and crushing associated with longer transport distances.

The importance of transport distance on matrix maturity was quantified in a benchmark study by Dreimanis and Vagners (1971), who assessed the grain-size changes in subglacial debris with distance from known bedrock sources (Figure 5.1c). They typically found that the coarse, residual clast mode was dominant close to the debris source (0–3 km transport) and that the finer modes increased in importance with increasing transport distances (75–500 km). Significantly, the data showed that there is an apparent lower size limit beyond which no further comminution takes place, regardless of the transport distance, a limit that they called the 'terminal grade'. This is equivalent to the 'limit of grindability' and can be partially explained by Griffith crack theory, in that the energy required for crack growth is inversely proportional to crack length; therefore, the fracture of small particles requires the growth of short cracks which requires greater energy than the growth of relatively long cracks in large particles. Unsurprisingly, the grain size of terminal grades varies according to bedrock type, well demonstrated by the early work of Elson (1961; Figure 5.1d), so that softer rock types will ultimately produce smaller terminal grades given sufficient transport distances.

It is important to understand at this juncture that grain-size distributions in tills often do not reflect fragment liberation from source outcrops and gradual comminution over long distances but rather

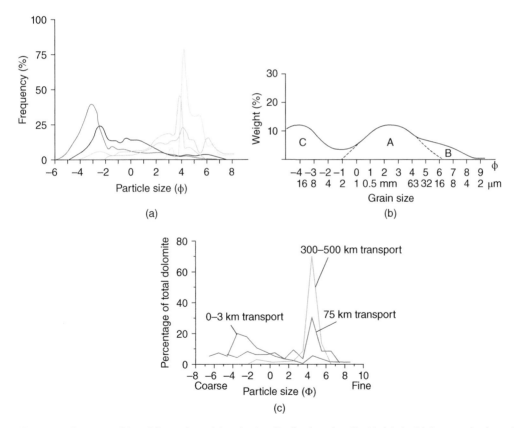

(a)

(b)

(c)

Figure 5.1 Diagnostic bimodal or polymodal grain-size distributions for tills; (a) debris-rich ice samples from the Matanuska Glacier, Alaska (Lawson, 1979a); (b) mean grain-size distribution of 150 till samples with areas of the graph classified as resistant crushed fraction (A), abrasion component (B) and residual component (C) by Haldorsen (1981); (c) frequency distribution of dolomite in till samples over transport distance (from Dreimanis and Vagners, 1971); (d) illustration of variation of terminal grades according to bedrock type (from Elson, 1961).

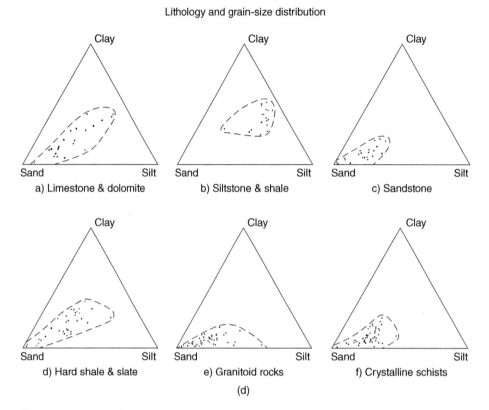

Figure 5.1 (*Continued*)

Figure 5.2 Details of quartz sand grain microscopy from glacial deposits: (i) schematic diagram showing styles of grain fracture (from Hiemstra and van der Meer, 1997, after Brzesowsky, 1995), where dashed lines define the centre lines of the grains in contact, arrows represent contact forces, and shaded areas are deformed contact regions. The styles include (a) diametrical loading at low compression rates with ring cracks passing into divergent cone cracks, (b) diametrical loading at high compression rates with radial fractures propagating as meridional cracks, (c) tangential loading at high angles of incidence with frictional sliding leading to the chipping of flakes and (d) tangential loading at low angles of incidence with attrition leading to crushing of all or part of a grain; (ii) photomicrograph illustrating crushing of a grain due to tangential loading at the centre of the image (from Hiemstra and van der Meer, 1997); (iii) typical glacially diagnostic quartz grain morphological features (from Sharp and Gomez, 1986) including (A) angular outline, (B) high-relief surface, (C) conchoidal breakage patterns, (D) stepped surface, (E) breakage blocks, (F) edge abrasion; (iv) typical grain surface textures from subglacial till (from Hart, 2006), including (A) pre-erosional surface (P) and smoothing and rounding, (B) pre-erosional surface (P) on one face with medium impacts, (C) pre-erosional surface (P) at base of 'hand-axe' form; (D) 'jagged' form, (E) smoothed conchoidal form, (F) conchoidal form. Mechanical features are numbered as: (1) large conchoidal fracture, (2) small conchoidal fracture, (3) arcuate steps, (4) straight steps, (5) crescentic gouges, (6) large breakage blocks, (7) fractured plates, (8) sub-parallel linear fractures, (9) curved grooves, (10) straight grooves.

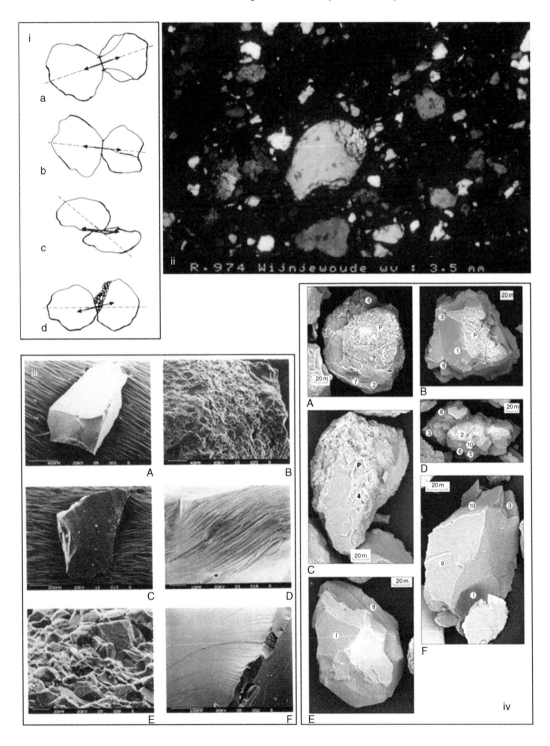

constitute a signature of cannibalisation of pre-existing sediments, a concept recognised by early workers such as Lamplugh (1911). In such cases, common in lowland settings where extensive and thick sediment piles accumulate between and during glacial advances, grain-size distributions are inherited from other glacial or non-glacial processes. Any subsequent shearing of such material can then modify the original grain-size distribution by particle comminution or sediment mixing.

Since the pioneering studies of Krinsley and Smalley (1972), Krinsley and Doornkamp (1973) and Whalley and Krinsley (1974), the impacts of crushing and abrasion within tills has been well documented at the microscale through observations on quartz grains using scanning electron microscopy (SEM) and thin-section micromorphology. SEM samples collected from subglacial tills (e.g. Eyles, 1978; Whalley, 1978; Whalley and Langway, 1980; Dowdeswell, 1982; Sharp and Gomez, 1986; Mahaney *et al.*, 1988, 2001; Mahaney, 1995; Mahaney and Kalm, 2000; Hart, 2006) contain quartz grains, which typically display features regarded as diagnostic of the comminution process, specifically related to crushing and abrasion. These include conchoidal fractures, sharp angular edges/outlines, stepped surfaces, edge rounding, breakage blocks, high-relief surfaces and pre-erosional surfaces between freshly removed fragments (Figure 5.2). Fractured quartz grains are also visible in thin sections collected from the matrix of subglacial tills, as illustrated by Hiemstra and van der Meer, 1997; Figure 5.2ii). They applied the studies of Brzesowsky (1995) to derive a range of fracture types that are observable within till matrices (Figure 5.2i).

6

Subglacial Sedimentary Processes: Modern Observations on Till Evolution

Without a solid empirical basis, even the most persuasive models are more akin to creation myths than products of scientific rigour.

Benn (2006, p. 437)

The three processes that are widely acknowledged as being involved in the development of subglacial till are lodgement, deformation and melt-out. Each of these processes has been observed in modern glacial systems and so it is pertinent at this stage to review the reconstructed process–form regimes derived from such observations and consequently for which we have confidence in moving forward in our process sedimentology for tills. Previous till classification schemes have strongly adhered to genetically related terms such as 'lodgement till', 'deformation till', 'deformed lodgement till' and 'melt-out', but it is important at this stage to resist such specific labelling because, as will be argued below, the processes can be neither separated operationally nor identified confidently in the geological record. Comprehensive reviews of our present state of knowledge of subglacial processes based on subglacial observation, laboratory experiment and physical theory are available elsewhere (van der Veen, 1999; Alley, 2000; Clarke, 2005; Benn and Evans, 2010; Iverson, 2010; Cuffey and Paterson, 2010), hence what follows is a brief overview of process observations as they pertain directly to till genesis and sedimentology.

6.1 Lodgement, Lee-Side Cavity Filling and Ploughing

As discussed in Chapter 2, the term 'lodge' was introduced by Upham (1892) and the term 'lodgement till' by Flint (1957), but the process was defined in detail by Boulton (1974) when he proposed a 'critical lodgement index' to describe the process whereby elevated effective pressures lead to increasing frictional resistance and, as a result, grain-by-grain lodgement at the sliding ice–bed interface. This was controlled by the frictional resistance acting against a particle in traction and moving over a hard or rigid bed, but Boulton (1975, 1976, 1982) also identified the process of clast ploughing (see also Boulton *et al.*, 1979; Clark and Hansel, 1989; Jørgensen and Piotrowski, 2003) and resulting lodgement and clast clustering on a soft or deformable bed (Figure 6.1; see Section 6.3). The lodgement process was further elaborated by Boulton (1982; Figure 6.1) when he proposed that whole sediment bodies could be plastered on to a rigid bed if they were part of the debris-rich ice facies of the overlying glacier; because the emplacement and release of the ice–debris mixture ultimately involves deformation, lodgement and the melt-out of the enclosing ice, potentially with the processes

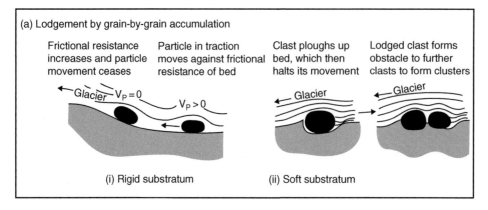

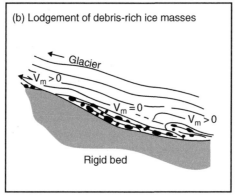

Figure 6.1 Boulton's (1982) depiction of the lodgement process: (a) particles lodging by (i) frictional retardation against a rigid bed and (ii) against obstacles ploughing on a soft bed; (b) lodgement of debris-rich ice masses, with a whole assemblage lodging against the bed and melting out in situ.

operating intermittently and in tandem prior to deposition, the till is in reality a product of lodgement, deformation and melt-out, an issue that we will revisit in Chapter 16.

Stemming from Boulton's (1974, 1975, 1979) subglacial observations, the critical role of friction in the lodgement of particles on a rigid bed has been further evaluated by the efforts of Bernard Hallet (1979, 1981) and Neil Iverson and colleagues (Iverson *et al.*, 2003; Cohen *et al.*, 2005; Iverson *et al.*, 2007). A Mohr–Coulomb equation was employed by Boulton (1974, 1975, 1979), in which the basal drag was assumed to be a function of the effective pressure and the fraction of the bed covered with debris:

$$\tau_b = \tan \Phi A_F N$$

where τ_b is basal drag, $\tan \Phi$ is the friction coefficient, N the effective pressure and A_F the fraction of the bed covered with debris, a modified version of which was developed by Schweizer and Iken (1992) and subsequently termed the 'sandpaper friction' model. The problem with such a Coulomb-type model is that it is strictly only applicable to the friction developed between rigid bodies rather than one rigid surface, like a rock bed, and a deformable surface like ice, which tends to flow around particles at the ice–bed interface; this enhanced deformation of ice around particles at the bed was acknowledged by Boulton (1975) when he proposed that larger (>10 cm diameter) clasts were preferentially subject to lodgement. Hence, Hallet (1979, 1981) argued that the contact forces at the bed will be independent of the effective pressure and therefore the contact force between a particle and the bed is the sum of the buoyant weight of the particle and the drag force resulting from ice flow towards the bed. Because the buoyant weight is the weight of the particle minus the weight of the same volume of ice, it is proportional to the mass of the particle and the difference between the densities of ice and rock. The ice flow towards the bed produces an additional contact force between bed and particle and the faster the ice flow towards the bed, the higher the contact force, and hence the higher the resulting drag. This 'Hallet model' accounts for the likelihood that ice flow towards the bed will result from a combination of ice melting due to geothermal and frictional heat and vertical ice strain and thereby predicts that high friction between a particle and a resistant bed will occur below large, heavy particles and/or where basal melting rates are high and ice is straining rapidly towards the bed. More recently, subglacial experiments at Svartisen, Norway, by Neil Iverson's team (Iverson *et al.*, 2003; Cohen *et al.*, 2005; Iverson *et al.*, 2007) have demonstrated that frictional drag or 'shear traction' is indeed strongly positively correlated with effective normal pressure, as proposed by Boulton (1974, 1975, 1979), which increases at times of low subglacial water pressure (Figure 6.2). Although these findings were consistent with the Coulomb-type model, the levels of frictional drag that were measured were an order of magnitude higher than those predicted by the Mohr–Coulomb equation above. This prompted Cohen *et al.* (2005) to propose a new theory of subglacial friction which accounts for the effects of both the effective pressure and ice flow towards the bed, a theory that also incorporates the important role of water-filled cavities on the lee side of particles which have been observed in laboratory experiments by Iverson (1993). We will return to the role of water-filled cavities and water films in till genesis in Section 6.3.

Following on from the subglacial observations reported by Boulton (1970a, b, 1974, 1975, 1976, 1982), *lodgement* was defined as 'the plastering of glacial debris from the base of a sliding glacier on to a rigid or semi-rigid bed by pressure melting and/or other mechanical processes' (Dreimanis, 1989).

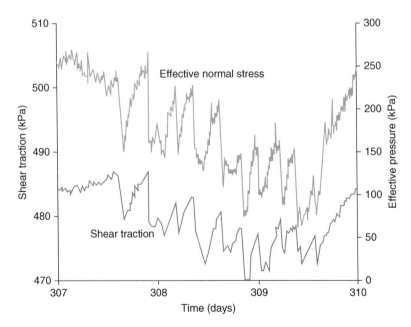

Figure 6.2 The relationship between shear traction and effective normal stress over time as measured during experiments in the Svartisen subglacial laboratory by Cohen *et al.* (2005).

Processes observed in subglacial cavities also alerted glaciologists to the intimate relationship between lodgement, melt-out and deformation in the infilling and smoothing of uneven hard glacier beds. For example, Boulton (1982) compiled a range of subglacial observations to summarise the processes of cavity filling and the subsequent development of streamlined landforms (Figures 6.3 and 6.4). Typically cavities filled up with fine debris slurries and individual particles falling from melting debris-rich ice (Boulton, 1970a), clasts derived from expulsion due to a sudden reduction in frictional retardation of the basal ice (Peterson, 1970; Vivian, 1976), debris-rich ice 'curls' created by the separation of debris-rich and debris-poor ice (Peterson, 1970; Rea and Whalley, 1994), and glacifluvial sediments washed in by meltwater migrating from the up-ice sides of the bedrock protrusions. Such crudely stratified deposits have been classified as 'flow tills', 'lee-side cavity fills' or 'lee-side tills' (Boulton, 1971; Hillefors, 1973; Haldorsen, 1982; see Section 9.4), but Boulton (1971) regarded the cavity filling as part of the broader process of bed smoothing by lodgement, a process association observed in a cavity beneath the Salieckna Glacier in Sweden (Boulton, 1982; Figure 6.4) where the glacier was in partial contact with the till-capped infill and had been streamlining its upper surface. Even when they do not produce cavities, undulatory glacier beds can initiate preferential till sedimentation in low areas and thereby reduce bed roughness by the processes of basal debris-rich ice thickening (Boulton, 1975; Figure 6.5a) and concentrated heat flow and concomitant increased basal ice melt rates in depressions (Nobles and Weertman, 1971; Figure 6.5b).

Boulton's (1974, 1975) observation-based theory on lodgement acknowledged that it was specific to individual particles and that the 'lodgement' of a wider range of grain sizes required the basal debris to be:

> a rigid frictional mass in which the tractive force produced by the ice velocity over its top surface is balanced by the frictional drag along the bottom surface.
>
> Boulton (1975, p. 22)

At this time, it was emerging that such a sediment body would be subjected to deformation (Boulton *et al.*, 1974; see Section 6.2), but its structureless appearance prompted Boulton (1975) to conclude that a deformational origin was difficult to infer, even though he clearly depicted crosscutting thrust faults that had been superimposed on 'lodged' till beneath Nordenskjoldbreen, Svalbard (Boulton, 1970a); this style of shearing at the ice–bed interface was later employed by Eyles and Boyce (1998) in their comparison of the subglacial deforming layer with fault gouge (cf. Pettijohn's (1957) 'extreme gouge' deposits; Figure 6.6). As outlined earlier in this section, deformation was also implicit in Boulton's (1975, 1976, 1982) clast ploughing and lodgement on to a soft bed, whereby the ploughing of the deformable substrate forms a prow that then arrests forward momentum and preferentially lodges the largest clasts (Boulton, 1982; Clark and Hansel, 1989; Jørgensen and Piotrowski, 2003). Schemes developed since the 1960s (Elson, 1961) that advocated the separation of subglacial till types based upon specific processes of lodgement versus deformation were therefore inherently untenable because all subglacial tills had been deformed and lodged (Ruszczynska-Sjenach, 2001; Evans *et al.*, 2006b).

Based upon these observations, the lodged component of a subglacial till comprises deforming sediment (matrix and smaller clasts) once it stops deforming and individual, larger clasts (cobbles and boulders) arrested at the bed by frictional retardation. Importantly, the subglacial observations made at Breiðamerkurjökull indicated to Boulton *et al.* (2001) that large flute-forming clasts were effectively lodged in the more stable lower till layers below the deforming bed. The interplay between ploughed/lodged larger clasts and a mixture of deforming matrix with smaller clasts beneath a soft-bedded glacier is well illustrated by the construction of cavity-infill type flutings where stoss

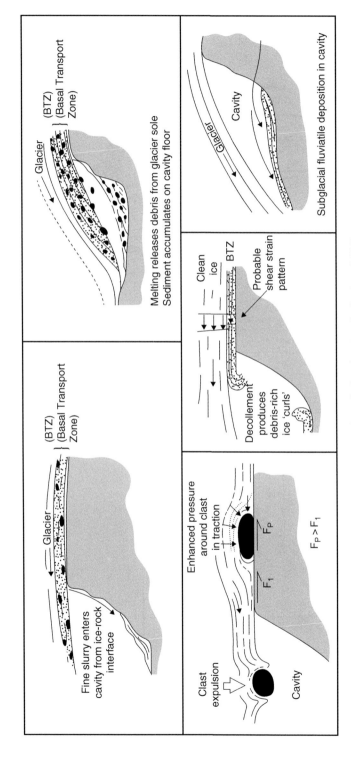

Figure 6.3 Mechanisms of debris release and accumulation observed in subglacial cavities by Boulton (1982).

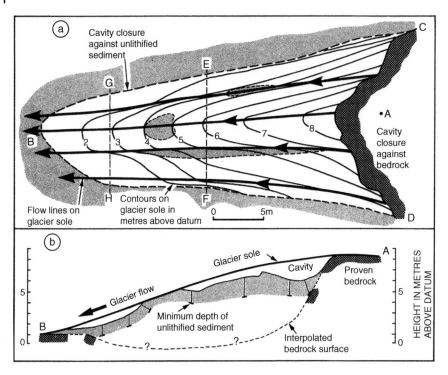

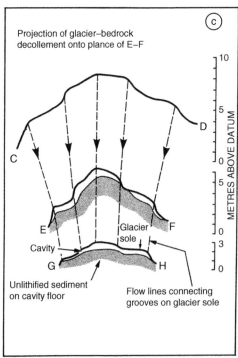

Figure 6.4 Observations on the till infilling of a subglacial cavity on hard bedrock beneath the Salieckna Glacier, Sweden by Boulton (1975).

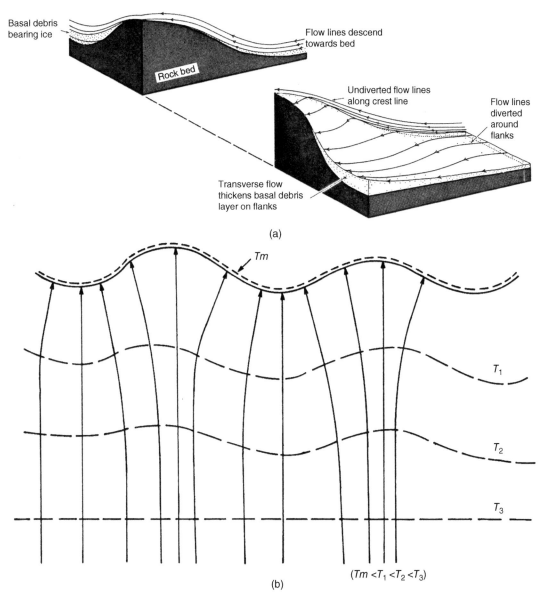

Figure 6.5 Models of preferential till sedimentation in low areas of a glacier bed by the processes of: (a) basal debris-rich ice thickening (Boulton, 1975); and (b) concentrated heat flow and elevated basal ice melt rates in depressions (Nobles and Weertman, 1971).

clasts lie at the ridge origin (Boulton, 1976; Benn, 1994b; Figure 6.7), the details of which we will investigate in Chapter 10. The occurrence of parallel flutings with and without stoss clasts suggests that at least some features are created by grooving of the substrate by the passage of large clasts embedded in the glacier sole before they become lodged, a process that is very likely significant in terms of subglacial till deformation styles and fabric development (see Section 6.2) and which Iverson (1999) proposes helps to couple the glacier to its bed by increasing roughness. The process

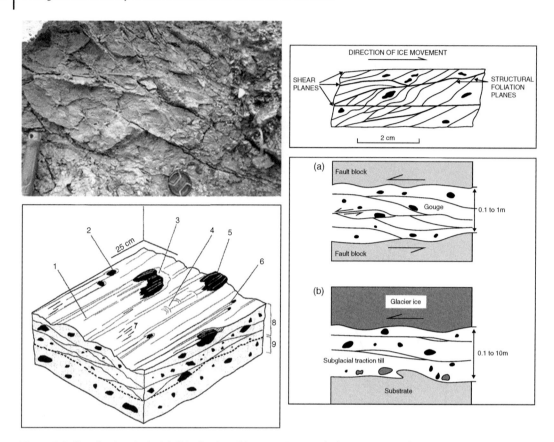

Figure 6.6 Shearing in subglacial till indicative of its operation as a fault gouge: top left – recent photograph and top right – Boulton's (1970a) sketch of cross-cutting thrust faults superimposed on 'lodged' till beneath Nordenskjoldbreen, Svalbard; lower left and right – depictions of the fault gouge as it pertains to subglacial materials (from Eyles and Boyce, 1998). Numbers in lower left are as follows: (1) flute ridge in lee of large clasts, (2) abraded clast, (3) crescentic groove on stoss side of striated clast, (4) crescentic fractures; (5) striated clast, (6) ridge-in-groove structure, (7) nail head striation, (8) gouge diamict with slip planes, (9) undisturbed footwall strata. The schematic comparison in lower right shows diamict production between fault blocks in (a) and between a glacier and its substrate in (b).

of clast ploughing has been well demonstrated inadvertently by Fischer *et al.* (2001) by their plough-meter experiments beneath Unteraargletscher, Switzerland. Even the design of ploughmeters has developed out of a subglacial experiment that went wrong, specifically the jamming of a drill stem in till below the Columbia Glacier, Alaska, and its subsequent dragging through the bed for five days before it was recovered in a bent and striated state (Humphrey *et al.*, 1993); this demonstrated that the uppermost 0.65 m of the bed was deformable sediment (see Section 6.2). Intentionally inserting a ploughmeter to investigate ice–sediment coupling, Fischer *et al.* (2001) found that the till in front of the ploughmeter was substantially weakened due its compression and concomitant increase in porewater pressure because the water could not drain away effectively (Figure 6.8a). They proposed that a glacier sliding over a soft bed would therefore move by ploughing so that the basal motion was concentrated near the ice–bed interface rather than at depth in the till (cf. Iverson *et al.*, 1994; Iverson, 1999). In terms of the lodgement process, it is clear that ploughing clasts need to be arrested by the deforming layer at some stage, presumably when water can drain away from the clast prow

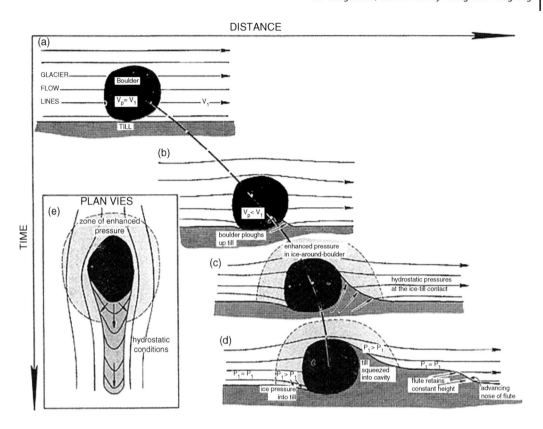

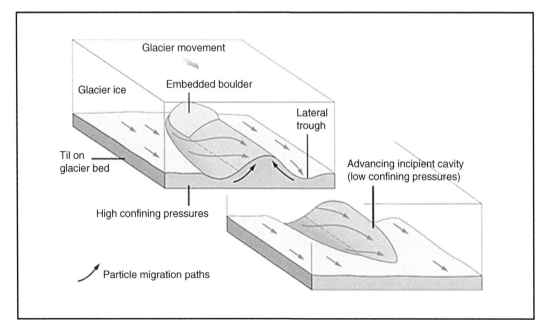

Figure 6.7 The relationships between a ploughed or lodged clast and the till matrix that propagates down flow to form a fluting. Upper diagram is Boulton's (1976) theoretical analysis of the process and lower diagram is from Benn (1994b).

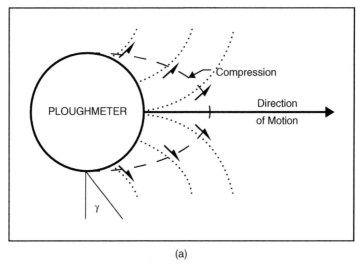

(a)

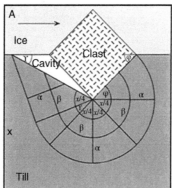

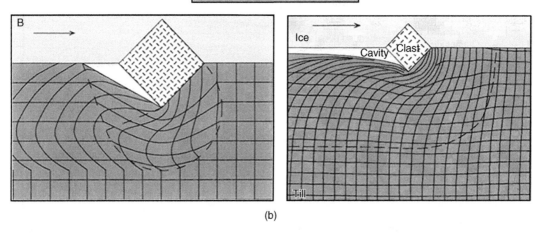

(b)

Figure 6.8 Observations from experiments on ploughing clasts: (a) the pattern of compressional impact of a ploughmeter on subglacial till (from Fischer *et al.*, 2001); (b) theoretical and experimental results from the assessment of clast ploughing. Box A shows Tulaczyk's (1999) application of Baligh's (1972) experiments on wedges dragged through soils, depicted here as an angular clast ploughing through perfectly plastic till. Symbols α and β are two slip lines, φ is the leading angle of the ploughing clast, and Υ is the trailing angle of the cavity developed behind the clast. In B, the deformation is measured by the distortions in an initially square grid, which is depicted in the right box from a photograph of the outcome by an experiment in which a wedge was ploughed through a homogenous clay.

and thereby reduce localised porewater pressures. The role of a ploughing clast in the deformation of tills has been modelled by Tulaczyk (1999) using previous investigations by Baligh (1972) into the impacts of wedges dragged along soil surfaces. The application of Baligh's (1972) experiments by Tulaczyk (1999) to a subglacial scenario (Figure 6.8b), indicates that a coarse-grained till will be subject to deformation that is distributed to depths several times greater than the depth to which the clast protrudes from the ice base. In thin, accreting till layers this disturbance will be considerable and has the potential to locally disturb or reorganise previously developed clast macrofabrics.

6.2 Deformation

As discussed in Chapter 2, it was the reporting in the late 1970s of the subglacial experiments undertaken at Breiðamerkurjökull, Iceland, by Geoffrey Boulton that placed till deformation at the centre of studies on till genesis. Since that time, glaciologists have been striving to deliver a flow law for till, the outcome of which has huge implications for the interpretation of ancient till genesis. Recognition that glacier beds could deform at relatively low stresses was not just down to the Breiðamerkurjökull work of Boulton (1979) Boulton and Jones (1979) and Boulton and Hindmarsh (1987) but also the discovery of saturated, weak sediment beneath Ice Stream B (Whillans Ice Stream) in West Antarctica by Alley *et al.* (1986, 1987a, b) and Blankenship *et al.* (1986, 1987). The Breiðamerkurjökull experiments, because they initiated the development of the 'Boulton–Hindmarsh model' of till deformation, deserve some detailed description here. In separate field campaigns in 1977, 1977/1978, 1980 and 1983, unfrozen till beneath the glacier snout was accessed via tunnels excavated into the ice margin in an attempt to understand the processes that produced a two-tiered till structure in the area, comprising an upper A horizon with a high void ratio overlying a more consolidated lower B horizon of denser, platy structure (Figure 6.9). The most celebrated results are those collected during the 1980 experiment when the till was accessed via boreholes through the tunnel floor and segmented rods and water pressure sensors (piezometers) were inserted into the till at various depths (Figure 6.10). Ice motion was measured at the same time using markers in the tunnel and on the glacier surface and basal motion was measured using a wire spool located in the tunnel and attached to an anchor emplaced in the deeper layers of the till. After 136 hours, water was pumped away from the bed via an access hole and a section excavated in the till to observe the new locations of the segmented rods. These had been displaced down glacier by varying amounts, and hence were effectively strain markers that recorded deformation of a layer of till approximately 0.5 m thick. The porewater pressure and velocity data were strongly positively correlated, but with velocity peaks lagging porewater peaks by a few hours; it was assumed that velocity peaks equate to peak strain rates in the deforming till. From these data came the first subglacial till flow law of Boulton and Hindmarsh (1987):

$$\dot{\varepsilon} = K(\tau - \tau_{\text{yield}})^a N^{-b} (\tau > \tau_{\text{yield}})$$

where K, a and b are empirically fitted constants. This states that the strain rate $\dot{\varepsilon}$ rises as the shear stress τ becomes increasingly greater than the yield strength τ_{yield} and decreases with the effective pressure N. The most important implication of these findings was that till viscosity was almost independent of strain rate and hence the till appeared to exhibit either a slightly non-linear viscous or Bingham-type viscous rheology, a flow law that was then widely applied to tills (e.g. Alley *et al.*, 1987a, b; Boulton, 1987; Hart *et al.*, 1990; Hicock and Dreimanis, 1992a, b; Hart, 1995; Benn and Evans, 1996; Hindmarsh, 1998a; Fowler, 2000, Hart and Rose, 2001).

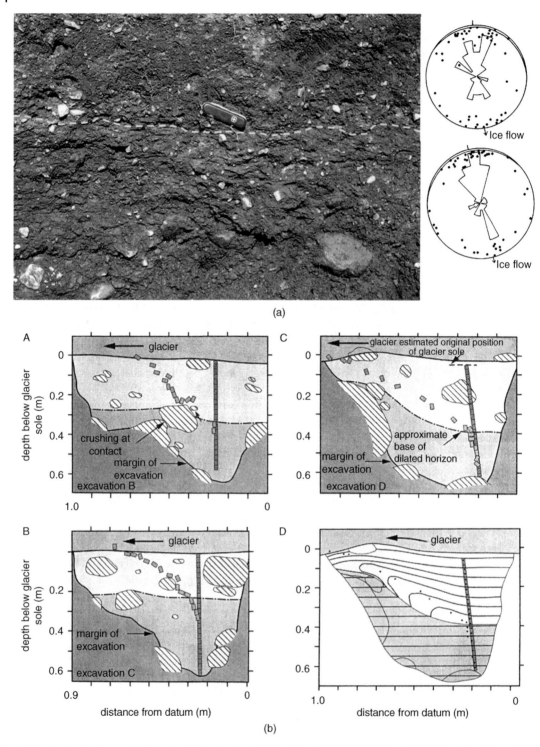

Figure 6.9 The subglacial till at Breiðamerkurjökull: (a) two-tiered till structure, comprising upper A horizon above pen knife overlying lower B horizon, with clast macrofabrics plotted on stereonets with rose plots (from Benn, 1995); (b) the pattern of strain as shown by the displacement of strain markers at three of the excavated locations (*A–C*) from the subglacial experiments (from Boulton and Dobbie, 1998). The sketch depicted in (*D*) shows how horizontally bedded sediments would have been deformed based upon the displacement patterns from the strain markers.

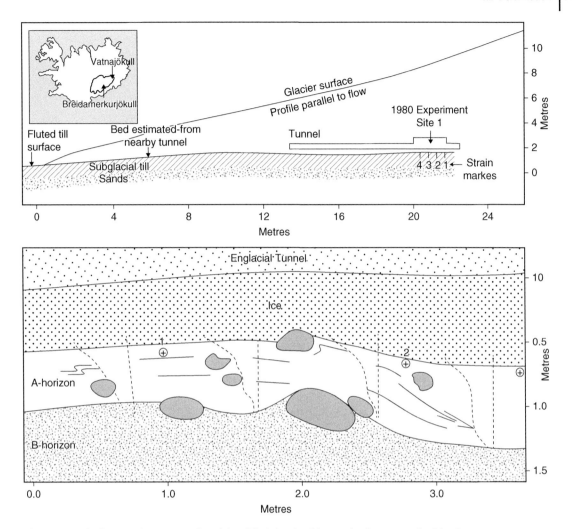

Figure 6.10 The location (upper panel) and simplified sketch of the results (lower panel) of the first Breiðamerkurjökull subglacial till experiment (from Boulton and Hindmarsh, 1987). The strain markers are depicted by broken lines and change from their original vertical alignment to a down flow displaced curve after 136 hours. The cross in circle symbols 1–3 are pore pressure gauge locations.

Although problems became evident with the representativeness of the Breiðamerkurjökull till flow law, specifically because the basal shear stress values at the experiment site near the glacier snout are not a simple function of local ice thickness and gradient, and also because later laboratory and borehole tests revealed a different and more complex till rheology (see below), the pattern of vertical increase in deformation through the upper till layers was replicated. Most significant in this respect were later experiments at Breiðamerkurjökull, where in 1988/89 Boulton and his colleagues instrumented a till prior to it being overrun by a mini-surge at the eastern glacier margin and emplaced an array of spool anchors beneath the glacier via a borehole, allowing the simultaneous measurement of strain rate profiles within the till and sliding at the ice–till interface (Boulton and Dobbie, 1998;

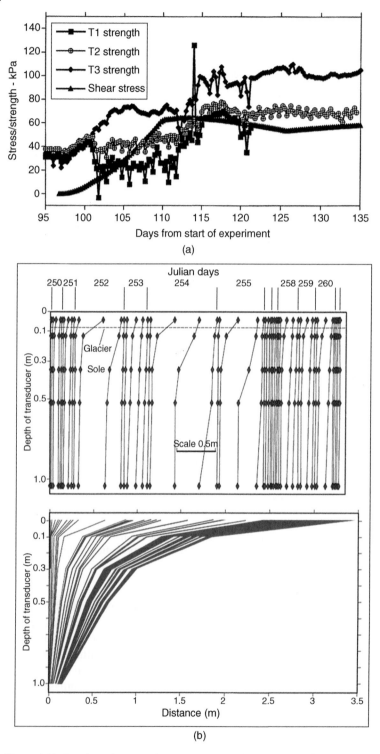

Figure 6.11 Data from the second Breiðamerkurjökull experiment of 1988/89, reported by Boulton and Dobbie (1998), Boulton *et al.* (2001) and Boulton (2006): (a) relationships between till shear strength at depths of 0.6 m (T1), 1.75 m (T2) and 2.5 m (T3) and shear stress; (b) time-dependent pattern of till strain inferred from the drag spool records; (c) proportion of movement of the glacier sole due to sliding versus till deformation.

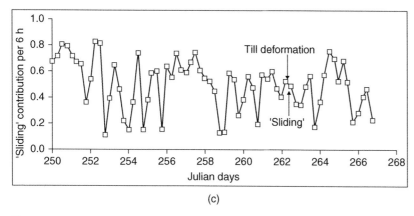

(c)

Figure 6.11 (*Continued*)

Boulton *et al.*, 2001; Boulton, 2006). In summary of these experiments, we can view the data depicted in Figure 6.11.

The relationships between till shear strength at depths of 0.6 m (T1), 1.75 m (T2) and 2.5 m (T3) and shear stress are depicted in Figure 6.11a, which is a record of changing till behaviour owing to loading during the mini-surge. Boulton *et al.* (2001) interpreted these trends as responses by groundwater flow over time, defined by four successive phases: (1) days 97–101, when there was a strong upward component of groundwater flow and hence a tendency for failure at the base of the till and top of the underlying gravels; (2) days 102–105, when groundwater flow reversed and failure tended to be at the top of the till; (3) days 106–114, when rapid ice advance created undrained loading and weakened the sediment so that it failed throughout to a depth of 2–3 m; (4) days 115–140, when the rate of loading decreased and the shear stress stabilised. In summary, the upper part of the till failed continuously from days 102 to 115 and sporadically from day 115 onwards, the lower part of the till failed from days 107 to 115, and the upper part of the sands and gravels failed during the period of high water pressure on days 112–113.

The time-dependent pattern of till strain inferred from the drag spool records is depicted in Figure 6.11b, interpreted by Boulton and Dobbie (1998), Boulton *et al.* (2001) and Boulton (2006) to be representative of most displacement over time taking place at or close to the ice–till interface, with intermittent deformation events at deeper levels so that cumulative displacement in the deforming till decreased with depth. This pattern of deformation is consistent with the predictions of groundwater behaviour indicated in Figure 6.11a.

The proportion of movement of the glacier sole due to sliding versus till deformation is depicted in Figure 6.11c, derived from Boulton and Dobbie's (1998) borehole experiments. Using strain markers at the glacier base (T_0) and at depths of 0.1 m ($T_{0.1}$), 0.3 m ($T_{0.3}$), 0.5 m ($T_{0.5}$) and 1.0 m (T_1) in the till, the proportion of sliding was calculated using $T_0 - T_{0.1}$ and the proportion of deformation by using $T_{0.1} - T_{0.3}$. From this, the variability in deformation in relation to changing effective pressure could be charted for every 6-hour period. The implications of this change in conditions at the ice–bed interface for till sedimentology will be explored further in Section 6.3 and Chapter 11.

In Section 6.1, results of subglacial experiments on deforming beds were briefly introduced through the discussion of Humphrey *et al.*'s (1993) discovery of till deformation down to a depth of 0.65 m when their drill stem was dragged along below the Columbia Glacier. A number of such observations, in addition to those reported above for Breiðamerkurjökull, are now available and provide a quantitative foundation upon which we can build an understanding of process–form regimes for

Table 6.1 Examples of the evidence reported from modern glaciers and ice sheets for deforming glacier beds.

Source	Location and nature of evidence	Nature of deformation
Boulton (1979), Boulton and Hindmarsh (1987)	Sub-marginal tunnel at Breiðamerkurjökull, Iceland – strain markers in two-tiered till with dilated upper A horizon (porosity 0.4) and compact, sheared lower B horizon (porosity 0.2)	Pervasive in A horizon (up to 0.45 m thick) and brittle shear in B horizon = non-linear strain distribution
Clarke *et al.* (1984), Blake (1992), Blake *et al.* (1992)	Boreholes in Trapridge Glacier, Yukon – double bend in flexible rod inserted into bed materials (effective pressure 78–292 kPa)	Pervasive(?) in top 0.3 m (subglacial sediment is 0.1–1.0 m thick)
Blankenship *et al.* (1986, 1987), Alley *et al.* (1986, 1987a), Engelhardt *et al.* (1990), Rooney *et al.* (1987), Atre and Bentley (1993, 1994)	Base of Ice Stream B – saturated and highly dilated sediment (porosity 0.4 and effective pressure 50 kPa)	Actively deforming but may not be pervasive? (5–6 m thick and continuous for 8.3 km across glacier and identified at a number of sites separated by 300 km)
Echelmeyer and Wang (1987)	Sub-marginal tunnel in Urumqi Glacier No. 1, China	Deformation of 36-cm-thick layer of frozen sediment = 60% of glacier motion
Engelhardt *et al.* (1990)	Ice Stream B – high porosity till	Till deforms down to 2 m
MacAyeal (1992), Alley (1993), Anandakrishnan and Alley (1994), MacAyeal *et al.* (1995)	Antarctic ice streams	Sticky spots = bedrock high, till discontinuity and/or till inhomogeneity
Humphrey *et al.* (1993)	Borehole through Columbia Glacier, Alaska – drill stem stuck in deforming till and bent	Viscous deformation in upper 0.65 m of till
Atre and Bentley (1993, 1994)	Ice Stream B.	Dilatant bed; deformation may not be pervasive
Fischer and Clarke (1994)	Ploughmeter at the base of Trapridge Glacier	Deformation varies with grain size/rheology (sliding also occurs)
Iverson *et al.* (1994, 1995)	Borehole through subglacial sediments at Storglaciären, Sweden – tilt cells show till deformation	Viscous deformation in upper 0.33 m of till
Engelhardt and Kamb (1998)	Ice Stream B (Whillans Ice Stream) – tethered stake	Shear dilation/shallow deformation and/or basal sliding in upper 0.05 m; large strain rate variability = failure of a non-linear, plastic material
Scherer *et al.* (1998), Tulaczyk *et al.* (1998, 2001)	Till cores from Ice Stream B – microfossils and lack of crushing and abrasion indicate mixing of pre-existing sediment and its deformation in its upper layers	Shallow plastic deformation by ice keel ploughing
Boulton and Dobbie (1998)	Breiðamerkurjökull – transducers emplaced via borehole	Sliding and deformation concentrated in upper 0.5 m of till but also diminishing in cumulative effect down to 1 m
Tulaczyk *et al.* (2000a,b)	Ice Stream B	Plastic deformation – failure plane migrating vertically on diurnal cycle

Table 6.1 (Continued)

Source	Location and nature of evidence	Nature of deformation
Truffer *et al.* (1999, 2000)	Black Rapids Glacier	Failure along a decollement at >2 m in a 7.5-m-thick till (later dismissed due to recognition of equipment malfunction)
Fuller and Murray (2000)	Continuous clay layers between tills in drumlin, Hagafellsjökull Vestari, Iceland	Shallow deformation (<16 cm) = sliding and deformation by ploughing clasts
Boulton *et al.* (2001)	Breiðamerkurjökull – drag spools and strain markers used over 12-day period in boreholes (porewater pressure measured in one borehole)	Stick–slip motion related to water pressure fluctuations – alternate sliding and deformation in descending dilating shear zone up to 1 m in depth (most displacement <0.3 m)
Porter and Murray (2001)	Bakaninbreen – tilt cells installed in the subglacial till	Deformation to at least 0.2 m depth in addition to glacier sliding; till is not behaving as a Coulomb-plastic
Kamb (2001)	Ice Stream D (Kamb Ice Stream) – drag spools	Significant deformation at depths >60 cm
Vaughan *et al.* (2003), Atre and Bentley (1993)	Antarctic ice streams – patterns in acoustic impedance of subglacial sediment correlated with deforming and lodged states	Spatial mosaic of bed processes related to material properties or sediment deformation history. Ice streams in stagnant phase underlain by lodged sediment
Mair *et al.* (2003)	Tilt cells below Haut Glacier d'Arolla	Intermittent shear strain at shallow depths (<7 cm)
Truffer and Harrison (2006)	Autonomous probes at 0.8 and 2.1 m in till beneath Black Rapids Glacier	Episodic deformation throughout the till but most concentrated together with sliding in upper 0.2 m
Kavanaugh and Clarke (2006)	Drag spool, tiltmeters and ploughmeter emplacement below Trapridge Glacier	Deformation throughout the diurnal cycle. Reduction of till strength and increased sliding speed with weak tendency towards lower strain rates at times of high water pressure = Coulomb-plastic rheology
Smith *et al.* (2007), Smith and Murray (2009), King *et al.* (2007, 2009)	Radar profiling of the bed of the Rutford Ice Stream	Bed elevation changes between 1991 and 2004 indicate small thickening or thinning of a deforming layer together with emerging drumlins

till (Table 6.1). Although the details of the findings summarised in Table 6.1 vary, recurrent patterns have emerged since the late 1970s. Specifically, deformation is typically confined to the uppermost few centimetres of the bed, although there is some evidence for deformation at greater depths (predominantly up to 1–2 m, with maximum depths of 5–6 m reported in the early work on Ice Stream B), which accounts for only small cumulative strain totals; Scherer *et al.* (1998) and Tulaczyk *et al.* (1998) regard the intra-till inhomogeneities in porosity, composition and microfossil content in Ice Stream B tills as inconsistent with pervasive deformation to depths any greater than 1.5 m. This pattern of deformation appears to be responsible for the exponential displacement curves displayed by the Breiðamerkurjökull data (Figures 6.10 and 6.11b). This is significant in till sedimentological terms because it clearly indicates that a process–form regime in which deformation, ploughing and lodgement (presumably also melt-out) emplace subglacial tills, cannot explain the deposition of thick tills (>2 m) by single accretionary events but rather by the incremental build-up of till layers over time. Moreover, a subglacial origin for deformation structures that span depths of tens of metres, for example, in glacitectonites (see Section 6.5 and Chapter 14), cannot be verified by observations on contemporary glacier beds; a proglacial origin and subsequent glacier overriding is the simplest interpretation of such features (e.g. Aber, 1982, 1985; van der Wateren, 1985, 1994, 2003; Aber *et al.*, 1989).

Further to Section 4.3 on the physics of material behaviour, it is relevant at this juncture to provide a brief overview of the changing views of glaciologists, geophysicists and glacial geologists on the rheology of tills and related materials. Further reflections on this debate, because they stem from laboratory experiments, will be provided in Chapter 7.

Even the initial interpretations of the Breiðamerkurjökull subglacial experiments revealed that glacial geologists were uncertain as to the style of rheological behaviour that was being observed. Boulton and Jones (1979) initially thought that the till had a perfect Coulomb-plastic rheology, but this was later revised by Boulton and Hindmarsh (1987), who considered the strain patterns to be more typical of a non-linearly viscous or Bingham material. We now understand that the characteristic exponential curves of the displacement profiles (Figures 6.10 and 6.11b) actually record the total or cumulative relative strain that has taken place in the till over the experimental period rather than mean strain, and hence it does not reflect a non-linearly viscous rheology in which strain rates increase non-linearly with applied stress (Figure 4.5). Indeed, the evidence that subsequently emerged from the instrumentation of glacier beds as well as laboratory experiments (see Chapter 7) was more consistent with a Coulomb-plastic rheology (Table 6.1). Importantly, it became clear that although till strength is reduced at times of high water pressure, the concomitant till strain rates do not increase but decrease (e.g. Iverson *et al.*, 1995; Hooke *et al.*, 1997; Fischer and Clarke, 2001; Kavanaugh and Clarke, 2006). The term introduced by Iverson *et al.* (1995) and Hooke *et al.* (1997) to refer to this situation is 'glacier–till coupling', specifically a process–response regime in which low water pressures in till create strong glacier–bed coupling through the process of ice infiltration into pore spaces; hence shear stresses are supported by the till (Iverson, 1993; Iverson and Semmens, 1995). When water pressure is high, ice infiltration is suppressed and the glacier is decoupled from its bed and traction across the ice–bed interface is reduced. In summary, high water pressures do create weaker tills, but their deformation is reduced at those high pressures, giving rise to glacier–till decoupling and rapid sliding along the ice–till interface. Critical to the mobility of the deforming till are short periods of negative strain rate at times of falling effective pressure and high sliding speed (Blake *et al.*, 1992; Iverson *et al.*, 1995, 2007; Fischer and Clarke, 1997, 2001; Hooke *et al.*, 1997). These are consistent with a Coulomb-plastic rheology because the reduced effective pressure renders the till as over consolidated with respect to the new pressure conditions, and hence decoupling and falling shear stress initiates till dilation.

Because these processes are active in the uppermost tens of centimetres of tills, where effective pressures and concomitant sediment strength are at their lowest, it is predicted that till deformation should not occur at depths where the sediment yield strength is higher than the shear stress. Hence, the evidence for deformation at depth has been regarded as problematic for the Coulomb-plastic model, specifically because it predicts confinement of deformation to a very thin layer just below the ice–till interface. Explanations have been forthcoming for this phenomena including, first, Tulaczyk *et al.*'s (2000a) and Kavanaugh and Clarke's (2006) idea of a meltwater pulse, which passes vertically through the till and thereby creates peak water pressures and concomitant minimum effective stresses at different depths during the diurnal cycle. This explains how the weakest part of the till varies in position through time. Second, a similar process but operating from the base of the till at Breiðamerkurjökull was proposed by Boulton *et al.* (2001) and Boulton (2006), where an underlying aquifer in glacifluvial outwash deposits drives water pressure increases and deformation at depth within the till (Figure 6.11a). Third, Iverson and Iverson (2001), developing ideas of granular aggregate flow conveyed by Boulton and Dobbie (1998), proposed that during shearing, intergranular forces are supported by chains of particles extending down into the till and this process gives rise to the distributed pattern of strain that is displayed in till displacement curves.

The understanding of till rheology in nature appears to be hindered by the very characteristics of the subglacial materials themselves. A Coulomb-plastic rheology is entirely predictable for a material that is homogeneous such as fine matrix-supported diamictons, but subglacial materials, as we shall see in Section 6.5 and Chapter 7, are commonly heterogeneous and hence we should perhaps be contemplating complex rheologies. This has been signposted by Hindmarsh (1997) in his accounting for deformation at depth in subglacial materials. He highlights that the bulk response to stress can be described by a viscous rheological model, even though the materials exhibit Coulomb-plastic behaviour at small scales. In geological terms, this can be equated to 'deformation partitioning', whereby brittle–ductile behaviour is often recognised in spatially variable material types. Hindmarsh (1996) has also proposed that the till itself may slide over the bed, giving rise to polished bedrock or even slickenside-covered soft sediment surfaces at a failure plane located below the 'normal' deformation profile, a process entirely consistent with a Coulomb-plastic rheology.

6.3 Soft-Bed Sliding (Ice Keel Ploughing), Meltwater Drainage and Ice–Bed Decoupling

The preceding section has presented the concept of alternating subglacial till deformation and soft-bed sliding (defined as ploughing by Brown *et al.*, 1987; Tulaczyk *et al.*, 2001; Clark *et al.*, 2003) as controlled by changing water pressures, even at diurnal temporal scales, giving rise to cycles of decoupling and coupling of the glacier from its bed and concomitant intermittent transmission of stress to the sedimentary substrate. This was observed at Breiðamerkurjökull by Boulton and Dobbie (1998; Figure 6.11c), where the proportion of glacier movement due to sliding versus till deformation was derived from the differential movement of strain markers at the glacier base and related to changing effective pressure every 6 hours. Similarly, Fischer and Clarke (1997) and Fischer *et al.* (1999) demonstrated a stick–slip sliding behaviour at the ice–bed interface due to decoupling of ice from the underlying till during periods of higher water pressures. In an assessment of the relative roles of sliding and deformation beneath ice streams, Thorsteinsson and Raymond (2000) proposed that sliding must be the dominant mechanism of basal motion, rather than till shearing unless: (a) the till can adhere to the ice sole; (b) short scale roughness develops at the ice–till interface; or (c) internal slip boundaries exist in the till. Engelhardt and Kamb (1998) also demonstrated that Ice Stream B was sliding over a till, which they interpreted as being controlled by the clay-rich nature of the till and its tendency to retard water migration. This lithological control on the deformation versus sliding regime was assessed by Boulton (1996a), who proposed that clay-rich tills, when compared to coarse-grained tills, not only preferentially decouple from the ice base but will also only deform to a shallow depth. Tulaczyk (1999) similarly suggested that ice–bed coupling is weaker in fine-grained tills, which facilitate sliding and clast ploughing. He proposed that lower hydraulic diffusivity and small numbers of clasts act to restrict strain distribution and hence produce a shear zone only 0.01 m thick, whereas coarse-grained tills enable strong ice–till coupling and deformation down to 0.1 m. Spatial variability in sliding versus deformation gives rise to the development of 'sticky spots' on the glacier bed, which can be controlled by a variety of features including bedrock bumps, lodged boulders and frozen patches within mainly thawed beds as well as well-drained areas of the bed (Alley, 1993; Stokes *et al.*, 2007). The implications of this spatial variability will be discussed further in Chapter 16 in relation to till mosaics.

In Section 6.1, it was demonstrated that ploughing clasts embedded in the glacier sole distribute shear to significant depths in subglacial till and then, through the production of a prow, retard their own movement and eventually lodge due to localised dewatering and increased consolidation of the till matrix. An alternative style of ice keel ploughing through soft substrates has been proposed by Tulaczyk *et al.* (2001) and Clark *et al.* (2003) in which subglacial materials up to a few metres may be disturbed and transported by the keels in a sliding, bumpy glacier sole. The ice keel ploughing model invokes till deformation around bumps in the ice sole as they are dragged through the substrate, helping to explain the streamlining of the substrate even though the basal ice is debris-poor and the underlying till is fine-grained and clast-poor. The fault gouge concept depicted in Figure 6.6 is advocated by Tulaczyk *et al.* (2001) as an analogue for this process in which the sliding ice base acts as the rigid upper fault plate, and asperities in the fault plate (clasts or keels in the ice sole) plough through the soft substrate and generate, or more specifically replenish, the deforming till layer (fault gouge layer) above the underlying, rigid strata (lower fault plate). Importantly, only the largest keels or bumps in the ice sole can generate new till, because they can protrude through the existing till layer into the underlying substrate. In addition to remobilising lower layers of potentially strain hardened till this process should be effective at liberating soft clasts (rafts) from the substrate while at the same time producing erosional contacts between the base of the deforming layer and the underlying non-deformed sediments (see Section 6.5). Meanwhile, smaller bumps are important in that they act to transport the existing till. This ploughing model is an attractive one in terms of till continuity (Alley, 2000; Iverson, 2010) because: (i) it requires no sediment input from the glacier base; (ii) it explains substrate deformation and streamlining of pre-existing sediments and hence can create or replenish deforming layers; and (iii) it explains extensive till units of relatively uniform thickness due to the operation of a stabilising feedback, whereby as a till thickens, fewer ice bumps can penetrate through it to the underlying substrate, thereby arresting and ultimately terminating the thickening rate. Alternatively, if a till thins, more ice bumps will protrude into and erode the substrate, thereby generating more till.

The results of the Breiðamerkurjökull experiments prompted Boulton *et al.* (2001) to propose a stick–slip cycle of sliding and deformation for the generation of the till in that setting, similar to the process reported by Fischer and Clarke (1997) and incorporating the Coulomb-plastic rheology proposed by Iverson *et al.* (1998; Figure 6.12a). The model proposed in Figure 6.12a accounts for the soft-bed sliding (ploughing) process when shear stress and water pressure build to the point where till dilates (effectively liquefies), ice–bed friction is reduced and ice–till decoupling takes place. The displacement curve to the left of each vertical profile diagram portrays the style of flow at the ice–bed interface, so that decoupling/sliding is depicted by decollement at the glacier–bed contact. Falling water pressures in the dilatant till will facilitate its deformation and ice–bed coupling as well as a reduction in the amount of sliding. Continued falling water pressures will consolidate the dilatant till and strengthen the interlocking of the ice–bed interface, causing the shear zone to descend through the till. Deformation stops once the water pressure falls below the critical level for failure and the bed then sticks. Diurnal changes in water pressure presumably then lead to repeat cycles of dilation and collapse so that the curvilinear displacement curve, representing cumulative strain, becomes more pronounced through time. In order to thicken a deforming till layer and to preserve evidence of meltwater deposits related to sliding bed phases, there must be net vertical accretion of the till, which Boulton *et al.* (2001) propose can only take place through loss of deforming till (cessation of its forward movement) from the bottom of the deforming layer, although liberation of sub-till materials can also thicken the deformed sub-stratum (see Section 6.5). Such processes are likely to accrete thick and multiple till and sliding bed deposits near glacier and ice sheet margins because only at such locations do shear stresses and meltwater discharges tail off enough to stack the materials advected from up ice (cf. Alley *et al.*, 1997; see Chapter 16, specifically related to incremental thickening and marginal till wedges). The net vertical accretion of deforming tills is critical to the preservation of any meltwater sediments created during the soft-bed sliding process.

Potentially also highly significant in the triggering of stick–slip styles of motion via soft-bed sliding are liquefaction events caused by ice quakes, as recently proposed by Phillips *et al.* (in press), who incorporate into their model the findings of recent studies in modern glacial environments (e.g. Ekstrom *et al.*, 2003, 2006; Bindschadler *et al.*, 2003; Tsai and Ekstrom, 2007; Wiens *et al.*, 2008), where seismic activity and ice quakes occur in response to movement on faults within the glacier

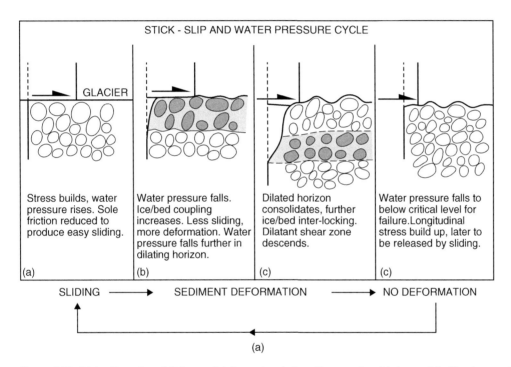

Figure 6.12 Stick–slip cycles of sliding and deformation during till generation: (a) the model of Boulton *et al.* (2001), emphasising the role of water pressure cycles in till deformation; (b) the model of Phillips *et al.* (in press) depicting the role of seismic activity in liquefaction and soft-bed sliding. Upper panel shows the conceptual model of the potential effects of seismic waves generated during an ice quake on unconsolidated subglacial sediments. The pulse of energy passes into the underlying saturated sediments and triggers transient liquefaction and soft-bed sliding. The short duration energy pulse causes individual clasts to vibrate, modifying the packing of the grains and leading to the pressurisation of the intergranular porewater. This then reduces the number of grain to grain contacts, allowing the individual clasts to move (slide or rotate) past one another, reducing sediment cohesion and leading to liquefaction and soft-bed sliding. The vibrating effect propagates away from the focus of the ice-quake as a pulse or series of pulses, so that areas of the bed initially undergo localised liquefaction and soft-bed sliding, followed by stabilisation outwards away from the ice quake focus. Liquefaction and soft-bed sliding probably occurs within discrete, laterally discontinuous patches or narrow zones in the order of only a few centimetres or even millimetres thick. Once the ice-quake energy has been dissipated (likely in a few minutes), the fall in intergranular porewater pressure and increase in sediment cohesive strength results in cessation of flow deformation. Lower panel shows a conceptual diagram of the microscale processes occurring during bed deformation, soft-bed sliding and basal sliding as a result of increasing porewater content within a soft glacier bed.

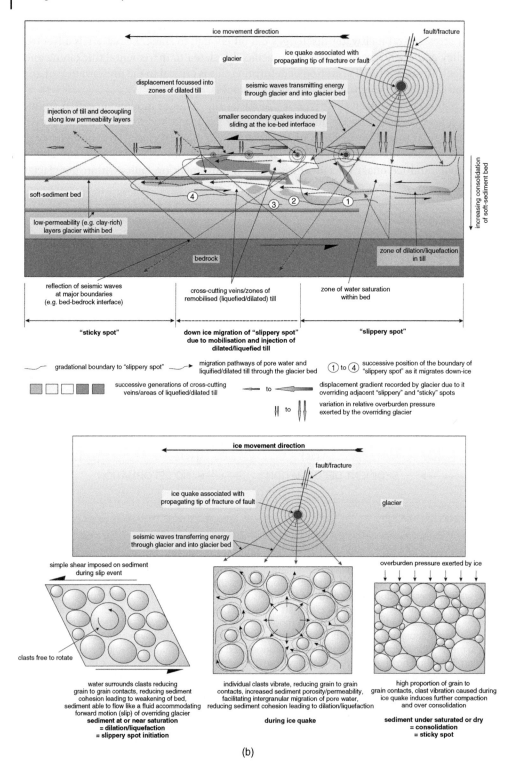

Figure 6.12 (Continued)

or underlying bed, crevasse/fracture propagation, iceberg calving, serac toppling, opening and closing of englacial drainage conduits and/or slip events at the ice–bed interface. Based upon detailed micromorphological evidence from subglacially deformed materials that we will discuss in more detail in later chapters, this model of soft-bed sliding is based upon explanations of the polydeformed nature of subglacial sediments; more specifically, the evidence for repeated phases of liquefaction followed by solid-state shear deformation. Localised liquefaction of tills is thought to lower the cohesive strength of the sediment and thereby create 'transient mobile zones' or soft-bed sliding zones where the shear imposed by the overriding ice is accommodated. Together with bed deformation and basal sliding, this soft-bed sliding forms a continuum that facilitates glacier movement, but rather than being a continuous uninterrupted process, glacier motion is regarded as a series of stick–slip events. Phillips *et al.* (in press) propose that the critical drivers of the slip events include not only the processes already widely acknowledged by the glacial research community, such as the introduction of pressurised meltwater into the bed (a process limited by till porosity and permeability) and the pressurisation of porewater as a result of subglacial deformation, but also episodic liquefaction of water-saturated till in response to glacier seismic activity or ice-quakes (Figure 6.12b).

Subglacial observations have facilitated an understanding of meltwater flow at the ice–bed interface and within subglacial tills, as summarised in Figure 6.13. A comprehensive review of the literature on modern subglacial drainage systems is provided elsewhere (Hubbard and Nienow, 1997; Fountain and Walder, 1998; Benn and Evans, 2010; Cuffey and Paterson, 2010) and hence only a brief review is presented here as it pertains to till sedimentology. Subglacial drainage systems are generally classified as either channelised systems, in which water is confined to relatively narrow conduits, or distributed systems which extend over larger proportions of the bed. The specific configuration and how it interacts with substrate deformation dictates the nature of the stratified sediments associated with tills. As discussed above, water may migrate through a soft glacier bed by bulk movement in till pore spaces or by Darcian porewater flow (1 and 2 in Figure 6.13; e.g. Stone and Clarke, 1993). Observations at the base of the Rutford Ice Stream, Antarctica, by King *et al.* (2004) indicate that deforming subglacial tills may drain meltwater internally by pipe flow (3 in Figure 6.13). Water may also flow at the ice–bed interface in dendritic channels, through linked cavities (more likely on hard beds), in braided canals or as thin films (4–7 in Figure 6.13). Much of our knowledge of subglacial meltwater networks and their associated deposits arises from theoretical models informed by observations on meltwater behaviour on modern glaciers, and in some cases retrodictively from ancient sediment characteristics, hence some circular reasoning is inevitable. The flow of water at the ice–bed interface is associated with periods of sliding and hence the sedimentological evidence for such events is likely to be preserved only in situations where till is accreting. Additionally, recoupling of the ice with its bed is likely to result in significant reworking of the stratified sediments by deformation, cannibalisation and re-ingestion into the deforming layer (see Section 6.5). At initiation, the meltwater deposits beneath soft-bedded glaciers and ice sheets are most likely to be created in irregular water layers (Creyts and Schoof, 2009) or braided canals (Walder and Fowler, 1994; Englehardt and Kamb, 1997; Ng, 2000), when discharges are too large to be evacuated through the bed or at times of ice–bed separation/soft-bed sliding. Pipe flow through the till, as observed by King *et al.* (2004) beneath the Rutford Ice Stream, is thought to be responsible for the construction of 'mini-eskers' (Alley, 1991). On hard or relatively impermeable beds, Evans *et al.* (2006b) speculated that Hindmarsh's (1996) till sliding could give rise to the generation of thin stratified beds by water films. Because many of our ideas on the processes and forms of till-related meltwater sedimentation have stemmed from interpretations of glacigenic deposits, genetic models of sliding bed deposits will be reviewed in Chapter 11.

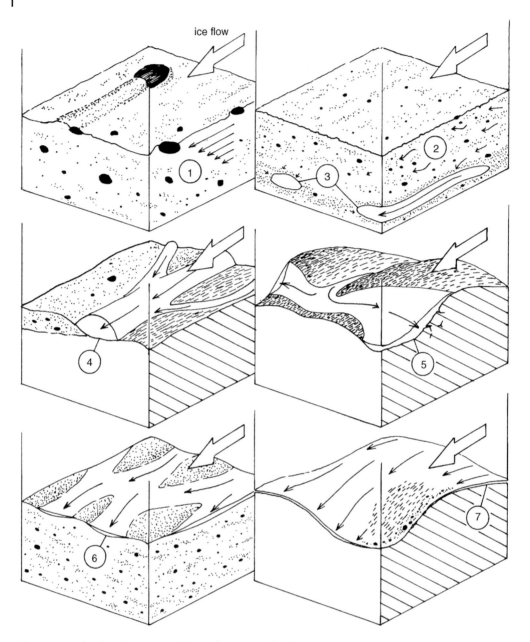

Figure 6.13 Sketches showing possible configurations of subglacial drainage networks (from Benn and Evans, 1996): (1) bulk water movement in deforming till; (2) Darcian porewater flow; (3) pipe flow; (4) dendritic channel network; (5) linked cavity system; (6) braided canal network; (7) thin film at ice–bed interface.

6.4   Melt-Out

The definition of a melt-out till as updated by Benn and Evans (1998) and Evans *et al.* (2006b) is:

> Sediment released by the melting or sublimation of stagnant or slowly moving debris-rich glacier ice, and directly deposited without subsequent transport or deformation.
>
> Evans *et al.* (2006b, p. 169)

This is a process–form regime that has traditionally been defined by theoretical postulates stemming from studies of ancient tills (e.g. Shaw, 1979; Haldorsen and Shaw, 1982; Shaw, 1982, 1983; Ham and Mickelson, 1994; Piotrowski, 1994; Munro-Stasiuk, 2000; Larson *et al.*, 2006; Möller, 2010) but supported by some observations on modern glaciers (Boulton, 1970a, b, 1971; Mickelson, 1973; Shaw, 1977; Lawson, 1979a, b, 1981; Shaw, 1982; Ronnert and Mickelson, 1992; Cook *et al.*, 2011a; Larson *et al.*, 2016; Figure 2.6c, d; Table 6.2) and ancient buried glacier ice/permafrost (Astakhov and Isayeva, 1988; Murton *et al.*, 2005). The problems of melt-out till production and preservation are discussed in Chapter 13, so this section deals specifically with the small number of relevant field observations of the melt-out process.

Subglacial melt-out (note that often the term 'supraglacial melt-out' is used when exposures are melting from the top downwards even though it is basal ice) is the slow and largely passive release of sediment from debris-rich, stagnant basal glacier ice and hence is a notoriously difficult process to observe and quantify. In situations where debris concentrations are high, the thaw consolidation ratio is low, and hence theoretically at least the delicate englacial debris structures could be preserved after melt-out of the encasing ice, if drainage is slow but efficient. This has been assessed by Carlson (2004), who has demonstrated that a till with low hydraulic conductivity can accommodate the transport of 1.3 m^3 water a^{-1} m^{-2}, which is up to three orders of magnitude more than is produced at the base of a glacier, without initiating dewatering and thereby destroying englacial structure. It has been proposed that even localised vigorous drainage of meltwater from the site can be accommodated in the sedimentology of potential melt-out tills in that inter-till stratified lenses and beds can be produced at the locations of debris-poor ice bands (Shaw, 1982; Munro-Stasiuk, 2000).

Because preservation potential of melt-out till is theoretically improved in situations where the thickness and debris content of basal ice facies are large, it is the relatively recent acknowledgment of the importance of supercooling in creating stratified basal ice that has given the melt-out process–form regime a greater degree of plausibility (cf. Alley *et al.*, 1998, 1999; Lawson *et al.*, 1998; Evenson *et al.*, 1999; Roberts *et al.*, 2002; Larson *et al.*, 2006; Cook *et al.*, 2007, 2010, 2011a, b). By definition, a subglacial melt-out till must inherit the foliation and structure of the parent ice, but during the melting process, at least some (it has been argued a lot) of modification takes place (see Chapter 13); the purest or least disturbed type of melt-out till should therefore be produced where the parent ice is removed by sublimation, thereby creating Shaw's (1977, 1989) 'sublimation till' (Figure 6.14), a till type recognised in arid, polar settings also by Lundqvist (1989) and Fitzsimons (1990). Based upon their observations at Matanuska Glacier, Larson *et al.* (2016) propose that preservation is enhanced by the burying of a debris-rich basal ice layer within a detached, stagnant part of a receding glacier snout, where supraglacially derived debris flows can accumulate quickly above the ice mass and thereby insulate it and slow the melt rate and release the meltwater by groundwater seepage. The characteristics of subglacial melt-out tills where they appear to be emerging from melting debris-rich basal ice (Table 6.2) will vary according to the mechanism by which the stratified basal ice was initially created, whether it be by regelation, supercooling, net

Table 6.2 Evidence from modern glaciers for melt-out till production.

Source	Nature of evidence	Process
Boulton (1970a, b, 1971), Lawson (1979a)	Similar fabrics to englacial debris but with reduced dips and reduced maxima.	Passive melt-out with lowered up-glacier dips due to ice removal and some re-orientation due to inter-clast contacts.
	Massive, fine-grained till with shear planes and structural foliation, whole shells and glacier ice lenses.	Passive release of englacial debris in stagnant basal ice, observed at late stage of melt-out. Shearing imparted by overriding active ice.
Shaw (1977), Fitzsimons (1990)	Foliated or densely horizontally jointed diamicton. 'Highly' and 'poorly attenuated till facies' relating to foliated ice and massive ice, respectively.	Removal of ice by sublimation preserves highly and poorly attenuated ice facies as foliated till – massive ice facies produces till that might be difficult to differentiate from lodged basal till.
Lawson (1979a, b, 1981a, b)	Three end members: (1) structureless, pebbly sandy silt (2) discontinuous laminae, stratified lenses and pods in massive pebbly silt (3) bands or layers of contrasting sediment.	Release of grains from debris-poor ice and melting of ice debris of similar texture both produce structureless till. Ice-poor debris lenses are intact but distorted. Ice-poor layers of mixed grain sizes produce blurred contacts.
Shaw (1982, 1983), Lawson (1981a)	'Stratified' appearance as a result of sorted intrabeds and draped layers over large clasts.	Volume loss during melting along ice folia.
	Infilled scours beneath clasts and between melt-out till and soft substrate.	Subglacial meltwater film erodes into substrate and around clast protruding from basal debris-rich ice.
	Infilled scours beneath clasts and in intra-till lens.	Meltwater flow in englacial cavity in debris-rich ice or between debris-rich ice and underlying melt-out till.
	Diapirs of till protruding into overlying stratified sediments.	Mass displacement in till due to thaw-consolidation.
	Sub-till sediment clasts (faulted and slightly deformed but with angular borders) in till.	Quarried in frozen state and melted out from debris-rich ice.
Cook *et al.* (2011b)	Preservation of sedimentary stratification in melted-out supercooled ice. Dominance of silt-sized material indicates supercooled origins.	Very rare preservation of stratification observed in supercooled glacier ice due to high ration of ice to sediment and hence a propensity to fail as sediment gravity flows during melt-out.
Larson *et al.* (2016)	Debris-rich (stratified) basal ice transitioning upwards into pseudo-stratified/laminated diamicton with strong macrofabrics and microscale porewater pathways.	In situ melting inherits stratified basal ice properties as laminated and bedded diamicton. Requires meltwater seepage from stagnant ice body protected from insulation beneath hummocky supraglacial debris flow cover.
	Linkage between stratified ice and overlying melt-out till is evidenced by transition zone 5–10 cm thick.	

A   Advancing glacier

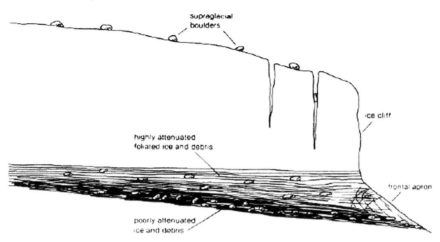

B   Retreating glacier

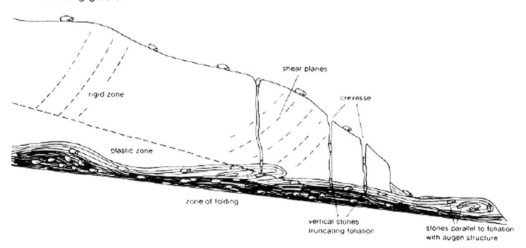

C   Depositional sequence

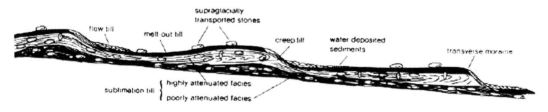

Figure 6.14 Shaw's (1977) genetic model for the production of sublimation till in arid, polar environments, where the englacial debris structures created by the folding of basal, debris-rich ice facies are preserved after passive removal of interstitial ice.

adfreezing or apron overriding (in polar settings) all of which can be thickened by englacial folding and thrusting.

What is emerging in settings like the Matanuska Glacier foreland is a deposit that displays 'pseudo-stratification', which includes discontinuous and contorted and sheet-like lamina, layers and lenses, textural and compositional banding with indistinct contacts and internal flow structures and undeformed, intra-till soft-sediment or soft-bedrock rafts (Lawson, 1979a, b; Shaw, 1977; Carlson, 2004; Larson *et al.*, 2016; Figure 6.15). The supercooled ice at the Matanuska Glacier in particular is creating a deposit that displays silt or sandy-silt lamina, *en echelon* silt lamina, irregular lenses of coarse to fine sand that offset or cross bedding planes, short lenses of gravel, sand and silt, thin lenses of silt and clay–silt aggregates, dispersed sub-rounded, but non-striated pebbles and cobbles around which lamina bend (Larson *et al.*, 2016; Figure 6.16). Similar features are seen at the microscale in addition to porewater pathways, compressed and lightly deformed clay–silt clasts and an absence of shear indicators (Larson *et al.*, 2016; Figure 6.17). Clast macrofabrics (see also Chapter 8) have been reported by Lawson (1979a, b) and Larson *et al.* (2016) for the Matanuska Glacier, Alaska (Figure 6.18), in which the sense of former glacier flow direction, reflected in the shear strain signature of the enclosing basal ice, is recorded by strong S_1 eigenvalues of $0.865–0.887$ (melt-out till) and $0.896–0.925$ (stratified basal ice), even though shear indicators are not otherwise visible. We will review the details of melt-out till, as it appears to be evolving at locations like the Matanuska Glacier, in Chapter 13, where problematic similarities with other glacigenic diamictons are analysed.

6.5 Glacitectonite Production, Rafting and Cannibalisation

Our knowledge of glacitectonite development is borne less out of modern subglacial process observations and more out of the discipline of structural geology and hence this section dwells more on that knowledge base. The concept of a subglacial shear zone and its manifestation in strain signature is illustrated in Figure 4.13, which conveys the principle of the strain ellipsoid and the increasing attenuation of materials towards the ice–substrate interface. In subglacial tills, we see this strain signature recorded in clast macrofabric strengths (see Section 4.5 and Chapter 8), but the structures developed within both tills and glacitectonites have been compared also with those within fault gouge settings (Figure 6.6; Eyles and Boyce, 1998). The subglacial shear zone is envisaged to be relatively thin, specifically a maximum of 10 m but most often significantly less.

It is well known that the depth of glacitectonic disturbance of strata can exceed 100 m, but this is accomplished only in proglacial settings where horizontal compressive stresses can elevate large masses of material above the height of the glacier margin (Aber *et al.*, 1989; van der Wateren, 1995a, b; Benn and Evans, 2010). Beneath thick ice, shear stresses are too low to produce such deep-seated failure and the normal stress component initiated by the ice load prevents the vertical growth by folding and thrust stacking that has been demonstrated by both experiment (e.g. Rotnicki, 1976; Mulugeta and Koyi, 1987) and observations on moraine construction (e.g. Sharp, 1985; Croot, 1988; Evans and England, 1991; Benediktsson *et al.*, 2009) to be related to gravity spreading. However, the development of the various brittle and ductile structures by glacitectonic disturbance, reviewed in Section 4.4 (Figures 4.6–4.9), is invariably initiated by compression in proglacial stress fields and then modified by later glacier overriding and simple shear (Kozarski, 1959; Aber, 1982, 1985) to produce glacigenic melanges (see Section 4.2; Figure 6.19). At this stage, pre-existing sediments and even bedrock can be converted into glacitectonite, the definition of which is:

Figure 6.15 Idealised vertical profile of the typical features observed in the melt-out till at the Matanuska Glacier, Alaska (from Lawson, 1981a). Zones include (a) structureless, pebbly, sandy silt, (b) discontinuous laminae, stratified lenses and pods of texturally distinct sediment in massive pebbly silt, and (c) layers of texturally, compositionally or colour-contrasted sediment. Laminae or layers may drape over large clasts.

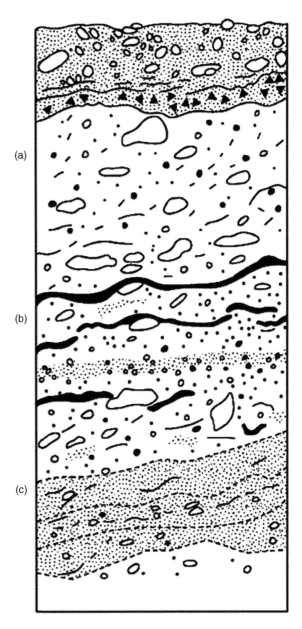

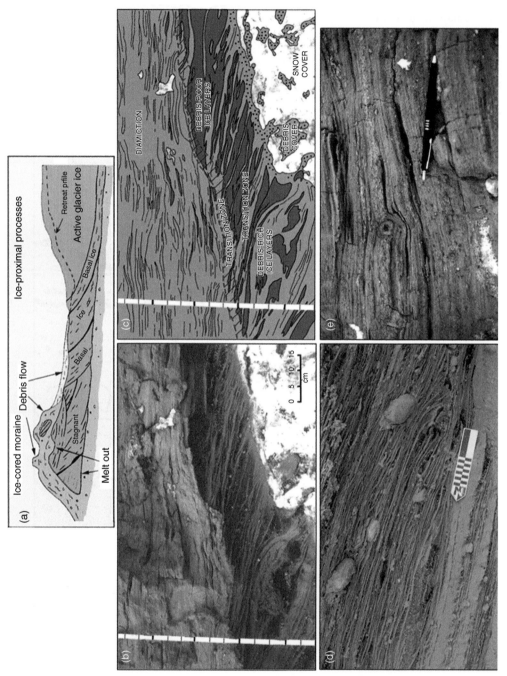

Figure 6.16 Details of the Matanuska Glacier melt-out sequence (from Larson *et al.*, 2016): (a) cross-sectional sketch through the glacier margin, showing the stratigraphy of the basal debris-rich ice sequence and its accumulating supraglacial debris cover; (b) exposure and (c) annotated sketch of exposure through the debris-rich basal ice and its overlying melt-out till. Light grey bands are debris-rich layers and dark bands are debris-poor layers and upper part of exposure is pseudo-stratified diamicton. Area marked with red diagonal lines is the approximate transition zone between ice and melt-out till; (d) basal debris-rich ice facies, showing discontinuous debris-rich (grey) and debris-poor (dark) laminations with a small fold and clear imbrication of elongate clasts; (e) detail of the pseudo-stratified diamicton created by melt-out, showing laminae of silt deformed around clasts.

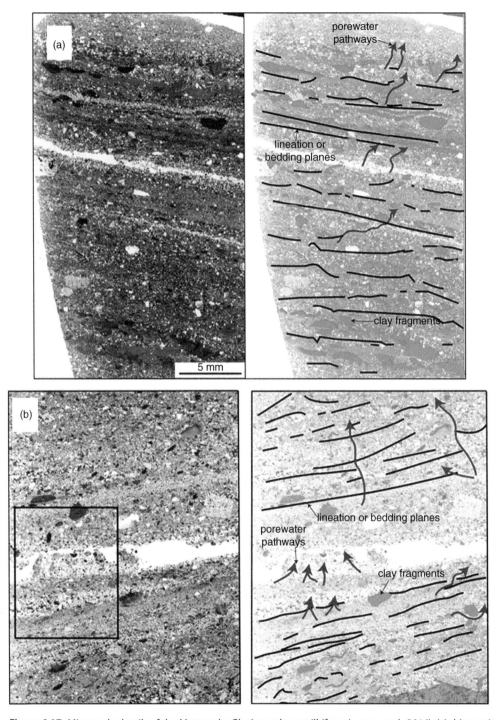

Figure 6.17 Microscale details of the Matanuska Glacier melt-out till (from Larson *et al.*, 2016): (a) thin section (left) and interpreted structures (right), showing silt-rich laminated diamicton with laminations (black lines), deformed aggregates of clayey silt (yellow fill) and porewater pathways (magenta lines). Aggregates occur in bands with their long axes generally subparallel to lamina; (b) thin section (left) and interpreted structures (right), showing silt-rich laminated diamicton with laminations (black lines), deformed aggregates of clayey silt (yellow fill), porewater pathways (magenta lines) and fine to coarse sand at base (pink fill). Aggregate long axes are generally subparallel to laminae.

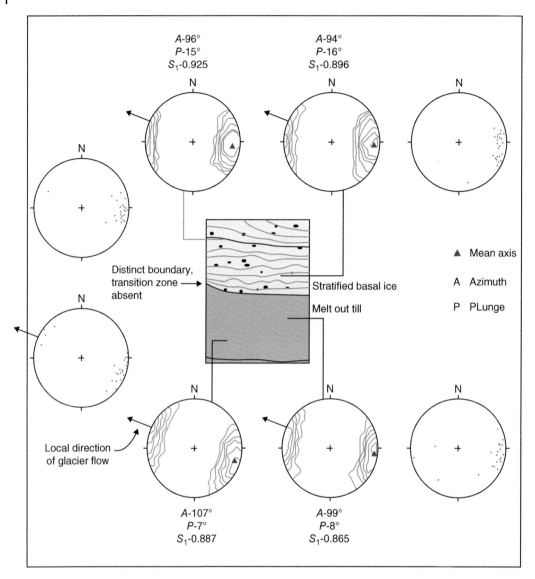

Figure 6.18 The stratigraphy and fabric characteristics of subglacial melt-out till and basal-debris-rich ice at the Matanuska Glacier, Alaska, showing virtually identical clast fabrics in the till and glacier ice (from Lawson, 1979b).

...rock or sediment that has been deformed by subglacial shearing (deformation) but retains some of the structural characteristics of the parent material, which may consist of igneous, metamorphic or sedimentary rock, or unlithified sediments.

Benn and Evans (2010, p. 375, after Benn and Evans, 1996).

Because of the similarities in tectonically induced structures observed in glacial deposits and metamorphic rocks, the process–form regimes and associated nomenclature of metamorphic petrology

Phase 1– proglacial thrusting & composite ridges

Phase 2 – overriding & cupola hill/glacitectonite & till carapace

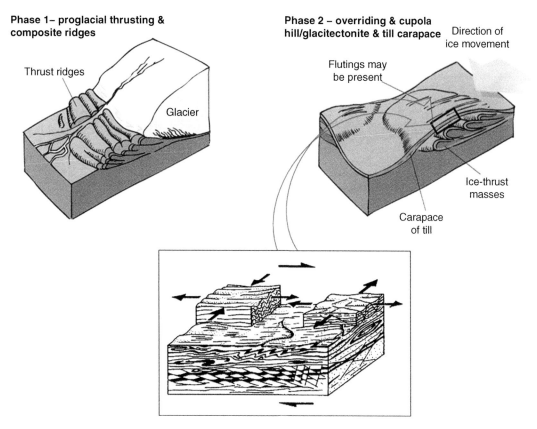

Figure 6.19 The production of a glacitectonite carapace during the second phase of glacitectonic disturbance of pre-existing deposits by an advancing glacier, when proglacially thrust ridges have been streamlined into a cupola hill (after Evans and Benn, 2001). The structural details of the carapace (inset diagram from van der Wateren, 1995a, b) displays the vertical continuum from S_r to S_b to S_h (see Figure 6.21).

are widely used in explanations of glacitectonites, along with process observations from modern sub-marginal to proglacial thrust moraines (Figures 6.19 and 6.20). The process–form details of proglacially thrust masses are discussed comprehensively elsewhere (e.g. Aber *et al.*, 1989; van der Wateren, 1995a, b, 2003; Benn and Evans, 2010), and hence we concentrate here on the details of sub-glacial glacitectonism. It is most common to see thick sequences of glacitectonite in former glacier sub-marginal locations, where they technically constitute the materials excavated from the footwalls of shear zones (Figure 6.20). As we saw in Chapter 4, the influence of simple shear in such locations is recorded by a vertical continuum from non-deformed to folded or faulted sediments at the base grading upwards to attenuated materials or melange and then to homogenised diamicton or till (Banham, 1977; Pedersen, 1989; Figures 2.8 and 4.6). In an assessment of such signatures in Polish glacial deposits, Kozarski and Kasprzak (1994, Table 6.3) devised a typically metamorphic process–form nomenclature, which recognised the change from less deformed glacitectonites (folds, faults, rafts) to increasingly deformed glaciocataclastites (breccia, boudinage) and glaciomylonites (gneissic and schistose styles of deformation). Because metamorphism involves quite different processes to modern

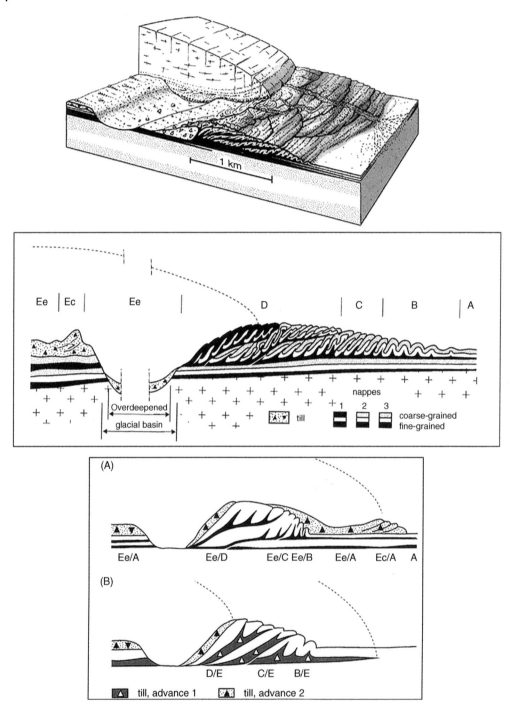

Figure 6.20 A model for the production of glacitectonite in glacier sub-marginal locations based on the margin of Holmströmbreen, Svalbard by van der Wateren (1995a, b, 2003). Upper panel: reconstruction of the late-nineteenth-century surge margin and thrust moraine of Holmströmbreen. Middle panel: glacitectonic styles as they relate to the ice margin (dashed line). A = undeformed foreland, B = steeply inclined structures, C = overturned and recumbent structures, D = nappes, E = extensional structures, including Ec (internally compressive structures), Ee (strong extension). Lower panel: examples of overprinted glacitectonic styles in (A) advance sequence and (B) readvance sequence during retreat.

glacitectonic regimes, such as pressure and heating, it is not particularly appropriate to use the terminology in this way, but the stress/strain continuum recognised in this scheme is nevertheless relevant to the present discussion.

Similarly, good analogues have been identified for the environment in which this takes place, specifically by van der Wateren (1995a, 2003), who likens the setting to accretionary wedges in structural geology. He views the glacier sole as the plate overriding and scraping sediments of variable shear strengths off their basement and accreting them to the advancing wedge. He uses the modern thrust and sheared masses around the margin of Holmströmbreen, Svalbard, to demonstrate the architecture of this process–form regime (Figure 6.20), subdividing the sub-marginal to proglacial area into five structural styles A–E, with styles A–D being representative of proglacial thrusting and composite moraine construction. In terms of melange/glacitectonite production, the subdivisions within zone E are instructive. Here the style of deformation is characterised by boudinage, boudinaged folds and folded boudins, sub-horizontal shear planes, transposed foliation, typical of extremely high simple shear strain and horizontal extension. The specific style is either compressional (Ec) and characterised by stacked till sheets and other compressive structures, or extensional (Ee), where the substratum is strongly eroded and overdeepened. Overprinted signatures are observed wherever advances or readvances result in the overriding of the thrust moraine (Figures 6.19 and 6.20). The former results in style E overprinting the compressive structures of zones B, C, D and the undeformed foreland of zone A. The latter results in styles B, C and D overprinting style E structures. Hence, the nature of glacitectonite can be complex and multi-generational, often including subglacial tills.

The shear zone operating beneath the ice in the scenarios depicted in Figures 6.19 and 6.20 has been summarised by van der Wateren (1995a) using the vertically increasingly deformed signatures identified in the models of Banham (1977), Pedersen (1989) and Kozarski and Kasprzak (1994; Figures 2.8 and 4.6; Table 6.3) and informed by shear zones developed at the base of nappes, both in hard rock geology and proglacial thrust moraines. It is subdivided (Figure 6.21) into a lower horizon (S_r) of

Table 6.3 A Process-form based nomenclature for glacitectonites based upon the application of structural terms from metamorphic petrology (from Kozarski & Kasprzak 1994).

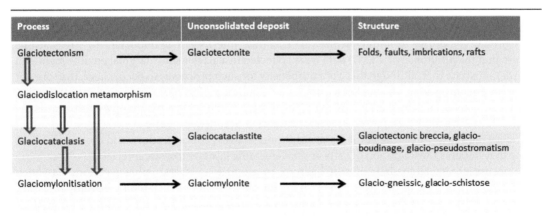

Process	Unconsolidated deposit	Structure
Glaciotectonism ⟶	Glaciotectonite ⟶	Folds, faults, imbrications, rafts
⇩ Glaciodislocation metamorphism		
⇩ ⇩ Glaciocataclasis ⟶	Glaciocataclastite ⟶	Glaciotectonic breccia, glacio-boudinage, glacio-pseudostromatism
⇩ ⇩ Glaciomylonitisation ⟶	Glaciomylonite ⟶	Glacio-gneissic, glacio-schistose

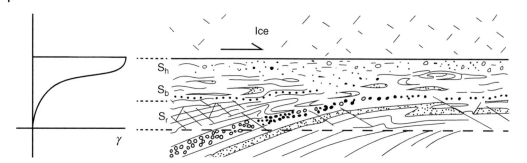

Figure 6.21 Schematic diagram to display the typical vertical continuum of structures in a subglacial shear zone (from van der Wateren, 1995a).

minimal strain, characterised by rooted structures such as drag folds, an intermediate zone (S_b) of strongly deformed material, characterised by transposed foliation of boudins, rootless folds and highly attenuated bands, and an upper zone (S_h) of maximum strain where materials are completely homogenised into diamictons.

This model of shear zone development illustrates that, despite the lack of deep-seated disturbance in deformable substrates, once ice advances over them and reaches a significant thickness, glacitectonic deformation appears to continue to play a significant role in subglacial till continuity (Alley, 2000; Iverson, 2010), likely adding to the processes of regelation and melting and ice keel ploughing in replenishing the mass lost to down-ice advection. For example, disturbance of materials below the base of the deforming layer has been explained by the ploughing theory as a product of the largest clasts or ice keels protruding through it to the underlying substrate and therein liberating soft clasts or rafts (see Section 6.3). Indeed, any perturbation in the sub-till materials can generate fresh material at the base of the deforming layer. This is illustrated by Boulton *et al.*'s (2001) model of 'tectonic/depositional slices' (Figure 6.22) wherein till layers are accreted incrementally from the base upwards as drag folds or by fault propagation folding (Brandes and Le Heron, 2010), each liberation event creating 'slices' comprising folded, attenuated and boudinaged substrate materials. Based upon their measurements of groundwater response to glacier overriding, Boulton *et al.* (2001) explain this type of deformation at depth as a response to varying water pressures in sub-till aquifers (see Figure 6.11a), giving rise to folding at the till–substrate interface; it is important to stress at this juncture that the Boulton *et al.* (2001) data were collected in a glacier sub-marginal environment and hence demonstrate that, although there is a propensity for such processes to occur there, they are not necessarily operating under the thicker ice up glacier. Boulton *et al.*'s (2001) findings demonstrate that there are two modes of deformation, involving the increase in shear strain both upwards towards the glacier sole (see Section 6.2) and downwards towards a decollement zone at or around the top of the underlying substratum. The latter produces localised stress concentrations and the concomitant folding that initiates the liberation of rafts at the deforming layer–substrate interface.

The occurrence of large slabs of soft bedrock in anomalous stratigraphic positions has been explained by the process of raft detachment along decollement zones dictated by bedding planes in the strata (Christiansen, 1971; Moran, 1971; Ruszczynska-Szenajch, 1976, 1987; Stalker, 1976; Ringberg *et al.*, 1984; Aber *et al.*, 1989; Hopson, 1995). For thick overlying ice to detach such megablocks

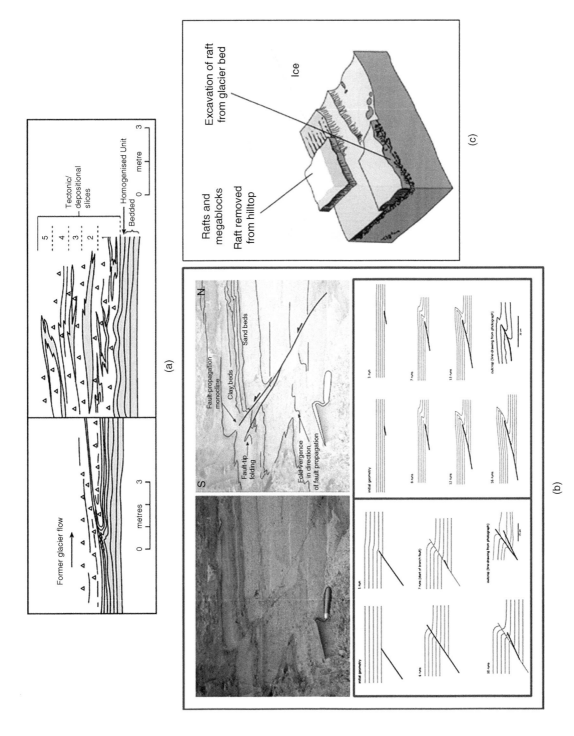

Figure 6.22 Modes of raft detachment or substrate cannibalisation in subglacial deforming materials: (a) Boulton *et al's* (2001) model of 'tectonic/depositional slices'; (b) fault propagation folding (from Brandes and Le Heron, 2010); (c) generalised sketch of detachment of large slabs of soft bedrock where decollement zones are dictated by horizontal strata (from Evans and Benn, 2001).

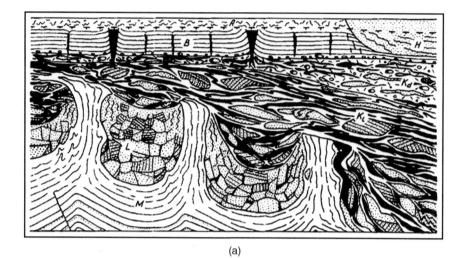

(a)

(b)

(c)

(d)

(e)

Figure 6.23 Examples of deformed ice masses entombed in permafrost: (a) section sketch of exposure in coastal bluff on the Yamal west coast, Siberia (from Astakhov *et al.,* 1996). M = clayey silt of the Marresale Formation, which is folded and occasionally faulted and displays diapirs, L = brecciated Labsuyaha sand, Kl = local Kara till, Kd = foreign diamict facies, B = Baydarata sandy silt with large ice wedges and minor syngenetic ice veins, H = Holocene limnic sediments with collapse structures, A = active layer. Thin lines are sedimentary and deformation structures and thick lines and blackened areas are ice. Dotted pattern represents sand. The Kl and Kd facies are heavily glacitectonised; (b)–(d) structures reported by Waller *et al.* (2009) as indicative of glacier–permafrost interactions at the former northwest margin of the Laurentide Ice Sheet in the Western Canadian Arctic; (b) frozen glacitectonite up to 8 m thick overlying massive ice; (c) raft of massive ice approximately 15 m long within frozen glacitectonite; (d) sand lenses and fold noses between silty clay indicative of glacier sub-marginal ductile deformation of ice-rich materials; (e) relict basal ice of the former Laurentide Ice Sheet, showing tight folds highlighted by layers of debris-poor ice.

or megarafts (Figure 6.22), it is likely that the basal freeze-on process identified in modern Antarctic ice streams (cf. Iverson, 2000; Bougamont *et al.*, 2003; Christoffersen and Tulaczyk, 2003; Vogel *et al.*, 2003) was effective in liberating the substrate in locations where deforming layer tills were temporarily thin or absent. These various processes of liberation of the substrate and its subsequent incorporation into the deforming layer to create first glacitectonite, then pseudo-laminated diamictons and ultimately subglacial till via gradual homogenisation, has been termed 'cannibalisation' (Hicock and Dreimanis, 1992a, b; Evans, 2000).

In a strict sense, the deformation of permafrost is also the development of a glacitectonite; but, as this material would couple to the overriding glacier ice, it would effectively constitute basal debris-rich or stratified ice facies in most instances and hence contribute to the production of melt-out till if the deglaciated landscape was completely de-iced. Numerous case studies have been reported in which such deformed ice masses, either debris-rich basal ice or permafrost glacitectonite, lie entombed in the permafrost (Mackay, 1956, 1959; Mackay *et al.*, 1972; Kaplyanskaya and Tarnogradskiy, 1986; Astakhov and Isayeva, 1988; Astakhov *et al.*, 1996; Waller, 2001; Dyke and Evans, 2003; Murton *et al.*, 2005; Waller *et al.*, 2009; Figure 6.23) and hence presumably merely become reconstituted as basal debris-rich ice facies or increasingly more mature glacitectonites during each glaciation. Continuous permafrost will respond in various ways to overriding glacier ice depending on grain size characteristics (Mathews and Mackay, 1960; Tsytovitch, 1975). In the case of coarse-grained sediments, higher ice contents raise peak shear strengths and hence they can resist deformation (Goughnour and Andersland, 1968; Nickling and Bennett, 1984). Conversely, fine-grained materials with large clay contents can contain large volumes of liquid water at sub-freezing temperatures and hence will deform in response to concomitant high porewater pressures (Andersland and Alnouri, 1970) and may even initiate glacier sliding at temperatures down to −17°C (Echelmeyer and Wang Zhongxiang, 1987; Cuffey *et al.*, 1999). Discontinuous, patchy or sporadic permafrost, because of its spatially highly variable water and ice content, would be ideal for the production of glacigenic melanges and rafts, especially if they were subject to gravitational mass wasting during thrust/composite moraine construction prior to glacier overriding.

7

Subglacial Sedimentary Processes: Laboratory and Modelling Experiments on Till Evolution

Experiments of this type are essential in helping field geologists distinguish fact from folklore.
Clarke (2005, p. 263)

The various ways in which till might behave in subglacial settings were introduced in Section 4.3, where the rheology of such materials was briefly explained in terms of viscous and plastic modes of failure. Details on the nature of this behaviour, specifically in relation to subglacial deformation, were discussed in Section 6.2, where the benchmark Breiðamerkurjökull experiments were seen to be crucial to the erection of a till flow law characterised by a non-linear viscous or Bingham-type viscous rheology. Problems with this flow law became apparent after further subglacial experiments appeared to predominantly suggest a Coulomb-plastic rheology (Table 6.1). Verification of such field observations has been delivered by laboratory experiments (Table 7.1), which until recently had been few in number, a situation that Iverson *et al.* (2008) view as a serious shortfall in the glacial research arena:

> Inferring depositional processes or dynamics of past glaciers from tills in the geologic record is difficult because the basis for interpretation of till properties is commonly weak. … Despite its clear utility, experimental reductionism is an approach that is nearly absent in the study of glacigenic sediments.
>
> Thomason and Iverson (2006, p. 1027)

Such experiments have evaluated the tendency for shearing materials like till to develop particle fabrics (see Section 4.5 and Chapter 8) and attempted to quantify fabric strengths in terms of the strain magnitude. This section reviews the details of laboratory experiments on the deformation of tills in the context of their rheology, and we will review the test results in terms of till fabrics in Chapter 8.

Earlier experiments, such as those by Engelhardt *et al.* (1990) and Kamb (1991) on till samples recovered from the bed of Ice Stream B (Whillans Ice Stream), involved rate-controlled and stress-controlled shear box tests and demonstrated a Coulomb-plastic rheology. Subsequent experiments (e.g. Iverson *et al.*, 1998; Hooyer and Iverson, 2000) employed a ring shear apparatus, allowing samples to be deformed to high cumulative strains. Using modern till from the bed of Storglaciären, Sweden, and ancient tills from the Late Wisconsinan glacial stratigraphy of the Superior and Michigan lobes of the southern Laurentide Ice Sheet, Iverson *et al.* (1998) found an almost Coulomb-plastic behaviour. Both ring shear and triaxial tests of undisturbed till cores and

Table 7.1 Examples of laboratory evidence for subglacial till behaviour.

Source	Experiment and nature of evidence	Till behaviour
Engelhardt *et al.* (1990)	Triaxial test on clay-rich till from Ice Stream B	Perfect plastic rheology
Kamb (1991)	Shear experiment on Ice Stream B till (<2 mm fraction) – single plane failure	Plastic deformation
Iverson *et al.* (1997, 1998)	Ring shear tests on modern Stoglaciaren and Pleistocene tills and linear viscous putty – strain localization in the tills	Perfect plastic rheology – failure along aligned clay particles
Tulaczyk (1999)	Ring shear tests and theoretical analysis	Coulomb-plastic rheology, fine-grained tills facilitate sliding and clast ploughing = shear zone of 0.01 m thickness. Coarse-grained tills = ice–till coupling and deformation down to 0.1 m.
Hooyer and Iverson (2000)	Ring shear tests on beads in putty and ancient tills	Diffusive mixing at boundaries of shearing granular layers. Constant clast rotation in putty but tendency to align a-axis permanently in direction of shear in till.
Iverson and Iverson (2001)	Modelling of ring shear tests and previous field data from Breiðamerkurjökull	Distributed shear via multiple failure planes (convex upward displacement profiles) = Coulomb-plastic response
Watts and Carr (2002)	Deformation tank using various materials	Brittle deformation along distinct shear planes (convex upward/S shape displacement curve) = plastic response
Moore and Iverson (2002)	Subjected Storglaciaren till and ancient Laurentide Ice Sheet, Des Moines lobe till to constant stress in ring shear device.	Slow shearing in ten slip episodes, involving pore dilation during shear and attendant porewater pressure decline and strengthening, followed by gradual pore-pressure recovery and weakening. At 'critical-state', porosity till failed catastrophically.
Hiemstra and Rijsdijk (2003)	Shear test on potter's clay	Microscale shear planes and grain rotations due to Coulomb-plastic failure
Thomason and Iverson (2006)	Ring-shear tests on water-saturated basal tills	Coulomb-plastic response: microscale shear planes and sand grain fabric developing
Rousselot and Fischer (2005, 2007)	Ploughing experiments using rotary ploughing device	Coulomb-plastic behaviour: variable pore pressure response related to location near the object creates either compaction or dilation.
Rathbun *et al.* (2008)	Deformation of Matanuska Glacier till in shearing apparatus	Nearly Coulomb-plastic deformation.
Thomason and Iverson (2008)	Ring-shear tests of Douglas Till and Batestown Till of south Laurentide Ice Sheet	Sand grain long-axis fabrics became progressively stronger with strain and reached steady state at strains of 7–39
Altuhafi *et al.* (2009)	Ring-shear tests on modern basal till from Langjökull, Iceland (secondary data from modern Storglaciaren and Whillans Ice Stream tills and ancient Two Rivers till)	Time-dependent (viscous) behaviour at low stresses

remoulded samples from below the Antarctic ice streams show convincingly that these tills also exhibit a Coulomb-plastic or nearly Coulomb-plastic rheology (Tulaczyk *et al.*, 2000a; Kamb, 2001). Importantly, Iverson and Iverson (2001) identified distributed shear, displayed as a convex upward displacement curve, in laboratory tests on Breiðamerkurjökull till. In contrast to Boulton and Hindmarsh (1987), Iverson and Iverson (2001) interpret this pattern of failure as an indication not of a viscous rheology but a Coulomb-plastic, specifically because failure occurred along numerous slip planes. A similar convex upward displacement curve, with failure taking place along discrete failure planes, was reproduced also by Watts and Carr (2002), based upon experiments in a deformation tank. These consistently reproduced failure patterns have led to a consensus (Fowler, 2002, 2003) that the rheology of subglacial till is typical of a Coulomb-plastic, but the spatially distributed deformation pattern of displacement is one resembling that of a viscous material.

Although they cannot replicate the exact nature of subglacial environments, experiments creating artificially induced strain on potter's clay provide clear evidence of the structural patterns associated with the deformation of fine-grained till matrixes. This has been illustrated at the microscale by shear tests undertaken by Hiemstra and Rijsdijk (2003), who identified the development of grain alignments and plasmic fabrics, albeit at a very low strain of <0.1, leading to the development of shear planes and ultimately to branching and merging unistrial plasmic fabrics (Figure 7.1).

Such visible structural patterns of deformation relate to the gradual compaction of materials during shear, an important process in the development of fine-grained compacted subglacial tills in the geological record. Ring shear tests have been instructive in understanding how tills may strengthen in response to the dilatency process that we understand as being critical to the subglacial deformation of materials (Clarke, 1987; Iverson *et al.*, 1998; see Section 4.3). For example, Moore and Iverson (2002) subjected modern Storglaciaren till and ancient Laurentide Ice Sheet, Des Moines lobe till to constant stresses and found that they sheared slowly without unstable acceleration in ten slip episodes. The process involved pore dilation during shear with attendant porewater-pressure decline and consequent strengthening, followed by gradual pore-pressure recovery and weakening. When the sediment had dilated to what is termed its 'critical-state' porosity (see Section 4.3), it could not dilate further and so the till failed catastrophically. From these experiments, it was concluded that the pore pressure decreases and consequent strengthening caused by shear-induced dilation can: (1) suppress rapid shear of subglacial till and fault gouge, and (2) result in slow episodic shear at rates controlled by till porosity and hydraulic diffusivity.

Also apparent in the ring shear experiments conducted by Hooyer and Iverson (2000) was mixing across the boundaries between shearing granular layers like tills. Employing the distribution of index lithologies across till boundaries, they identify that the mixing coefficient (D) decreases with strain and an upper limit on bed shear strain can be calculated. This mixing can effectively give rise to the development of gradational contacts between till units or layers and has been used to propose pervasive deformation by some field researchers (e.g. Kemmis, 1981; Clayton *et al.*, 1989). This contrasts with predictions of sharper boundaries, specifically by Alley (1991), as the product of a downward increase in effective pressure; this involves till deforming too slowly at a critical depth to maintain its high porosity and therefore being compact and unable to shear. The resulting sharp boundary between the upper deforming layer and the lower rigid layer is compatible with the A and B horizons identified at Breiðamerkurjökull. Whether or not such boundaries are sharp or diffuse, the occurrence of distinctly different lithological signatures above and below them indicate that the provenance of the till certainly changes and that this reflects the importation of different matrixes, which potentially introduces different rheologies and deformation signatures in till sequences.

In an attempt to address both laboratory and field observations, Damsgaard *et al.* (2013, 2015) have developed a numerical model of a two-way coupled particle–fluid mixture undergoing shear

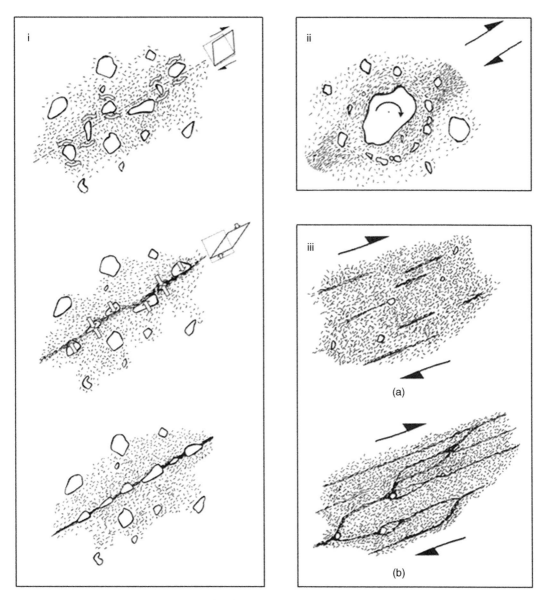

Figure 7.1 Micromorphological evidence of artificially induced strain developed during a shearing experiment on potter's clay (from Hiemstra and Rijsdijk, 2003): (i) grain alignments and plasmic fabrics that develop sequentially (from top to bottom) of grain lineaments and unidirectional plasmic fabrics. Dashed line in upper sketch represents the position of a developing shear plane. Elongate grains near the shear plane rotate until they are aligned plane-parallel. They also move towards the shear plane due to contraction of the sediment (see strain boxes); (ii) sketch to illustrate the relationship between unidirectional plasmic fabrics, skelsepic plasmic fabrics and turbate structures; (iii) schematic diagram to show the development of branching and merging characteristics in unistrial plasmic fabrics. Short, discontinuous unistrials in (a) grow to form continuous features in (b), and where unistrials meet they split or bifurcate.

deformation. The results reveal vertical displacement curves for fluid and particle behaviour with progressive shear strain (Figure 7.2) and confirm the findings of previous investigations (e.g. Iverson *et al.*, 1998; Moore and Iverson, 2002) that subglacial materials undergo transient strengthening. This demonstrates that the deformation of dense granular material is governed by inter-grain contact mechanics, so that the porosity evolves asymptotically towards a critical-state value with increasing shear strain. The model also confirms a plastic rheology but dilative strengthening appears to contribute to the material strength during the early stages of fast deformation. Spatial and temporal variability in subglacial till forming processes (triggering variable glacier movement) is also acknowledged in this study, whereby: (a) subglacial mosaics of deforming and stable spots will create localised sinks for meltwater in patches of deforming materials; (b) sediment dilation can result in substantial thinning of water films at the ice–bed interface; and (c) halts in subglacial shearing movement initiates gradual sediment weakening due to the fluid pressure readjusting to hydrostatic. When combined with the replenishment of till forming materials by whatever process appropriate to the particular setting (i.e. regelation and melt-out, ploughing, raft liberation), these modifications of the deforming layer by shear-induced dilation have the potential to vertically accrete multiple subglacial till units.

Although subglacial observations, laboratory experiments and modelling exercises have predominantly highlighted a Coulomb-plastic rheology for tills, the potential for viscous behaviour has certainly not been entirely dismissed and has even been identified as significant in some case studies (cf. Jensen *et al.*, 1995, 1996; Sane *et al.*, 2008; Altuhafi *et al.*, 2009). Indeed, geotechnical engineering recognises that plastic and viscous behaviours are not mutually exclusive, and pseudo-viscous till behaviour is acknowledged specifically in assessments of dilatancy (see Section 4.3). Moreover, a pressure-dependent viscosity has been assumed in a number of modelling exercises (e.g. Hindmarsh, 1998a, b; Fowler, 2000, 2003; Schoof, 2007), especially those pursuing a universal explanation for till-cored subglacial bedforms like drumlins (Stokes *et al.*, 2013). Criticisms of laboratory-based tests inevitably highlight the non-representative conditions, such as the selection of only the fine-grained matrix and the tendency for ring shear devices to shear only within a narrow band in the centre of the sample, despite repeated replication of exponential displacement curves indicative of distributed shear increasing in magnitude towards the ice–till interface. For researchers who are unconvinced that a viscous rheology is appropriate for subglacial till (e.g. Iverson and Iverson, 2001), such curves have been regarded as merely 'apparently viscous' when in fact they represent distributed Coulomb-plastic slip events. In Section 4.3, a viscoplastic rheology was briefly entertained for materials in subglacial environments, because meltwater volumes and porewater pressures vary and there is temporal and spatial variability in strain, however the experimental reductionism that is represented by laboratory ring shear tests forms the foundation for understanding the fundamental relationships between subglacial deformation and till properties through the direct measurement of strain magnitude. It is important, therefore, that we heed the sage advice of Thomason and Iverson (2006) quoted at the beginning of this chapter and move forward to the measurement of strain in Chapter 8, but at the same time bear in mind that till sedimentology is not entirely an exercise in determining strain magnitude, for fabric alone does not deliver the answer on depositional genesis.

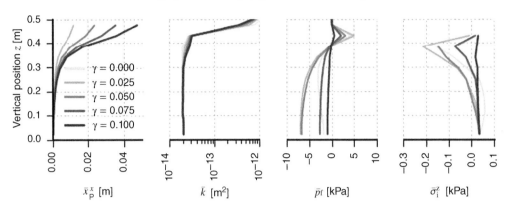

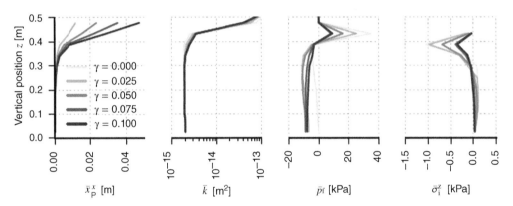

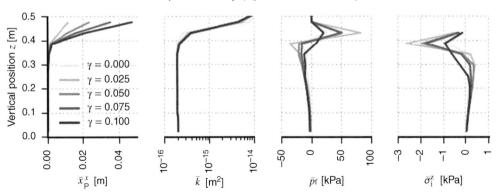

Figure 7.2 Numerical modelling results for a particle–fluid mixture undergoing shear deformation (from Damsgaard *et al.*, 2013, 2015).

8

Measuring Strain Signatures in Glacial Deposits

No section, however carefully drawn, is of more than ephemeral interest.
Sir H.H. Howorth (cited with disdain in Slater, 1926, p. 395)

The assessment of strain signature using macrofabric analysis has a long history in glacial geology (see Section 4.5), but microscale applications and laboratory experiments in particular have evolved more recently, especially in relation to attempts at strain quantification (Table 7.1). The nature of clast behaviour in a shearing medium like till has also been a subject for scrutiny, the results of which have culminated in an emerging consensus that March rotation (see Section 4.5; Figure 4.12) is the dominant mechanism for orientating particles in a subglacial deforming layer. The widespread acknowledgment by the glacial geology community in the late 1970s through to the 1990s that glaciers could be moving over deformable tills led to the routine application of clast macrofabrics in assessing the strain signature of that deformation (Hart, 1994). Some of the literature arising from these applications proposed that thicker deforming layers produced weaker macrofabrics and this was explained as a potential product of what was then regarded as a viscous till rheology (e.g. Hicock, 1992; Hart, 1994, 1997; Benn, 1995; Benn and Evans, 1996), an interpretation that was to be labelled as disciplinary belief or folklore (cf. Clarke, 2005) for glacial sedimentologists.

Although we now understand that Coulomb-plastic behaviour is more likely for deforming tills, numerous observations of macrofabrics weakening in A horizon deforming layers remain problematic, especially as laboratory experiments, as we shall see below, deliver strong fabric signatures at low strains. Glacial sedimentologists have speculated on the cause of such weak clast macrofabrics in tills, proposing ductile intergranular shear, localised dilatancy or even sediment flowage as explanations of this characteristic, potentially controlled by the spatial and temporal history of solid state deformation (e.g. Roberts and Hart, 2005; Evans *et al.*, 2006b). Nevertheless, studies of till sedimentology have consistently acknowledged the role of increasing cumulative strain in the development of stronger clast fabrics (e.g. Hicock *et al.*, 1996; Evans *et al.*, 1998, 1999a, b; Hiemstra and Rijsdijk, 2003) despite Iverson *et al.*'s (2008) criticism that they have done quite the opposite. Problematic in terms of reconciling the remarkably strong fabrics attained in laboratory shear tests and the range of clast macrofabrics (more notably the relatively weak examples) discovered in geological outcrops is the incompatibility of the material properties in the two environments (Table 8.1). However, Iverson and co-workers (cf. Iverson and Hooyer, 2002; Iverson *et al.*, 2008) confidently charge that: (a) no systematic relationship between fabric development and shear-strain magnitude has emerged from field studies due to the lack of an independent means to determine strain magnitude; (b) neither till thickness (Hart, 1994, 1995; Hart and Rose, 2001) nor fluidity (Dowdeswell *et al.*, 1985; Benn, 1995;

Till: A Glacial Process Sedimentology, First Edition. David J A Evans.
© 2018 John Wiley & Sons Ltd. Published 2018 by John Wiley & Sons Ltd.

Table 8.1 Clast macrofabric strengths (S_1 eigenvalues) for tills from a range of laboratory and field case studies (brown = subglacial till and glacitectonites; blue = melt-out till; A axis data unless listed otherwise).

Case study	Till/glacitectonite type	S_1 eigenvalue range	S_1 eigenvalue mean
Dowdeswell *et al.* (1985)	Lodgement till = subglacial traction till	0.63–0.79	0.69
Dowdeswell and Sharp (1986) – inc. Kruger (1982)	Lodgement till (deformed and undeformed) = subglacial traction till	0.45–0.79	0.60
Catto (1990)	Subglacial till on roches moutonées	0.73–0.93	0.81
Catto (1998)	Basal till (including melt-out till in which eigenvalues are at the upper and lower extreme values)	0.47–0.92	0.62
Hicock (1991)	Clast pavement	0.56–0.79	0.64
Hicock *et al.* (1996) – inc. Little (1995), Hicock (1987, 1991, 1992), Goff (1993), Hicock and Dreimanis (1995), Hicock and Fuller (1995)	Lodgement and ductile flow with minor melt-out = subglacial traction till	0.47–0.95	0.62
Benn (1994)	Subglacial till	0.65–0.79	0.72
Benn (1995)	Subglacial till (upper till/A horizon) fluted	0.60–0.71	0.67
	Subglacial till (upper till/A horizon) unfluted	0.48–0.56	0.52
	Subglacial till (lower till/B horizon)	0.61–0.72	0.66
Gordon *et al.* (1992)	Fluted till	0.50–0.85	0.66
Evans *et al.* (1998)	Glacitectonite continuum	0.46–0.61	0.54
Evans and Rea (2003)	Fluted till	0.41–0.62	0.51
	Ploughed boulder prow (unpublished)	0.37–0.63	0.52
Evans (2000)	Subglacial traction till (A/B planes):		
	A horizon	0.50–0.52	0.51
	B horizon	0.47–0.60	0.53
Evans and Twigg (2002)	Subglacial traction till (A/B planes):		
	A horizon	0.46–0.50	0.47
	B horizon	0.56–0.67	0.61
	Glacitectonite/amalgamation zone and subglacial traction till carapace over thrust moraine (A/B planes):		
	Amalgamation zone	0.43–0.68	0.56
	Subglacial traction till (overprinted A and B horizons)	0.48–0.67	0.55
Carr and Rose (2003)			
Larsen and Piotrowski (2003)	Subglacial tills	0.65–0.97	0.88
Evans and Hiemstra (2005)	Lodged clasts	0.83–0.94	0.89
Nelson *et al.* (2005)	Fluted till (Bruarjokull surge)	0.52–0.75	0.63
Kjaer *et al.* (2006)	Fluted till (Bruarjokull surge)	0.45–0.84	0.68

(*continued*)

Table 8.1 (Continued)

Case study	Till/glacitectonite type	S_1 eigenvalue range	S_1 eigenvalue mean
Carr and Goddard (2007)	Subglacial till (16–32 mm)	0.60–0.77	0.70
	Subglacial till (8–16 mm)	0.55–0.71	0.64
Iverson *et al.* (2008)	Experimentally sheared tills (consolidation fabric to strain of 30)	0.51–0.94	0.94
Davies *et al.* (2009)	Clast pavement	0.72	
Shumway and Iverson (2009)	Douglas till (basal till)		
	AMS	0.72–0.98	0.85
	Sand particles	0.51–0.71	0.59
Evans *et al.* (2012)	Clast pavement	0.79	0.79
Gentoso *et al.* (2012)	Drumlin till (AMS)	0.33–0.90	0.70
	(Pebble)	0.44–0.71	0.61
	Fluting till (AMS)	0.61–0.84	0.74
	(Pebble)	0.68–0.81	0.74
Ó Cofaigh *et al.* (2013)	MSGL glacitectonite/hybrid till	0.43–0.82	0.57
Spagnolo *et al.* (2015)	MSGL till	0.47–0.94	0.75
Eyles *et al.* (2015)	Sub-marginal fluting till	0.45–0.68	0.60
	Ploughed boulder prow	0.43–0.58	0.51
Evans *et al.* (2016)	Lodged clasts		
	A axes	0.77–0.81	0.79
	A/B planes	0.56–0.59	0.58
Evans *et al.* (2016)	Subglacial traction till		
	Massive (A horizon) – A axes	0.47–0.59	0.53
	– A/B planes	0.40–0.51	0.46
	Fissile (B horizon) – A axes	0.44–0.63	0.55
	– A/B planes	0.44–0.52	0.48
Johnson *et al.* (2010)	Subglacial traction till	0.51–0.82	0.65
Tylmann *et al.* (2013)	Basal till/amalgamation zone and deformed lenses	0.76–0.88	0.81
Lawson (1979)	Melt-out till (modern, Alaska)	0.87–0.89	0.88
Fitzsimons (1990)	Melt-out till (modern, Antarctica)	0.45–0.77	0.58
Hicock *et al.* (1996) – inc. Hicock (1987) and Goff (1993)	Melt-out till (ancient)	0.65–0.83	0.74
Larson *et al.* (2016)	Melt-out till (modern, Alaska)	0.79–0.80	0.795

Benn and Evans, 1996; Evans *et al.*, 2006b), but rather shear-strain magnitude is the variable most closely correlated to fabric development; and (c) fabric variability in field data can be attributed most reasonably to strain heterogeneity. Hence, Iverson and co-workers have embarked upon the task of exploring whether or not there is a simple relationship between fabric development and shear-strain magnitude, the results of which we will now review.

A benchmark paper on the fabric results of ring-shear experiments is that of Hooyer and Iverson (2000), who show that particle long (A) axes rapidly rotate into parallelism with the principal strain axis and remain there, with stable cluster fabrics being attained at cumulative strains of around 2.0 and then no further modification up to strains of at least 370. Subsequently, Thomason and Iverson (2006) identified a clear strengthening of sand grain microfabric with increased strain in ancient tills when they were placed in the ring shear device, with fabric strength attaining S_1 values of 0.71–0.74 at shear strains of 7–39 but then remaining steady up to shear strains of 108 (Figure 8.1). Clearly, such shear strains are entirely realistic beneath ice sheets. In order to generate some compatibility between the analytical interpretations of the results gleaned from laboratory experiment and those collected from field settings, Iverson *et al.* (2008) use the standard clast fabric shape ternary plot

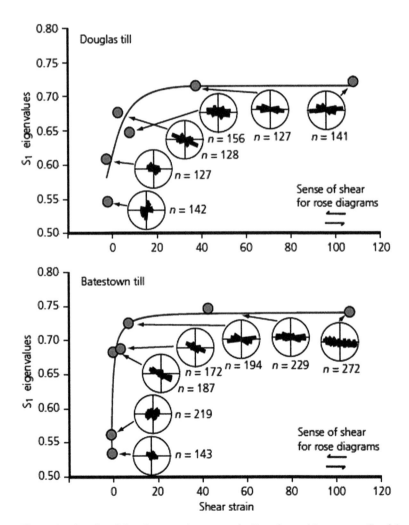

Figure 8.1 Results of shearing experiments on the Douglas and Batestown tills of the Laurentide Ice Sheet, showing microfabric strength related to strains up to 108. The rose diagrams record the alignments of sand grain long axes in two dimensions and therefore S_1 eigenvalues will not fall below 0.5 (Thomason and Iverson, 2006).

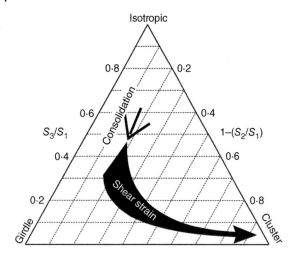

Figure 8.2 The relationship between fabric shape development and consolidation and shear strain, plotted on the standard clast fabric shape ternary plot, based on AMS fabrics produced during shearing experiments (form Iverson *et al.*, 2008).

(Figure 4.14) to indicate the influence of initial consolidation of a till (the imparting of normal stress of 65–85 kPa prior to shearing) and then increasing shear strain on AMS fabric shapes (Figure 8.2); this is represented by the arrow on Figure 8.2 depicting the changing fabric shape with increasing shear strain magnitude, from isotropic to girdle to cluster.

Field macrofabric data, as highlighted by Hooyer and Iverson (2000) and Iverson and Hooyer (2002), although routinely collected from tills, cannot be related quantitatively to shear strain at the ice–bed interface. However, Benn's (1995) study of the freshly exposed subglacial tills at Breiðamerkurjökull, at a site located close to the location where Boulton and Hindmarsh (1987) reported measurements of subglacial till deformation in the 1980s (see Section 6.2), is nonetheless instructive as it provides us with the field fabric signature of that deforming till (Figure 6.9a; cf. Benn, 2002). The till had a two-tiered structure, with a low-strength, high-porosity upper layer (A horizon) overlying a stronger, higher-density lower till (B horizon), and clasts in both horizons had A–B planes lying close to horizontal, with striated, polished facets on their upper and lower surfaces, indicative of clast 'gliding' in quasi-stable orientations. Clast A-axes were strongly clustered, parallel with former ice-flow direction in the lower till and the fluted parts of the upper till. In contrast the unfluted parts of the upper till revealed weaker fabrics (Figure 4.14b), a characteristic interpreted by Benn (1995) as evidence of unsteady strain conditions in response to porewater pressure fluctuations and clast interactions. Such variable and some relatively weak clast macrofabrics in a till that demonstrably had been subject to subglacial deformation was explained by Iverson and Hooyer (2002) as a product of strain heterogeneities (Iverson and Hooyer, 2002; cf. Jaeger and Nagel, 1992; Beeler *et al.*, 1996). Spatially and temporally unsteady patterns of deformation have been proposed by Benn (1995) and Benn and Evans (1996) as possible ways to 'weaken' fabrics, so that March rotation need not necessarily cause monotonic development of strong cluster fabrics.

The many cases of massive ancient and modern subglacial tills with clast macrofabrics that are relatively weak when compared to laboratory shearing experiments have been explained as the product of thick, subglacial deforming layers, in which clasts underwent continuous Jeffery type rotation (Figure 4.12; e.g. Hicock and Dreimanis, 1992a, b; Hart, 1994; Hicock *et al.*, 1996). This has been countered by those who argue that only March type rotation can occur in subglacial deforming layers, and hence any till with a 'weak' fabric cannot have formed by subglacial deformation (e.g. Hooyer and Iverson, 2000; Piotrowski *et al.*, 2001; Thomason and Iverson, 2006; Iverson *et al.*,

2008). Although March type rotation is now firmly established as the most likely mechanism for clast alignment in tills, there is a clear dichotomy still surrounding clast macrofabrics in tills, wherein the same evidence, weak clast macrofabrics, is being used as evidence both for and against subglacial deformation.

In contrast to the variable and sometimes weak fabric signatures collected from subglacially deformed tills, both in ancient deposits and those on the Breiðamerkurjökull foreland, consistently strongly oriented clast macrofabrics in till sequences have been reported and attest to the accretion of strongly sheared subglacial materials, presumably with a Coulomb-plastic rheology. An excellent example is that of Larsen and Piotrowski (2003) and Piotrowski *et al.* (2006) in their study of stacked till units in Poland (Figure 8.3), which reveals both vertically and horizontally consistent strong clast fabric cluster orientations through three till units (A–C) overlying glacitectonised sand over a thickness of 2 m, wherein each till is up to 0.2, 1.45 and 1.1 m thick, respectively, and S_1 eigenvalues range from 0.65 to 0.96. Similarly, Spagnolo *et al.* (2016) report macro- and microfabric data, including

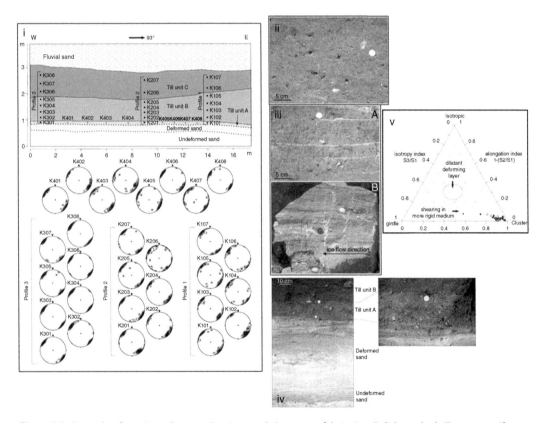

Figure 8.3 Example of consistently strongly orientated clast macrofabrics in a Polish stacked till sequence (from Larsen and Piotrowski, 2003): (i) section sketch showing main facies and locations of fabric samples. Macrofabrics are displayed below as stereonets; (ii) the structureless appearance of Till C; (iii) till unit B showing the characteristic millimetre-thick sand stringers which give the till a stratified appearance (A), and a view on to the top of a decollement surface within the till (B); (iv) transition zone between outwash sands and till, showing gradational contacts between component units which relate to the upwards-increasing strain rate; (v) clast macrofabrics plotted on clast fabric shape ternary diagram.

the results of X-ray tomography, from a homogenous, ≥1.4-m-thick, single massive silty–sandy diamicton within megascale glacial lineations (MSGL), which reveal remarkable consistency in orientation, both vertically and horizontally, aligned parallel with MSGL crests (Figure 8.4). The alignment patterns of clasts at microscale reflect the development of Riedel shears in the shearing zone immediately beneath the ice. Spagnolo *et al.* (2016) conclude that the range of consistently orientated data demonstrates a continuously accreting, shallow but pervasively deforming, bed under consistent basal conditions where a plastic style of deformation was facilitated by continuous sediment supply and an inefficient drainage system. The consistency of the various fabric signatures in such studies is often used to argue against the ploughing model for till deformation, either by ice keels or large clasts, specifically because ploughing is envisaged to displace material sideways away from the keel or clast and would thereby disturb ice-flow parallel fabric orientations. However, the basal shear stresses acting upon both groove and ridge would likely constrain clast deflection equally (i.e. both would be subject to simple shear), and hence it could be argued that spatially consistent, strongly aligned fabrics are not necessarily incompatible with groove-ploughing.

Case studies on clast fabrics in tills appear to be somewhat inconclusive on the subject of variations in fabric strength with clast size, although it has been proposed that abnormally weak fabrics could be identified in certain grain size ranges. For example, Larsen and Piotrowski (2003) found no correlation between fabric strength and grain size for the 7–56 mm grain size range, and laboratory experiments reported by Iverson *et al.* (2008) revealed no correlation below 8 mm. However, Spagnolo *et al.* (2016) isolated variable signatures for different grain size ranges below 2 mm (Figure 8.4), wherein some grain sizes appear to adopt transverse alignments in equal measure to ice-flow parallel orientations. Others have identified a tendency for smaller particles to display weaker or transverse A-axis fabrics than larger particles in the same till (Kjær and Krüger, 1998; Carr and Rose, 2003; Carr and Goddard, 2007), which may be due to smaller particles being more susceptible to collisions with equal-sized or larger neighbours (Thomason and Iverson, 2006). This tendency for fabric strength to be affected by clast collisions in coarser-grained tills with large numbers of variably sized clasts has a precedent in the experimental study of Ildefonse *et al.* (1992). Inhomogeneous or unsteady deformation may therefore produce a wide range of clast fabric strengths, with localised fabric patterns reflecting local strain conditions, but the variability appears to be demonstrated in different grain size ranges depending on the nature of sorting of the sediments. For example, the tendency for clasts to align transverse to ice flow is recognised only in the <2 mm grain sizes in the relatively fine-grained, massive silty–sandy diamicton of Spagnolo *et al.* (2016), whereas coarser-grained tills with abundant cobbles and boulders display localised fabric variability at larger grain sizes.

The latter pattern is well illustrated in the coarser-grained tills of Icelandic glacier forelands where lodged boulders, which are identified using sedimentological criteria that are independent of fabric signature (Sharp, 1982; Kruger, 1984; Evans and Hiemstra, 2005; Evans *et al.* 2016), tend to display very strong preferred orientations, reflecting tightly constrained shear at the ice–till interface, but smaller neighbouring particles do not. This is well illustrated by the tendency for clast fabrics to display herringbone patterns or weak fabrics on the down-flow sides of stoss boulders in flutings (e.g. Rose, 1989, 1992; Benn, 1994a; Evans *et al.*, 2010; Eyles *et al.*, 2015; Figures 6.7 and 8.5). Lodged particles form a bridge between ice and till, and the movement of the ice will tend to rotate the particle until its A-axis and A–B plane are parallel to the plane of shear; an up-glacier imbrication will also develop due to the tendency for the till matrix to plough up in front of the particle (Figures 6.7 and 6.8). Evans *et al.* (2007) presented A–B plane fabrics of boulders lodged on the forelands of Icelandic glaciers, where lodgement was independently determined for those boulders that were partially buried in a fluted till surface, with their upper exposed faces striated parallel to the surrounding flutings and often forming stoss boulders to the flutings. The sample exhibited a strong cluster

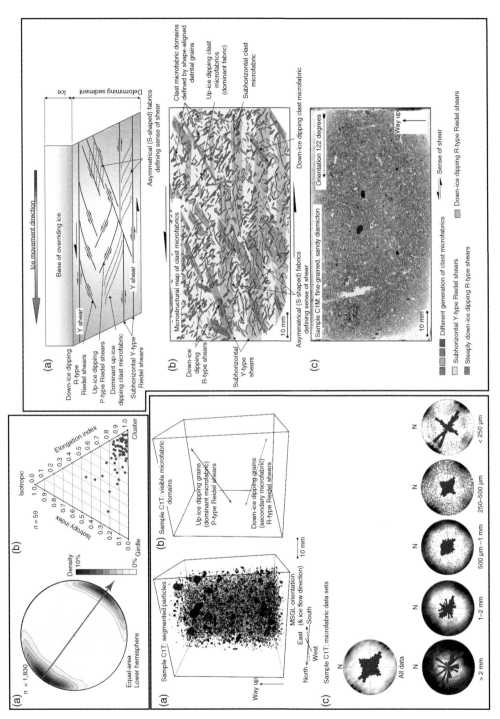

Figure 8.4 Macro- and microfabric data from a homogenous, ≥1.4-m-thick, single massive silty–sandy diamicton within megascale glacial lineations (from Spagnolo *et al.*, 2016): (top left) composite stereonet (a) and clast fabric shape ternary plot (b); (lower left) X-ray tomography image and microfabric subdivided according to grain size; (right) example of thin section mapping, showing: (a) summary of different sets of Riedel shears developed within the diamicton; (b) microstructural map with colours representing different generations of microfabrics which define the Riedel shears, sub-horizontal shear fabric and up-ice dipping foliation; (c) image showing massive, fine-grained sandy nature of the diamicton.

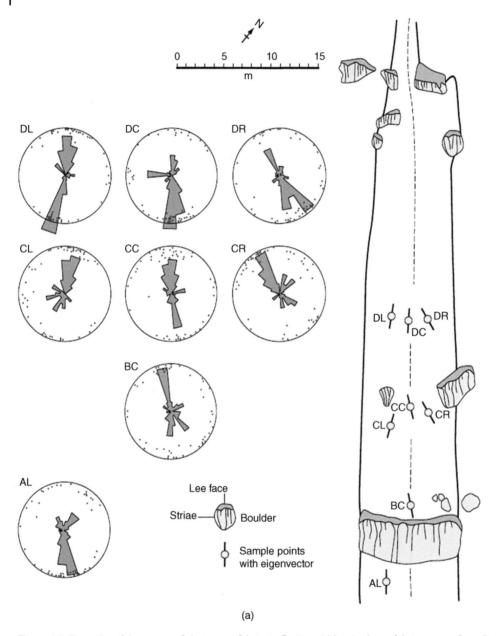

(a)

Figure 8.5 Examples of the nature of clast macrofabrics in flutings: (a) herringbone fabric pattern from fluting at Slettmarkbreen, Norway (from Benn, 1994a); (b) selected flutings and their clast macrofabrics from Sandfellsjökull, Iceland (from Evans *et al.*, 2010); (c) stratigraphy, morphology and fabrics from flutings at Austre Okstindbreen, Norway, displaying herringbone patterns (from Rose, 1989).

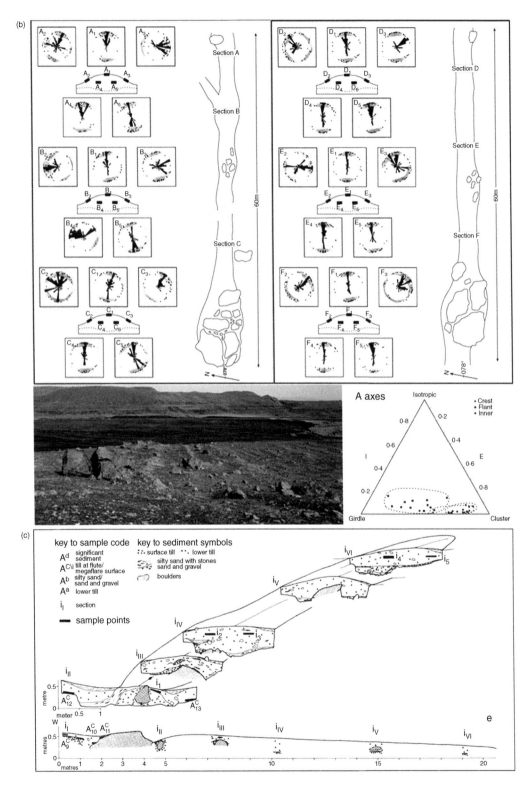

Figure 8.5 (*Continued*)

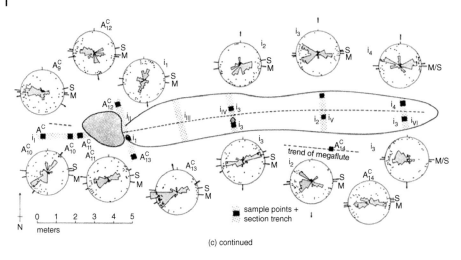

(c) continued

Figure 8.5 (*Continued*)

fabric with near horizontal A–B planes (Figure 4.14c), which was clearly in contrast to the weaker macrofabrics on both A–B planes and A-axes from the A and B horizon tills associated with the lodged boulders and flutings (Figure 4.14b). A similar contrast between the fabrics of unequivocally lodged boulders and other smaller clasts in coarse-grained tills has been identified in a case study of multiple, stacked tills at þorisjökull, Iceland by Evans *et al.* (2016; Figure 8.6). The relatively weak S_1 eigenvalues (A axes = 0.44–0.62; A/B planes = 0.39–0.52), hence higher isotropy, of the Þorisjökull tills are not unlike those previously reported for Icelandic tills (e.g. A axes = 0.51–0.74, A/B planes 0.46–0.67; cf. Sharp, 1984; Dowdeswell and Sharp, 1986; Benn, 1995; Benn and Evans, 1996; Evans, 2000a; Evans and Twigg, 2002; Evans and Hiemstra, 2005), and difficult to reconcile with the shear strains of >100, regarded as typical of modern glaciers and ice sheets. Moreover, laboratory experiments demonstrate that strong 'steady-state' fabrics ($S_1 > 0.78$) are developed at lower strains of only 7–30 (Hooyer and Iverson, 2000; Thomason and Iverson, 2006; Hooyer *et al.*, 2008; Iverson *et al.*, 2008). Hence, the strong lodged boulder fabrics, especially on A-axes, and their contrast with the relatively weak till fabrics at þorisjökull are interpreted as the products of rapid boulder orientation during subglacial shear followed by interruption of clast alignments in the sub-boulder fraction as the till is deformed and deposited in between and around lodged boulders. The þorisjökull till fabric data are not unusual in that the S_1 eigenvalues on A-axis fabrics reported from ancient tills are wide-ranging even though they can be strong (e.g. 0.47–0.95, Hicock *et al.*, 1996; 0.44–0.83, Gentoso *et al.*, 2012; 0.65–0.97, Larsen and Piotrowski, 2003). So many deposits like the Þorisjökull tills record strains too low to represent a steady-state strain signature, despite displaying the sedimentological and structural characteristics of subglacial till emplacement. For example, generally well-orientated till fabrics in recently exposed flutings on the forelands of Sandfellsjökull, Iceland (Figure 8.5b; Evans *et al.*, 2010) and Athabasca Glacier, Canada (Figure 8.7; Eyles *et al.*, 2015) deliver A-axis S_1 eigenvalues of 0.55–0.82 and 0.43–0.68, respectively. Moreover, a 1.5-m-thick till sequence beneath the Sandfellsjökull fluting field (Figure 8.8) displays predominantly weak spread-bimodal and multi-modal fabrics. This indicates that steady-state strains are more difficult to achieve under field conditions than in laboratory experiments, and this appears to be related to grain size characteristics and associated clast collisions and/or large clast interruptions of deformation patterns in the till matrix.

In direct contrast to the strong macrofabrics reported by Spagnolo *et al.* (2016) for MSGL, and hence for subglacial rather than sub-marginal tills, Ó Cofaigh *et al.* (2013) found relatively weak signatures in the Dubawnt Lake palaeo-ice stream MSGL tills (Figure 8.9; cf. Figures 4.14c and 4.15),

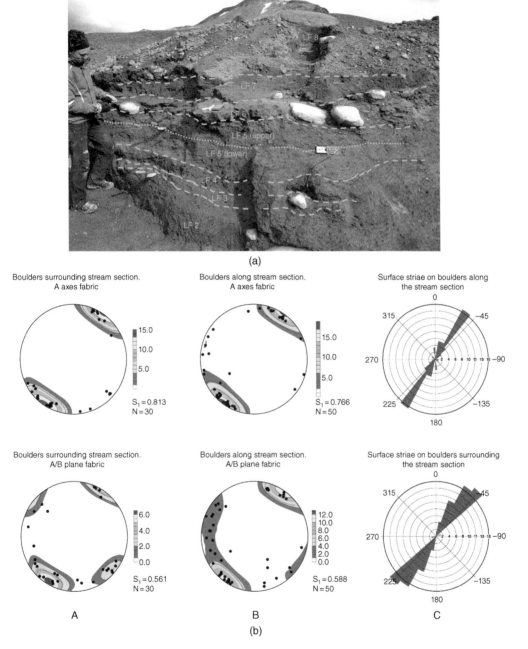

Figure 8.6 Clast macrofabrics from a case study of multiple, stacked tills at þorisjökull, Iceland (from Evans *et al.* 2016): (a) annotated photographic log showing one exposure through the stacked till units and discontinuous clast lines comprising lodged clasts; (b) macrofabric and striae data for lodged boulders from the till sequences; (c) clast macrofabric shape ternary diagrams, depicting the positioning of glacial deposits of known origin as envelopes and the samples from the stacked till units at þorisjökull. The indications of fabric shape development in relation to consolidation and shear strain, after Iverson *et al.* (2008), are also shown.

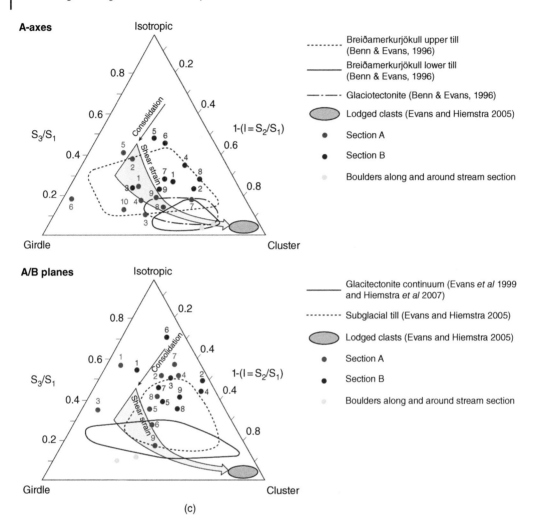

Figure 8.6 (*Continued*)

with S_1 eigenvalues ranging from 0.43 to 0.82. The occurrence of numerous stratified bodies in the till are used to infer a glacitectonite/hybrid till origin, whereby the MSGL were constructed from pre-existing outwash deposits that were subject to non-pervasive deformation and localised lodgement of larger clasts. The short transport distances and incomplete mixing of the material indicate that groove ploughing was likely responsible for the construction of the MSGL during a short period of ice sheet margin readvance and rapid ice flow. A significant difference between the subglacial materials (tills/glacitectonites) at the Dubawnt Lake and Spagnolo *et al.*'s (2016) Polish MSGL sites is the grain size, with the former constituting a more poorly sorted, coarser deposit with significantly more larger clasts of variable size. If we are to accept laboratory findings that strong macrofabrics are achieved at very low strains (Figure 8.1), then the tills in both MSGL studies should tend towards S_1 eigenvalues of 0.70. This is the case for the Polish MSGL with a mean S_1 of 0.75 (range 0.47–0.94; compatible with Larsen and Piotrowski's 2003 data for Polish tills with S_1 mean of 0.88), but not for

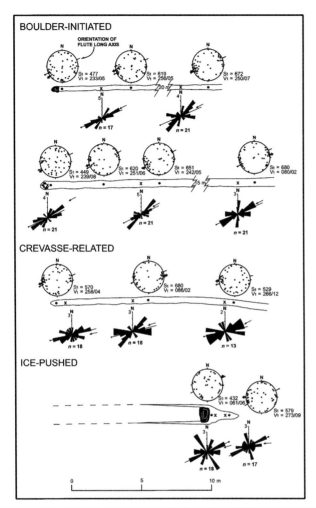

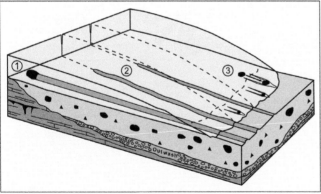

Figure 8.7 Clast macrofabrics and subglacial bedforms on the foreland of the Saskatchewan Glacier, Canada (from Eyles *et al.*, 2015): Upper panel shows macrofabrics measured in boulder-initiated, crevasse-related and ice-pushed flute ridges (ice flow direction from left to right). Lower panel is a schematic reconstruction of the process–form regime in which three principal flute types are created, including (1) boulder-initiated flutes, (2) flute ridge formed by till squeezing into crevasses, and (3) ice-pushed flutes and ploughed grooves.

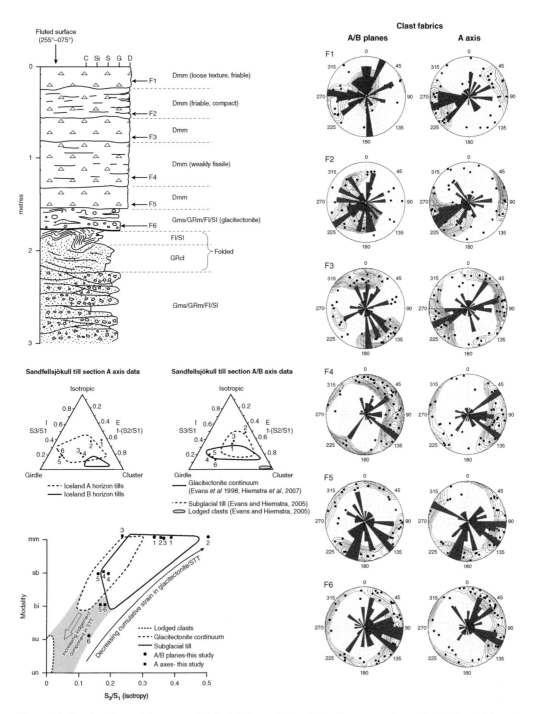

Figure 8.8 Stratigraphy and clast macrofabric data from a 1.5-m-thick till sequence beneath a fluting field on the Sandfellsjökull foreland, Iceland (from Evans *et al.*, 2010).

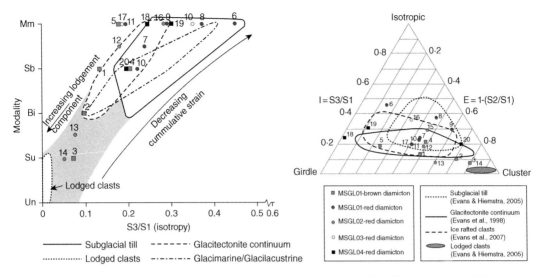

Figure 8.9 Clast macrofabric data from the Dubawnt Lake palaeo-ice stream MSGL tills, plotted on modality-isotropy and fabric shape ternary diagrams (from Ó Cofaigh *et al.*, 2013).

the Dubawnt examples with a mean S_1 of 0.57. It appears likely, therefore, that grain size variability is significant in determining clast macrofabric strengths, due to the increased clast collisions and concomitant disruption of consistent clast alignments that take place in coarse tills.

The variable grain-size distribution and large proportion of cobble and large-sized clasts in Icelandic tills also appear to generate more variable clast fabric strengths within flutings, a direct contrast with the patterns identified in MSGL by Spagnolo *et al.* (2016). This has been demonstrated by Boulton (1976) with clast fabric strengths that vary vertically and horizontally across flutings and which deviate in orientation quite significantly from fluting long axes (Figure 8.10). The occurrence in the clast macrofabrics of bi-modal to spread bi-modal signatures (see Figure 4.15) and high dip angles in the cores of the ridges are instructive in that they suggest clast mobility, potentially even vertical alignment due to till squeezing or flowage. A similar scenario has been identified by Evans and Rea (2003) in the parallel-sided flutings that extend for hundreds of metres on the foreland of the surging glacier Brúarjökull in Iceland, which are characterised by relatively weak clast macrofabrics ($S_1 = 0.41 - 0.62$) particularly in their cores ($S_1 = 0.41 - 0.56$; Table 8.1; Figure 8.11a), prompting Evans and Rea (2003) to propose till flowage into grooves in the ice base created by the numerous lodged boulders; this appears to have been followed by slight strengthening of surface fabrics ($S_1 = 0.49 - 0.62$) as the ice streamlined the fluting. In contrast, Kjær *et al.* (2006) discovered stronger and ice-flow-aligned clast fabrics ($S_1 = 0.45 - 0.84$) in one fluting on the same foreland, similar to fabric strengths reported for the same area, but largely below fluting cores, by Nelson *et al.* (2005; $S_1 = 0.52 - 0.75$), indicating that strain patterns vary between flutings and with till depth. The range of macrofabrics discovered in this area of rapidly fluted bed are potentially related to the various stages of subglacial deformation that are captured in the subglacial landform–sediment associations, with the weaker signatures relating to early stage till squeezing or flowage into newly created cavities, which stopped evolving because the snout surge ceased; good illustrations of this are the weak fabric signatures recorded in the prows developed in front of ploughing boulders ($S_1 = 0.37 - 0.63$; Figure 8.11b), which would be at the core of later stage flutings whose surface clast alignments would be stronger due to better ice–bed coupling.

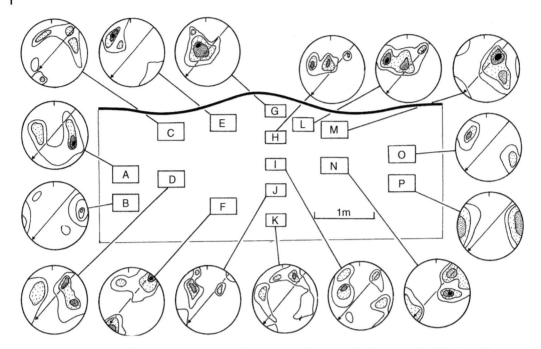

Figure 8.10 Clast macrofabrics from various depths within a fluting on the Breiðamerkurjökull foreland (from Boulton, 1976). Arrows on stereonets show ice-flow direction.

The liquefaction required to initiate till flowage at the ice–bed interface and allow particles to freely rotate (Evans *et al.*, 2006b) has been questioned as a viable subglacial process by Iverson *et al.* (2008), because this scenario entails a material that has zero strength and would lead to catastrophic acceleration of the glacier. Although at Brúarjökull this is exactly what did take place during the fluting-forming surge of 1964 (Evans and Rea, 1999, 2003; Kjær *et al.*, 2008; Rea and Evans, 2011), in the absence of surges the flowage process would be a localised one and more likely to take place under the lower basal shear stresses of sub-marginal settings, as demonstrated by the close association of flutings, marginal crevasse-squeeze ridges and the elongate limbs of sawtooth push/squeeze moraines, as well as 'till eskers', in active temperate settings (cf. Price, 1970; Christoffersen *et al.*, 2005; Evans *et al.*, 2010, 2016; Eyles *et al.*, 2015). Weak ice–bed coupling in the surge and active temperate sub-marginal settings is potentially creating the till liquefaction and concomitant particle momentum exchange (cf. Iverson, 1997) necessary for greater clast mobility than that replicated in the highly confined shearing environments of ring shear devices. Verification of this weak coupling appears to be manifest also in the macrofabric signatures of sub-marginal till sequences.

The clast macrofabric and microstructural signatures within sub-marginal till wedges or the ice-proximal ramps of push moraines, created by incremental till thickening (Evans and Hiemstra, 2005; see Chapter 16), allow us to analyse the strain signatures of sub-marginal deposits. Whereas subglacial tills should be expected to be highly strained to the levels replicated in laboratory shearing experiments, even if tills are thinner or repeatedly overprinted, sub-marginal till stacks are more likely to display a range of clast macrofabric and microfabric strengths. This is due to the changing rheological characteristics of the tills driven by changing environmental temperatures and concomitant porewater pressures beneath a seasonally oscillating glacier snout. Seasonally

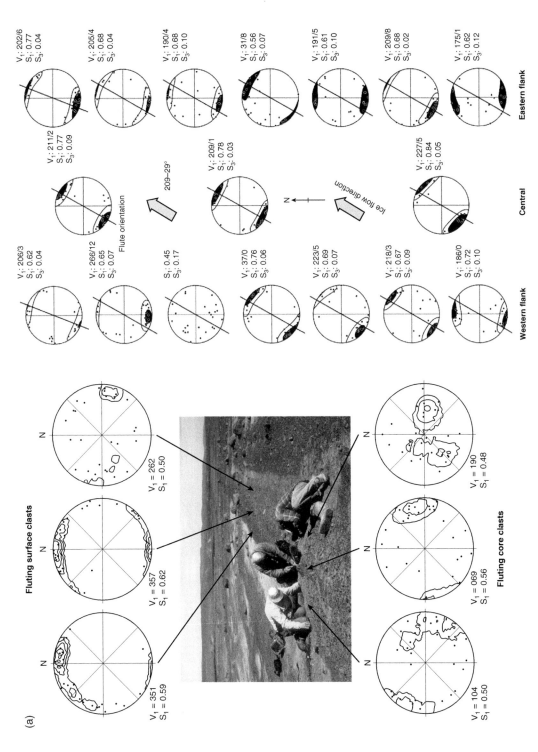

Figure 8.11 Clast macrofabric data from the surge flutings on the Brúarjökull foreland, Iceland; (a) data from parallel-sided flutings. Left panel shows data from the flanks and centre of a single fluting through two separate cross-sections. Right panel shows clast macrofabrics from both flanks and the centre of a 100-m-long fluting (from Kjær *et al*., 2006); (b) data from two separate examples of prows or incipient flutings developed in front of ploughing boulders.

driven sub-marginal processes of deformation, squeezing and freeze-on and melt-out give rise to the advection of subglacial till to the glacier snout and the construction of push moraines in which the macrofabrics are moderately strong and conform to the ice flow directions indicated by surface flutings and strongly orientated lodged clasts, but microstructures indicate that the matrix was subject to water escape and sediment flowage (Figure 8.12a, b). This late-stage modification of the microfabrics of subglacial tills has been identified also by Neudorf *et al.* (2013), who propose that this should not be unexpected for coarse-grained tills. Hence, till fabric strength appears to vary in some situations not just because of clast collisions and localised perturbations set up by larger clasts, but also due to smaller size fractions becoming relatively more mobile in the liquefaction and dewatering of matrixes; in the sub-marginal till wedges this appears to be conditioned by lower basal shear stresses in the outer wedge during the ablation season when push/squeeze moraines

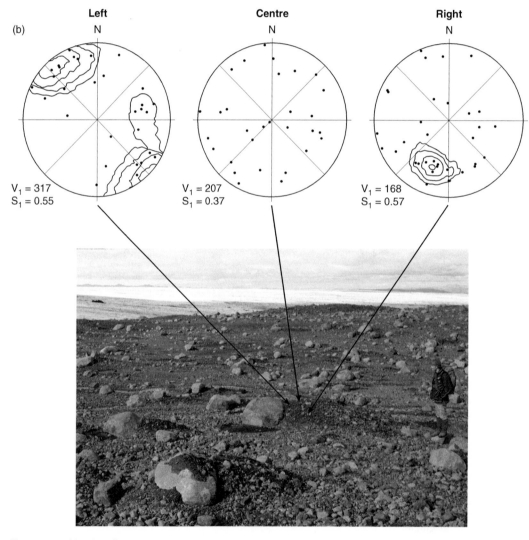

Figure 8.11 *(Continued)*

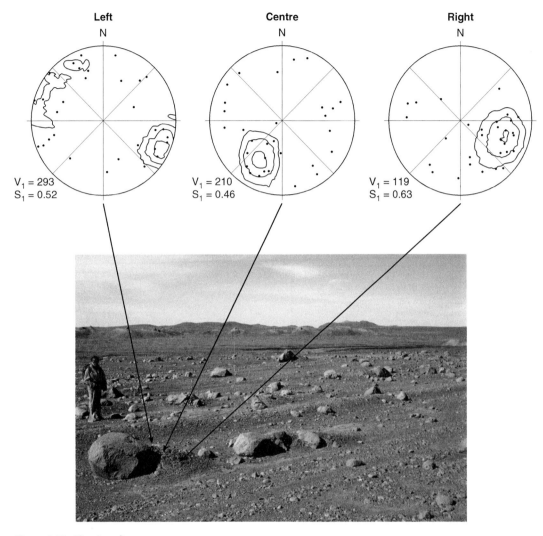

Figure 8.11 (*Continued*)

are constructed (Price, 1970). At the thin end of such sub-marginal till wedges, the combined subglacial processes of lodgement, deformation and ice keel and clast ploughing repeatedly rework and advect the subglacial till to produce overprinted strain signatures and clast pavements; this has been equated to the 'erodent layer hypothesis' (ELH) by Eyles *et al.* (2016), a concept that we will revisit in Chapter 16. Hence, the ice-proximal ramps should contain subglacial till fabric signatures including those of typical A and B horizons but also superimposed horizons with strong cumulative strain signatures (Figure 8.12c). Nevertheless, even here, the S_1 eigenvalues (0.43–0.79; Table 8.1) indicate low shear strain magnitudes (cf. Figure 8.2) for the tills, despite the associated lodged clasts displaying evidence of high strains ($S_1 = 0.83$–0.94; Table 8.1).

Also apparent at glacier sub-marginal settings are down-ice thickening wedges of glacitectonised materials, displaying a range of deformation structures but often containing evidence of deep-seated low strain signatures. Subglacial deformation of such materials creates glacitectonite over cupola hills (*sensu* Aber *et al.*, 1989; Figure 6.19), the sedimentology and broader architecture of which are reviewed in Chapters 14 and 16. The extent of deformation in such glacitectonites and the related patterns of strain that exist within the overlying amalgamation zone of 'tectonic/depositional slices' (cf. Boulton *et al.*, 2001; Figure 6.22) and the capping till are well illustrated by the Brennhola-alda

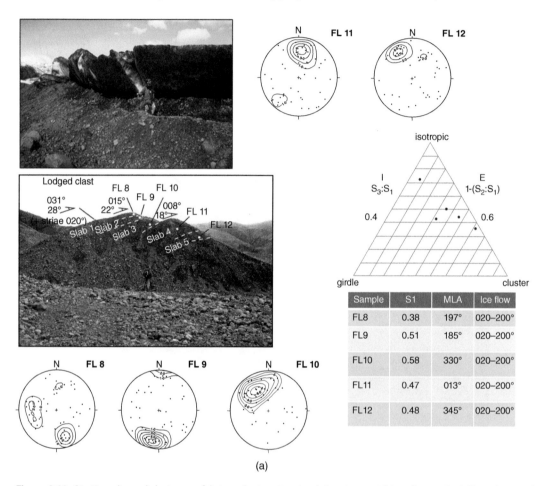

(a)

Figure 8.12 Stratigraphy and clast macrofabric and microstructural signatures within sub-marginal till wedges or the ice-proximal ramps of push moraines: (a) multiple till wedges/moraines constructed at the margin of Fláajökull, Iceland in the early- to mid-1990s (after Evans and Hiemstra, 2005); (b) multiple till units emplaced on the ice-proximal side of a glacially overridden ice-contact subaqueous outwash fan and delta sequence at south Loch Lomond, Scotland (after Benn and Evans, 1996; Phillips *et al.*, 2002; Benn *et al.*, 2004; Evans and Hiemstra, 2005). Upper panel shows the main lithofacies associations, of which LFA 3 includes the stacked till units. Middle photograph and sketch shows the till details and the locations of macrofabrics (data at lower right) and thin section samples (images at lower left). The thin sections reveal: (A) silt-rich till with linear arrangement of silt and sand grains (arrow) and diffusely bounded intraclast (circle); (B) turbate structure with associated lineament 'tail' to the right; (C) anastomosing silt and clay laminae with many silt grains aligned parallel to the lamination; (D) muddy intraclasts (arrows) and casings around individual grains (circles); (E) irregularly shaped mammilated void, composed of intergrown more rounded pore spaces; (F) water escape structure (arrows); (c) data from overprinted till sequences within the proximal ramps of sub-marginal till wedges, including independently collected lodged surface clast data, from Breiðarlon on the Breiðamerkurjökull foreland and Skalafellsjökull, Iceland.

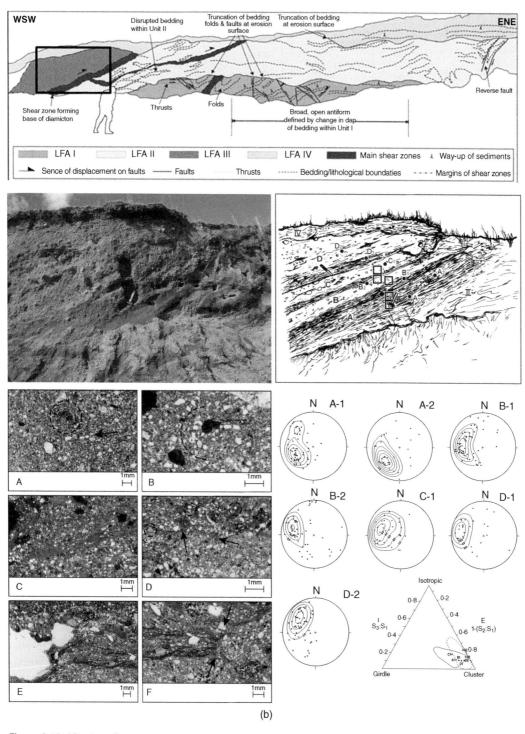

(b)

Figure 8.12 (Continued)

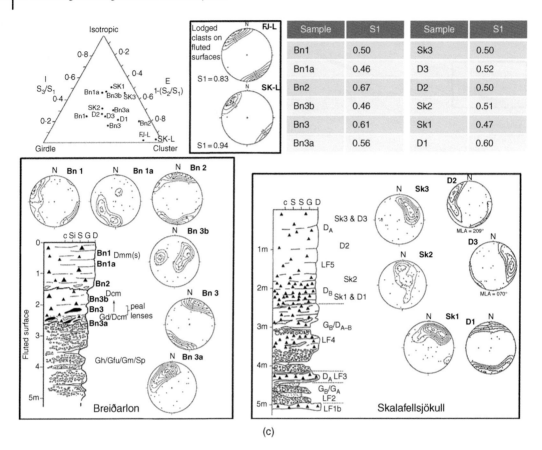

(c)

Figure 8.12 *(Continued)*

overridden thrust moraine at Breiðamerkurjökull (Evans and Twigg, 2002; Evans and Hiemstra, 2005; Figure 8.13). The stratified sand and minor gravel glacitectonite with its numerous Riedel shears, clastic dykes and water escape structures is abruptly truncated by a matrix-supported to pseudo-laminated diamicton with attenuated sand lenses (tectonic slices), which passes vertically into a capping massive, matrix-supported diamicton with localised fissility and sheared sand wisps. The macrofabric signature reveals a remarkably consistent northwest orientation for each 30-cm-thick sampling band (the direction from which shear stresses were imparted during thrust block construction and overriding), but S_1 eigenvalues ranging from 0.43 to 0.68 are indicative of low strains when compared to laboratory tests on ancient tills (Figure 8.1).

Clasts in glacitectonites can display a range of macrofabrics from strong clusters, reflecting the strongly constrained shear that often takes place in such materials (Benn and Evans, 1996), to girdles (Figure 4.14c), depending on the degree of modification of the parent material. Hence, a strain signature continuum has been proposed for glacitectonites using the clast macrofabric strengths of materials whose relative degree of deformation can be assessed independently by their structural appearance (Evans *et al.*, 1998, 2007; Figure 8.14; see Sections 4.2 and 4.4 and Chapter 14). Weaker fabrics, typically girdles, are observed in the more immature tills/tectonic amalgamation zones and

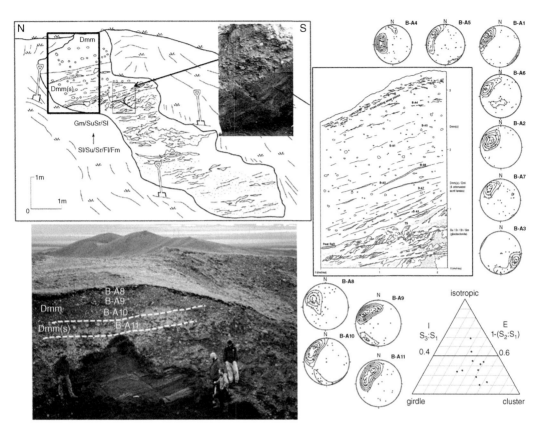

Figure 8.13 Glacitectonite and till carapace over the Brennhola-alda overridden thrust moraine at Breiðamerkurjökull, Iceland (after Evans and Twigg, 2002; Evans and Hiemstra, 2005). Photograph shows stratified sands, silts, clays and fine gravels with peat layers which have been glacitectonically thrust and cross-cut by a sub-vertical clastic dyke (hydrofracture fill) and sheared into an amalgamation zone (clast fabric BA-11) at the base of an overlying till sequence (clast fabrics BA8–10). Top left section sketch shows the details of the upper glacitectonically deformed part of the stratified deposits and the overlying amalgamation zone (Dmm with sheared rafts) and capping till. Inset photograph shows a vertically descending till-filled dyke injected into the stratified deposits during subglacial shearing. Top right section sketch shows the details in the box on the top left sketch and the location of clast macrofabric samples BA1–7.

glacitectonites that have been subject to lower cumulative strain. Fabric strengthening occurs as glacitectonites are moved increasingly further from their source and attenuated.

Strong clast macrofabrics are often predicted for melt-out tills but only a few studies have actually collected data from materials that have unequivocally evolved directly from the passive melting of debris-rich ice (Lawson, 1979a, b; Larson *et al.*, 2016; Figures 4.14d and 6.18; see Section 6.4). This is thought to reflect the original englacial fabrics, but with some overprinting during dewatering and consolidation (Boulton, 1970a; Lawson, 1979a, b; Lundqvist, 1989; Murton *et al.*, 2005; Figure 2.6d). Englacial clast fabrics often have strong preferred orientations parallel to the direction of shear, due to the rotation of clasts by the surrounding deforming ice (Lawson, 1979a; Ham and Mickelson, 1994) but may be transverse in zones of compressive flow (Boulton, 1970b). Although the Matanuska Glacier data set of Lawson (1979a, b; Figures 4.14d and 6.18) has been widely used as a reference for the interpretation of ancient melt-tills, there are a number of reasons why melt-out may not

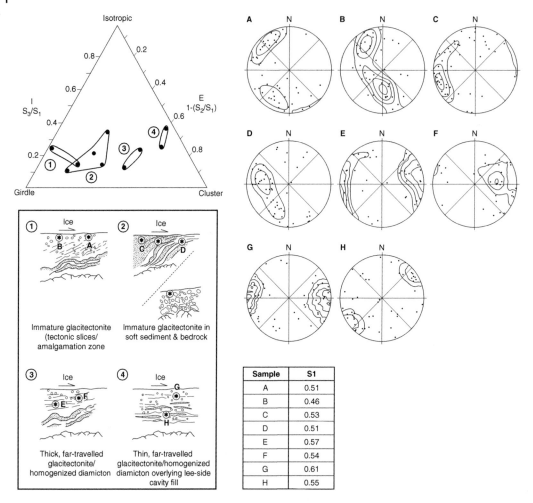

Figure 8.14 Clasts macrofabrics from a selection of glacitectonites (from Evans *et al.*, 1998), displaying an apparent strain signature continuum. Note that the glacitectonite maturity was assessed independently based upon their structural appearance and degree of mixing/homogenisation compared to their parent materials.

create such strong fabrics due to varying degrees of remoulding during and following deposition. For example, the reduction in dip values and an increase in dispersion relative to englacial fabrics may take place, and clast interactions during settling may weaken the preferred orientation.

Significant advances have taken place in recent years in the identification and quantification of microscale shearing indicators using thin section analysis and scanning electron microscopes (Figures 4.9, 7.1 and 8.4) as well as microtomography (Tarplee *et al.*, 2010). As discussed in Chapter 7, microstructural studies of artificially deformed clays by Hiemstra and Rijsdijk (2003) have shown that distinctive grain arrangements develop during shearing. Focused strain along shear zones produces grain lineaments, and the rotation of particles or particle aggregates within shear zones produces 'turbate structures' (Figure 7.1). These structures consist of small particles arranged concentrically around a larger mass, and have been recognised in many ancient tills (van der Meer, 1993, 1997; van der Meer *et al.*, 2003). Hiemstra and Rijsdijk (2003) showed that both grain lineaments and turbate structures increase in frequency at higher cumulative strains, and pointed out that planar movements are necessary to create the torques required to rotate the grains. Similarly, Thomason and Iverson (2006) showed that most strain was accommodated by microshears (Riedel shears) aligned at low [R1] and high [R2] angles to the shearing direction (Figures 8.1 and 8.15a), which could be quantified using the dimensionless I_L index:

$$I_L = \frac{\sum_{i=1}^{n} L_i \cos \psi_i}{\sum_{i=1}^{m} H_i \cos \theta_i + \sum_{i=1}^{n} L_i \cos \psi_i}$$

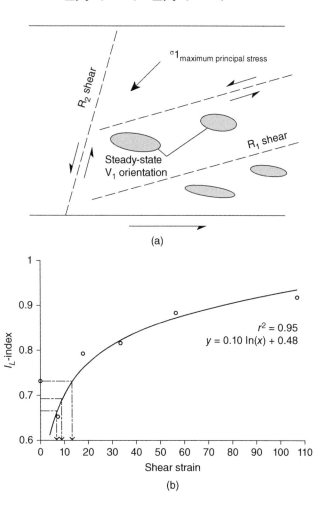

(a)

(b)

Figure 8.15 Parameters and data relating to the identification and quantification of microscale shearing indicators (from Larsen *et al.*, 2006; Thomason and Iverson, 2006): (a) summary diagram to show how most strain is accommodated by Riedel shears aligned at low [R1] and high [R2] angles to the shearing direction; (b) graphic plot that demonstrates a progressive I_L index increase during shearing.

where n and m are the numbers of low-angle (L) and high-angle (H) shears, respectively, and ψ and θ are the acute angles of low- and high-angle shears, respectively (the division between high- and low-angle shears is 25°). The I_L index varies from 0 to 1, with high values being indicative of a predominance of shears parallel to the shearing direction. The principle here is that microshears become aligned more in parallel to the principal zone of displacement under progressively higher strain. Further tests by Larsen *et al.* (2006) involved subjecting 40 mm of diamicton to shear strains up to 107, the result of which was displacement, measured from S-matrix microstructures in thin sections, only in a 14–20-mm-thick zone within the diamicton. The plotting of I_L indexes from ancient tills together with those from the laboratory experiments allowed Larsen *et al.* (2006) to identify a progressive I_L index increase during shearing (Figure 8.15b). Although this demonstrates the potential of using the I_L index for estimating strain in ancient tills, the homogenised ancient tills recorded strains of only 8–13, indicating that most of the microstructural changes, like microfabric development (Figure 8.1), take place at low strains.

The incompatibility of clast (micro- and macro-) fabric strengths collected during many field-based studies and from laboratory experiments (Table 8.1) poses some significant problems for glacial sedimentologists, and indicates that clast and/or matrix behaviour beneath glaciers is likely to be more complex than simple, universal models would suggest. The studies reviewed in this chapter can be summarised as follows:

- Fabrics in subglacial tills in field settings vary from relatively weak ($S_1 \leq 0.50$) to strong (>0.70).
- Fabrics in field settings appear to be very strong ($S_1 > 0.80$) for unequivocally lodged clasts when assessed independently.
- Some variability in fabric strength with grain size is evident in field settings, with more poorly sorted and coarser-grained tills displaying relatively weaker fabrics, likely reflecting greater clast collisions and interruptions of alignments.
- Laboratory experiments on sheared materials, derived from both artificial and field samples, reveal the development of strong ($S_1 =\sim 0.73$) fabrics at low strains but no further strengthening above strains of around 37. Additionally, there is no significant variability in fabric strength with particle size, but materials do not replicate the coarseness or poor sorting of field conditions.
- Late-stage depositional changes appear to be recorded in microscale structures, where dewatering remobilises only the finer-grained matrix of many tills leaving macrofabrics undisturbed.
- Strain heterogeneities appear to be important in the localised development of weak fabrics, especially in glacitectonites or immature homogenised tills, which may relate to late-stage development (i.e. immaturity) or substrate ploughing (ice-keel- or clast-induced) of coarser-grained materials.
- Replicating till deformation and concomitant strong fabric development in a ring shear device does not necessarily disprove the Hart (1994) hypothesis that weaker fabrics are created in thicker deforming layers, because the same hypothesis implies strong fabrics in thin and constrained shear zones such as those created in a shear box! Hence, we must seriously consider the notion that laboratory experiments at present cannot replicate all the stress/strain relationships that occur in spatially and temporally variable subglacial deforming materials.

Solutions to the problems arising from such observations can be delivered only through till deformation experiments at a scale large enough to replicate true field conditions. This is a tall order but has been initiated by the pioneering experiments of Hart *et al.* (2006, 2009, 2011), in which autonomous probes are inserted into modern subglacial tills via boreholes and then monitored over time. In one such experiment at the bed of Briksdalsbreen in Norway, data from the tilt sensors in the probes transmitted to the glacier surface indicated that the probes tended to exhibit progressive reduction of dip (i.e. March rotation), on which was superimposed short-term dip oscillations and rotation about the a-axis. This indicates that particle rolling could be taking place in conjunction with March rotation, thereby exerting some influence on fabric development. The complexity of subglacial processes, with respect to porewater pressure changes and concomitant brittle versus ductile responses by the till, appear to be reflected in the signals returned by the probes (Hart *et al.*, 2011), although linking such data to till sedimentological properties remains challenging and often unavoidably speculative.

9

The Geological Record: Products of Lodgement, Cavity Fill and the Boulder Pavement Problem

No one, I presume, will for a moment entertain the idea, that the one hundred and forty one pieces composing this bed were transported to this spot, having been striated elsewhere, and accidently deposited with their surfaces in the same plane, and their grooves substantially parallel. The chances against such an occurrence are so enormous, that we might with safety say, it could not happen except by miracle.

Stoddart (1859, p. 227)

9.1 Introduction – Repositioning Field Studies and Experimental Reductionism

A number of criticisms of glacial sedimentology and sedimentologists have been levelled by Neil Iverson and his co-workers since the beginning of their efforts to isolate a specific till rheology and measure strain magnitude through till fabric. They rightly highlight the impoverished nature of experimentation in the study of glacigenic deposits, particularly tills, but are all the criticisms valid? Some fundamental questions have not been fully addressed by the laboratory shear tests and predictably they relate to the problems of representativeness of the spatial and temporal complexity of real glacier and ice sheet beds. Thomason and Iverson (2006) correctly point out that this should not be used as an excuse to avoid simulating deforming beds at laboratory ring-shear scale, as simulating complexity is not the prime aim of such experiments but rather it is isolating the variables that are notoriously difficult to measure in the field. This chapter and Chapters 10–15 are critical reviews of our present state of knowledge on till deposits, importing the invaluable observations and data derived from the subglacial exploration, laboratory experiment and modelling reviewed in preceding chapters and reconciling them with the characteristics and associated process–form interpretations derived from field-based assessments. With this in mind, we first need to reflect on both the 'answers' and the unanswered questions that carry over from the laboratory experiments. The most significant of the unanswered questions is: how do we interpret all the low-strain signatures in subglacial materials?

The overarching and critically important aim of the laboratory till shearing experiments as set out by Iverson and co-workers is the testing of the 'bed deformation model', the definition of which is instructive in terms of how wide-ranging the laboratory test results are purported to be for assessing till process sedimentology. Specifically, Iverson *et al.* (2008) summarise the 'bed deformation model' as the assertion that 'a glacier can move primarily by shearing its bed, usually assumed to consist of

Till: A Glacial Process Sedimentology, First Edition. David J A Evans.
© 2018 John Wiley & Sons Ltd. Published 2018 by John Wiley & Sons Ltd.

till'; the notion that this was widespread and overwhelmingly dominant as a subglacial process was developed by Boulton's (1996a, b) model of regional till architecture (Figure 9.1), which is instructive in terms of subglacial sediment advection (cf. Alley *et al.*, 1997; Figure 1.4) but regarded by some as an overestimate of how much advection is accomplished by deformation alone (e.g. Piotrowski *et al.*, 2001, 2002, 2004), as we shall explore in the following chapters. The important corollary quantified by Iverson *et al.* (2008) is that shear strains (ratio of glacier displacement to shearing bed thickness) must exceed 100 if most basal motion was through bed deformation. Hence, in order to verify the 'bed deformation model', it is necessary to find till fabrics indicative of the very high shear strains predicted by the model. We will see in the following chapters that subglacial materials have most definitely been deformed and indeed that deformation is central to the process–form-based nomenclatures, which have been developed for tills and associated deposits, but the low strains indicated

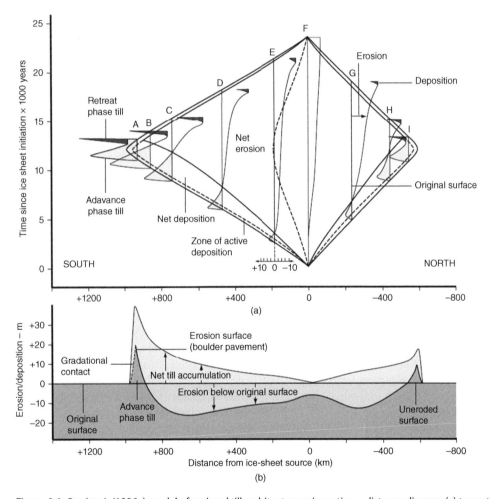

Figure 9.1 Boulton's (1996a) model of regional till architecture using a time–distance diagram (a) to portray the shifting patterns of erosional and depositional zones. The curves A-I represent the changing trends of erosion and deposition at selected locations through time, with advance and retreat tills identified. The lower cross-section (b) displays the pattern of till deposition at the end of a glacial cycle, emphasising the net ice sheet marginal-thickening of such deposits.

by a range of common features such as till heterogeneities and weak fabrics (Table 8.1) are regarded as incompatible with the 'deforming bed model' by Thomason and Iverson (2006) and Iverson *et al.* (2008). Problematic for reconciliation between the results of shearing experiments and the field characteristics of tills is the undoubted occurrence of low-strain signatures in the latter. Therefore, in terms of developing a process sedimentology for tills and associated deposits we have to acknowledge that glacier and ice sheet beds have been deformed but not always everywhere (temporally and spatially) to the high shear strains predicted by the 'bed deformation model', specifically at the stage when till is finally deposited. The term 'bed deformation model' is an unfortunate label for what should probably be more precisely termed the 'high shear strain, fully ice–till coupled, pervasive bed deformation model'. Otherwise the deformable bed problem is an all or nothing issue wherein its either all very highly strained or its not and so our deposits with low-strain signatures are either non-deformed tills (an oxymoron) or not tills at all? If they are tills, and a range of sedimentological criteria indicate that they are, how are relatively low-strain signatures being preserved in till stratigraphies? Clearly, Boulton's (1996a, b) advection process cannot be driven predominantly by pervasive subglacial deformation, and/or strain heterogeneities (Iverson and Hooyer, 2002) are widespread in the deforming bed, especially at the time when some tills are finally deposited?

9.2 Lodgement

In previous chapters, the traditional genetic label of 'lodgement till' was critically reviewed, leading to the realisation that the sedimentological signatures of none of the till forming processes, including lodgement, melt-out, deformation and undermelt, were sufficiently unequivocally diagnostic to be used confidently in discerning a specific process–form regime for glacigenic deposits. Indeed, many researchers have recommended a more broadly genetically defined classification such as 'subglacial till' (e.g. Anderson *et al.*, 1980, 1986; Kemmis, 1981; Bergersen and Garnes, 1983; Dreimanis, 1983; Lundqvist, 1983; Stephan and Ehlers, 1983; Ringberg *et al.*, 1984; van der Meer *et al.*, 1985; Rappol, 1985; Hansel and Johnson, 1987) or 'subglacial traction till' (Evans *et al.*, 2006b). The lodgement process was defined first by Chamberlin (1895) and then later by Dreimanis (1989) as the plastering of glacial debris from the base of a sliding glacier on to a rigid or semi-rigid bed by pressure melting and/or other mechanical processes, concepts developed by Boulton (1974, 1975, 1976, 1982; Figure 6.1) based upon process observations. Rather than label a whole till deposit according to this process, it has become common to identify specific characteristics of diamictons that could be related to lodgement (see Table 2.1) and hence link them to subglacial till production. The isolation of clast macrofabric signatures of lodgement from those of matrix deformation has also been attempted using the process-specific criteria on individual clasts prior to measurement (Evans and Hiemstra, 2005; Evans *et al.*, 2007, 2016; Figure 4.15; Table 8.1). This approach acknowledges that, in soft-bedded settings, clast lodgement involves the ploughing of the deformable substrate to form a prow (Boulton, 1982; Clark and Hansel, 1989; Evans and Rea, 2003; Jørgensen and Piotrowski, 2003; Eyles *et al.*, 2015; Figure 8.7 and 8.11b), which then arrests the forward momentum of the clast and gives rise to a final deposit that contains lodgement and deformation signatures. Several researchers have proposed simultaneous lodging, ploughing and shear deformation at the glacier bed, whereby large particles lodge first due to their higher drag force and smaller particles remain in traction until strain rates fall (Boulton *et al.*, 1974; Boulton and Hindmarsh, 1987; Benn, 1994a; Krüger, 1994; Evans and Twigg, 2002; Evans *et al.* 2016). Hence, it has become clear to glacial sedimentologists that lodgement and deformation act in combination to create subglacial tills, as we shall discuss in Chapter 16.

The lodgement process is manifest sedimentologically in both clast morphology (form) and sediment characteristics. Modifications to clasts take place in the subglacial traction zone as they make their way through the glacial debris cascade (Figure 1.2b), specifically in terms of shape, roundness and surface texture and wear patterns (e.g. Holmes, 1960; Boulton, 1978; Dowdeswell *et al.*, 1985; Benn, 1994a, 1995, 2004a; Benn and Ballantyne, 1993, 1994; Spedding and Evans, 2002; Hambrey and Ehrmann, 2004; Lukas *et al.*, 2013). In terms of shape, active glacial transportation for most lithologies tends to result in the production of compact, blocky shapes, which reflects both the initial shapes of subglacially plucked fragments and preferential clast breakage across their long axes. Clast roundness is modified by fracture and abrasion, whereby fracture creates new, sharp edges and fresh faces and abrasion increases edge rounding and creates polished facets; the net effect on these two processes in subglacial till clasts is that both angular and well-rounded forms tend to be rare and most clasts have intermediate roundness characteristics or are sub-angular and sub-rounded (Figure 1.2b). The surface texture and wear patterns of subglacial clasts are very distinctive, especially on fine-grained lithologies, and comprise polished and striated faces or facets. Striae can be straight and parallel if the clast has been held in a stable position but can also be cross-cutting if it has moved position in the traction zone (Hicock, 1991; Benn, 1995). Lodged clasts tend to display straight striae on their upper surfaces that are aligned parallel to their long (A) axes (Benn, 1994a; Krüger, 1994; Clark and Hansel, 1989).

The combined effects of these shape, roundness and surface wear changes during lodgement tends to produce clasts that have 'stoss-and-lee' or bullet-shaped forms with striated surfaces (Krüger, 1979; Sharp, 1982) and aligned with their A axes and/or A/B planes ice flow-parallel (Figures 1.2b, 4.11 and 9.2a). This reflects in situ abrasion of the up-glacier (stoss) sides and plucking of the down-glacier (lee) sides of clasts in a similar way to the erosion of a roche moutonnée (Boulton, 1978; Krüger, 1984; Benn, 1994a). Striated upper and lower surfaces of clast A/B planes, referred to as a double stoss-and-lee form, is regarded as particularly diagnostic of lodgement by Krüger (1984), who attributes this to a two-stage process of clast wear at the ice−till interface. Specifically, this involves clast ploughing through the till prior to deposition, causing abrasion of the lower leading edge and fracture of the lower trailing edge. After lodgement, the stoss side is abraded while the lee side is fractured. Potentially, a double stoss-and-lee form could also develop on a clast being rafted along within deforming till, as long as the clast maintains a quasi-stable orientation so that abrasion will be focused at the lower leading and upper trailing edges, and fracture at the lower trailing and upper leading edges (Benn, 1995). The nature of all these asymmetric subglacial wear patterns and their association with lodgement and till deformation are summarised in Figure 9.2b, after the assessments of Krüger (1984) and Benn and Evans (1996). Cases (a) to (c) reflect the lodgement process specifically, whereby a stoss-and-lee form is created by sliding ice over a lodged clast (a), and a double stoss-and-lee form results from ploughing (b) followed by lodgement (c). Continued modification takes place even within deforming layers, as indicated by (d) to (f), in which the smaller clasts in either the deforming matrix (d) or along internal slip (shear) planes (e, f) modify the facets.

The ploughing of large clasts is clearly manifest in prow construction and incipient flute formation on recently deglaciated glacier forelands, where clasts had only just made contact with the bed or whose momentum had just been arrested by frictional retardation before ice flow ceased (e.g. Boulton, 1982; Benn, 1995; Evans and Rea, 2003; Eyles *et al.*, 2015; Figure 8.11b). Similarly, clast ploughing to produce flutes and lodged clasts, as discussed above and in earlier chapters, is widely acknowledged (e.g. Boulton, 1975, 1976, 1982; Boulton *et al.*, 1979; Tulaczyk, 1999; Fischer *et al.*, 2001). Sedimentological evidence for ploughing by clasts dragged along by sliding ice or by deforming till (erodent layer of Eyles *et al.*, 2016) is manifest as linear grooves or 'sole casts' and sediment prows in front of clasts in some till exposures (e.g. Westgate, 1968; Ehlers and Stephan, 1979; Clark and Hansel, 1989;

(a)

Figure 9.2 Products of the lodgement and ploughing processes: (a) photographs of lodged clasts exposed on till surfaces and displaying prominent upper striated facets, stoss-and-lee forms, clast clustering and ice flow-parallel A-axis alignments; (b) schematic diagrams to explain the development of asymmetric clast wear patterns associated with sliding ice and deforming till, showing principal locations of abrasion (A) and fracture (F). Arrow lengths show relative velocities. (a) lodged clast with stoss-lee form due to stoss-side abrasion and lee-side fracture. (b, c) double stoss-lee morphology due to a two-stage process of ploughing and lodgement. (d) double stoss-lee clast due to a single-stage process within a deforming layer. Low-pressure zones are shaded. (e, f) flat, polished facets eroded on the upper and lower surfaces of clasts, where there is significant slip between the clast and the adjacent shear plane. (Modified from Krüger, 1984; Benn and Evans, 1996); (c) sketches of cross-sections through sediment slab containing a ploughed clast and prow (from Clark and Hansel, 1989). The long axes of elongate grains are depicted as lines; (d) left panel shows sketches of examples of plough marks on the basal contacts of tills (from Ehlers and Stephan, 1979): (1) ribs (with one clast in place), (2) wedge, (3) edge, (4) slickenside, and (5) undulation. Right photograph shows slickensides, at end of pencil, developed at the boundary between laminated clays and till (former ice flow from left to right).

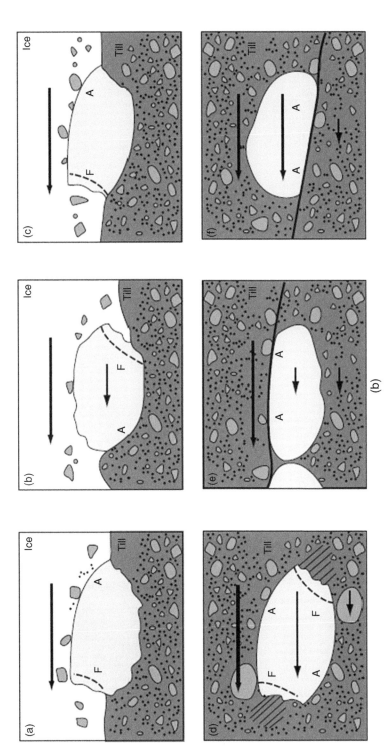

Figure 9.2 (*Continued*)

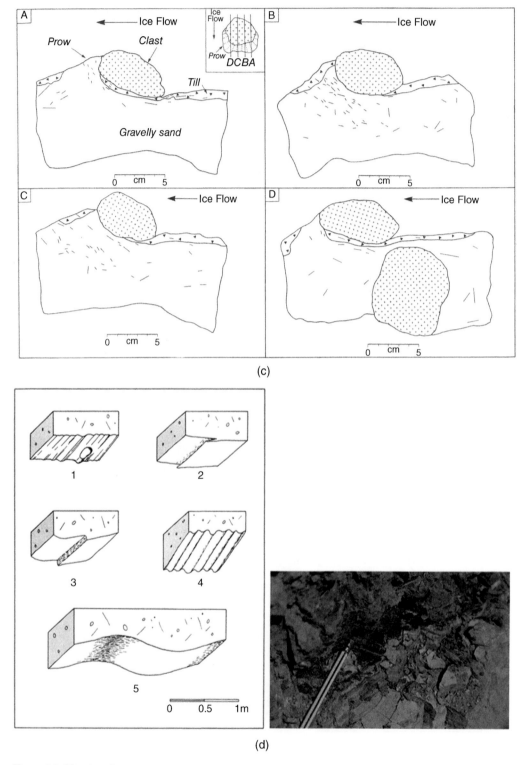

Figure 9.2 (*Continued*)

Piotrowski *et al.*, 2001; Jørgensen and Piotrowski, 2003; Figure 9.2c, d). Sole casts are ice flow-parallel grooves at the contacts between till and underlying deposits, usually best preserved in sandy materials. Occasionally these are associated with clasts and slickensides and hence linked to the sliding, ploughing and lodgement of clasts in the substrate, similar to the development of lineations or streamlined features on fault slip planes (cf. Tija, 1967; Means, 1987; Lin and Williams, 1992; Eyles and Boyce, 1998). Sediment prows are evident where underlying sediments are clearly deformed and displaced vertically on one side of lodged clasts, clearly impacting upon the alignment of fabrics or strata in the substrate (Figure 8.11b). Individual clast lodgement is known to initiate the further lodgement of other clasts following up from behind, forming clast clusters (Boulton, 1975; Krüger, 1979; Figure 6.1). Unlike clast pavements (see section 9.3), clast clusters will form linear strings of two or potentially more clasts within tills (Figures 8.6a and 9.2a).

9.3 Clast (Boulder) Pavements

The occurrence of distinct horizontal lines of large clasts within some till successions, usually with striated upper facets and often lying at the contacts between different tills or between a till and its underlying deposits (Figures 1.3e and 9.3) has been the subject of some debate in till sedimentology (cf. Miller, 1884; Gilbert, 1898; Holmes, 1944; Dreimanis and Reavely, 1953; Meneley, 1964; Christiansen, 1968; Sauer, 1974; Dreimanis, 1976; Dreimanis *et al.*, 1987; Clark, 1991; Mickelsen *et al.*, 1992 and Clark, 1992). These clast or boulder pavements are not dissimilar to the lags produced by other geomorphic agents such as aeolian (Hobbs, 1931) or fluvial erosion (Kay, 1931), nearshore fast ice (Hansom, 1983; Dionne, 1985) or subaqueous winnowing (Eyles, 1988; Eyles and Lagoe, 1990) and indeed subglacial processes may inherit such lags and accentuate them (cf. Gilbert, 1898; Dionne and Poitras, 1998). Certainly, the case for a subaqueous genesis in the glacimarine setting to produce a palimpsest lag (Powell, 1984; Eyles, 1988) is an important one, because massive iceberg rafted diamictons overlying pavements potentially could be misinterpreted as subglacial till products. However, the clasts in such settings would lack the diagnostic subglacial traction zone signatures such as stoss-and-lee forms and strong A-axis fabrics, even if they were created at the oscillating grounding line of a glacier as envisaged by Eyles and Lagoe (1990).

A lag interpretation is not incompatible with a subglacial origin for clast pavements, because they could demarcate a former position of the ice–till interface where preferential removal of the matrix of the underlying till has been affected by the combined effects of subglacial meltwater flushing and glacier sliding (see Chapter 11), a concept introduced by Miller (1884) and revisited by Gilbert (1898) and Holmes (1944). This would isolate larger clasts at the tops of till sheets, which could then be buried by renewed till deposition (e.g. Boyce and Eyles, 2000; Jørgensen and Piotrowski, 2003; Davies *et al.*, 2009). Indeed, an in-depth study of clast macrofabrics in boulder pavements by Hicock (1991) has shown that some have likely been disturbed by till deformation after their initial formation; their S_1 eigenvalues (Table 8.1) are certainly not always particularly diagnostic of lodgement even though lodgement has been widely associated with pavement formation (e.g. Boulton and Paul, 1976; Hicock, 1991; Jørgensen and Piotrowski, 2003; Menzies *et al.*, 2006). A model for the development of a subglacial lag by an erodent deformation layer (cf. Eyles *et al.*, 2016; see Chapter 10) has been proposed by Boulton (1996a; Figure 9.4), wherein a sub-till clast pavement is formed during the downward excavation of the deforming layer into an older till unit. This involves the preferential mobilisation of fine material and the accumulation of large particles which resist entrainment during the development

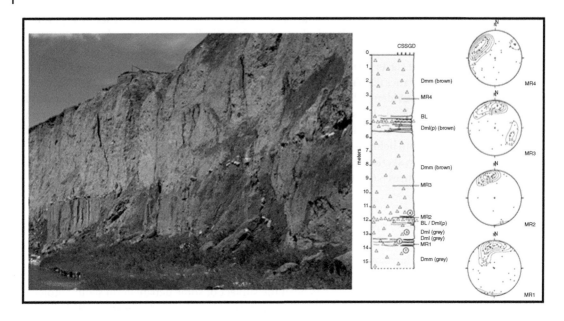

Figure 9.3 Examples of clast lines or pavements. Upper panel shows clast pavement developed between tills at Milk River, Alberta (from Evans *et al.*, 2012a). A macrofabric from the lower clast pavement is numbered MR2, which can be compared with those from the matrix of the tills. Lower panel shows a discontinuous clast pavement (clast line) between tills at Whitburn, northeast England, together with macrofabric (from Davies *et al.*, 2009). Inset photograph shows striated nature of upper facets.

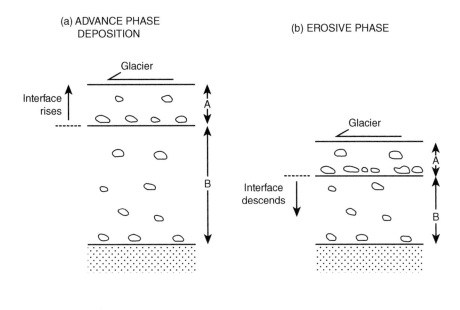

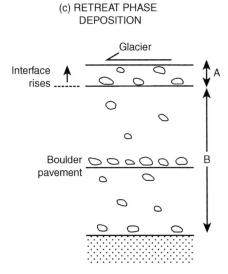

Figure 9.4 Boulton's (1996a) model for clast pavement development in relation to the rising and descending interface between A and B horizons in tills, whereby clasts concentrate at the interface and can be isolated at depth when tills thicken in the depositional zone beneath ice sheets.

of A and B type deforming horizons. A novel explanation presented by Clark (1991) involves the sinking of large clasts to the base of a dilatant till layer, similar to the process observed in debris flows. However, quite apart from the criticism of this theory aired at the time (cf. Mickleson *et al.*, 1992 and Clark, 1992), the likelihood of a subglacial till being in the necessary fluid state for clast sinking over tens of square kilometres of the ice sheet bed are unrealistic, as we have discussed in previous chapters.

In addition to the clast morphologies and stratigraphic and sedimentological characteristics of clast pavements, the nature of clast lithology and surface wear patterns can also provide important information on potential origins. Clast lithologies have been reported to be similar to those in overlying tills by a number of workers, a trend used by Clark (1991) to support his model of clast sinking, but this is by no means universal and some pavements comprise lithologies from materials and/or bedrock from both above and below (e.g. Davies *et al.*, 2009). Such lithological trends would be expected in erosional lag models of formation like that of Boulton (1996a), which predicts that the boulders in sub-till pavements will be lithologically similar to those in the underlying sediment but clearly the erosional till layer also introduces more far-travelled materials.

Clast wear patterns that appear particularly diagnostic of clast pavements in till successions are the upper faceted and striated surfaces. Striae orientations have been reported as similar to the clast fabrics in overlying till by, for example, Flint (1955) and Hicock and Dreimanis (1989), but cross-cutting striae also record multiple ice flow directions operating over some pavements during one glaciation (e.g. Davies *et al.*, 2009). This indicates that their development may relate to persistent subglacial ice and/or till sliding with concomitant preferential removal of finer materials from the same horizon over time. This likely requires a resistant substrate down to which the preferential winnowing has progressed, a good example being the Davies *et al.* (2009) case study at Whitburn, northeast England, where the pavement seals in a patchy and mostly thin lower till and continues locally onto bedrock high points and the most prominent striae and the macrofabric data record one dominant ice flow direction (Figure 9.3b). The occurrence of crudely stratified sand a gravel lenses and channel fills within the pavement at Whitburn attest to the likelihood of meltwater flow and sediment flushing, and hence sliding, at the ice–bed interface (see Chapter 11). As many clast pavements occur in depositional settings where proglacial lakes operated during glaciation, it is not unreasonable to entertain the notion that they may have originated by subaqueous winnowing (cf. Powell, 1984; Eyles, 1988) but were then 'inherited' by subglacial erosional processes during glacier readvance.

9.4 Lee-side Cavity Fills

The processes associated with the deposition of sediments in the lee-sides of bedrock protuberances on the glacier substrate were discussed in section 6.1, where the subglacial observations of Boulton (1982) in particular were used to support this as a viable process–form regime in till sedimentology. The sedimentological evidence for former lee-side cavity fills ('lee-side tills' of Hillefors, 1973; Figure 9.5a) is best preserved in hard bed settings, especially mountain environments, where the bed roughness manifest in roches moutonnées and bedrock steps is sufficiently great to initiate numerous cavities at the ice–bed interface. In such a setting in the Canadian Rocky Mountains, Levson and Rutter (1989a, b) report massive diamictons with rare, steeply dipping sand and minor gravel lenses, located on the lee-side of bed obstructions. Clasts within the diamicton display typical subglacial traction zone characteristics such as striations and strong fabrics, but the characteristics of ploughing and lodgement are absent. Moreover, clast A-axes dip predominantly downvalley, parallel to the sand lenses. These characteristics are regarded as indicative of cavity infill by mass flows, melt-out and debris fall and occasional meltwater pulses, resulting in down ice-thinning, wedge-shaped masses of massive to crudely stratified diamictons with steeply dipping lenses of water-sorted sediments. Similar to the contemporary observations of Boulton (1982; Figures 6.3 and 6.4), these sediment wedges are truncated by an overlying subglacial till, indicative of glacier–bed coupling once the cavity was full. In addition to hard bed settings, cavity fill origins have been proposed for stratified sediments in the lee sides of drumlins (e.g. Dardis *et al.*, 1984; McCabe and Dardis, 1989) where more vigorous subglacial meltwater discharges are envisaged. A particular type of lee-side cavity fill has been proposed

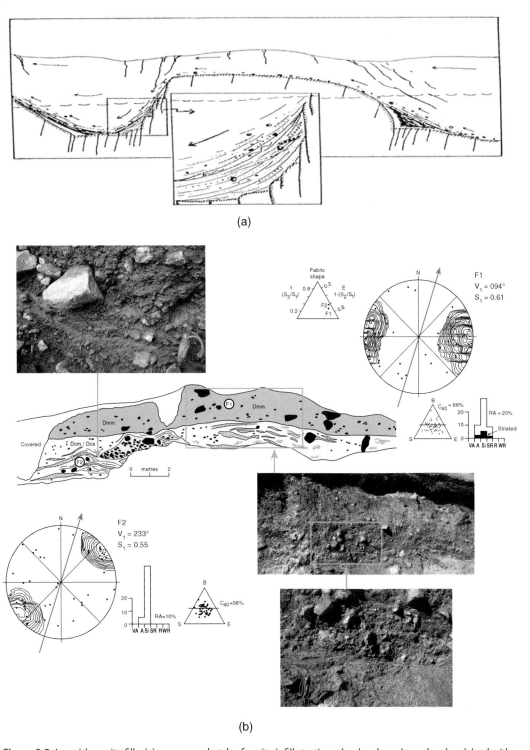

(a)

(b)

Figure 9.5 Lee-side cavity fills: (a) summary sketch of cavity infill stratigraphy developed on a hard rock bed with roches moutonnées (from Hillefors, 1973); (b) stratigraphy and clast form and macrofabric data from a lee-side cavity fill on the floor of a Norwegian mountain valley (after Evans *et al.*, 1998). Valley axis and downstream trend is marked by blue arrow.

also for the margins of tuyas (subglacial volcanic edifices) by Bennett *et al.* (2006), where geothermal activity is particularly effective in mobilising large volumes of sediment at the ice–bed interface.

Ice–bed coupling with lee-side cavity fills results in the reworking and/or overprinting of their upper layers by the subglacial processes of ploughing, lodgement, glacitectonic disruption and deformation, and hence Evans *et al.* (2006b) predict that the apparent stratification in some glacitectonites and subglacial tills may be imparted by the structures in the source materials. The stratigraphy of a lee-side cavity fill coupled with an overlying subglacial till is reported by Evans *et al.* (1998) from central Norway (Figure 9.5b), where a lower laminated to massive diamicton with lenses of sand and gravel and clast-supported diamicton (cavity fill) is glacitectonically attenuated and partially incorporated into a capping massive matrix-supported diamicton (till). Clast A-axis macrofabrics strengthen between the two units ($S_1 = 0.55–0.61$; see also Figure 8.14), with a northeast–southwest alignment in the cavity fill reflecting the valley axis and a W-E orientation in the overlying till recording an ice-flow transverse fabric probably developed because of the dominant slab and elongate clast shapes. Clast forms reveal the inclusion of freshly plucked fragments in both units but an increase in such material, as well as the introduction of striated clasts, in the overlying till.

10

The Geological Record: Deforming Bed Deposits

Detritus carried along in and under the ice … are subangular and rounded, with abundant scratched and polished pebbles and boulders stuck in a fine tough clay. This matrix is sometimes laminated, and … may be well stratified, or in other cases entirely without any definite arrangement.

A. Geikie (1903, p. 547)

Such is the perceived critical importance of deformation as a subglacial process that tectonic terminology, such as 'deforming bed till', 'deformation till' and 'tectomict', have been proposed for tills (e.g. Elson, 1961; Alley, 1991; Hart, 1995; Menzies *et al.*, 2006). A wide range of ancient glacigenic deposits have been interpreted as the products of former subglacial deforming beds, including those materials that have previously been called 'deformed till' but are discussed below as glacitectonites. In this chapter, we will restrict our discussions to those materials that we are confident represent subglacial deforming till layers, constrained by the observations and case studies reviewed earlier in Section 6.2 and Chapters 7 and 8. Most significantly, we are constrained by the process observations of subglacial deformation depths being restricted to only a few metres. Hence, greater thicknesses of deformed materials (glacitectonites) must be equated to proglacial tectonics followed by glacier overriding and till carapace development or, in the case of deforming bed tills, the thickening, through advection and stacking, of homogenised subglacial materials.

Interpretations of ancient till sequences must be informed by modern analogues for the envisaged glacier process–form regimes. This has been largely delivered through studies on the sedimentology of subglacial materials on recently deglaciated forelands, which can be reasonably confidently linked to known processes operating at the ice–bed interface of the glacier responsible for their deposition. In Section 6.2 and Chapter 8, we discussed the importance of the Breiðamerkurjökull till studies to our understanding of what a subglacially deformed till should look like and how its strain signature should appear in clast macrofabrics. The studies of tills on the Icelandic glacier forelands are just some of a number that have been undertaken on recently exposed glacier beds, which in combination provide us with a clear sedimentological signature for deformation as well as ploughing and lodgement. Case studies reveal that individual till layers are commonly thin (<2 m) and can display the characteristics of the typical two-tiered structure (A and B horizons) of the Breiðamerkurjökull sequence, which evolve as a response to vertically migrating porewater pressures and dilation and associated modifications to the till matrix framework, reflected in the ability of the material to take up solid state deformation (Figure 10.1). In reality, the A and B horizon classification relates specifically to their different textural and structural characteristics, which could develop independently or could

Till: A Glacial Process Sedimentology, First Edition. David J A Evans.
© 2018 John Wiley & Sons Ltd. Published 2018 by John Wiley & Sons Ltd.

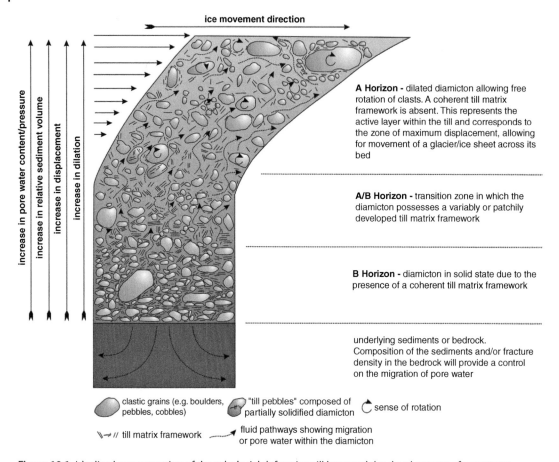

A Horizon - dilated diamicton allowing free rotation of clasts. A coherent till matrix framework is absent. This represents the active layer within the till and corresponds to the zone of maximum displacement, allowing for movement of a glacier/ice sheet across its bed

A/B Horizon - transition zone in which the diamicton possesses a variably or patchily developed till matrix framework

B Horizon - diamicton in solid state due to the presence of a coherent till matrix framework

underlying sediments or bedrock. Composition of the sediments and/or fracture density in the bedrock will provide a control on the migration of pore water

clastic grains (e.g. boulders, pebbles, cobbles)

"till pebbles" composed of partially solidified diamicton

sense of rotation

till matrix framework

fluid pathways showing migration or pore water within the diamicton

Figure 10.1 Idealised reconstruction of the subglacial deforming till layer and the development of porewater migration pathways, till–matrix framework and till pebbles and their relationship to potential A and B horizon development and geotechnical properties relevant to dilation and solid state deformation (after Evans *et al.*, 2006b).

even be superimposed in accreting till sequences. For example, Benn and Evans (1996) identify two scenarios, including Breiðamerkurjökull, where a low-lying, poorly drained bed promotes the pervasive deformation of upper till layers to produce A and B horizons (Figure 6.9a), and Slettmarkbreen, Norway, where the steep, well-drained bed maintains low porewater pressures and a stiffer B horizon type till (Figure 10.2). The bases of such tills can also display diagnostic features of subglacial deformation, for example, mélange zones comprising a mix of till and underlying materials. An excavation of a fluting at Isfallsglaciaren, Sweden, by Eklund and Hart (1996) demonstrated that the deformation involved in fluting construction created a shear zone comprising an amalgamation of the till base and the cannibalised top of the underlying, glacitectonically folded, stratified sediments (Figure 10.3). The tills and underlying deposits on the Icelandic forelands display the same vertical sequence of glacitectonised sands and gravels, amalgamation zone and capping till, the latter displaying A and B horizon characteristics and vertically decreasing numbers of attenuated peat rafts, sand lenses and wisps cannibalised from the substrate (Figures 8.12c and 8.13; Evans, 2000a; Evans and Twigg, 2002; Evans and Hiemstra, 2005); clast forms may also be inherited from underlying strata, especially from glacifluvial deposits (e.g. Evans, 2000a).

Figure 10.2 Two scenarios for subglacial till formation where either an A and B horizon stratigraphy arises (e.g. Breiðamerkurjökull) or only a B type horizon (e.g. Slettmarkbreen, Norway; after Benn and Evans, 1996). TRS = total relative strain or differential horizontal displacement. Photographs show structures typical of A (upper) and B (lower) horizons.

Assessments such as that by Eklund and Hart (1996) of tills that make up flutings provide us with a clear set of unequivocal subglacial deforming layer characteristics, because each fluting is related to deformation of the substrate, either by lee-side cavity infilling (Boulton, 1976; Gordon *et al.*, 1992; Benn, 1994a; Eyles *et al.*, 2015; Figures 6.7 and 8.7) or substrate grooving by erodent layers or ice keels (e.g. Evans and Rea, 2003; Eyles *et al.*, 2016). The wide range of clast macrofabric strengths in fluting tills (Table 8.1) likely reflects not just the mode of landform streamlining but also the clast interactions associated with the ploughing and lodging of larger clasts and the coeval deformation of matrix and smaller clasts, especially in the construction of cavity-infill type flutings wherein clast fabrics tend to display herringbone patterns or weak fabrics on the down-flow sides of stoss boulders (e.g. Rose, 1989, 1992; Benn, 1994a; Evans *et al.*, 2010; Eyles *et al.*, 2015; Figures 6.7 and 8.5; see Chapter 8). Early stages of lodgement on a soft bed also demonstrate the importance of clast interactions in fabric development, illustrated by the up-glacier imbrication that develops in till matrix ploughed up in front of embedded boulders (Figures 6.7, 6.8, 8.7, 9.2c and 10.4; Boulton, 1976; Evans and Rea, 2003; Eyles *et al.*, 2015) and the contrast between the strong fabrics of unequivocally lodged boulders and other smaller clasts in coarse-grained tills on the forelands of Icelandic glaciers (Evans *et al.*, 2007, 2016; Figures 4.14b, c and 8.6). Boulton's (1976) exhaustive treatment of clast macrofabrics in

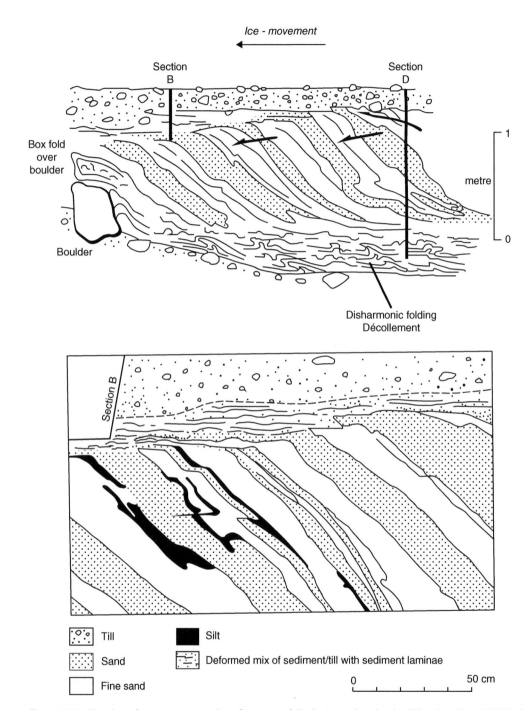

Figure 10.3 Sketches of sections exposed in a fluting at Isfallsglaciaren, Sweden, by Eklund and Hart (1996), showing a shear zone created by the amalgamation of the till base and the underlying stratified sediments.

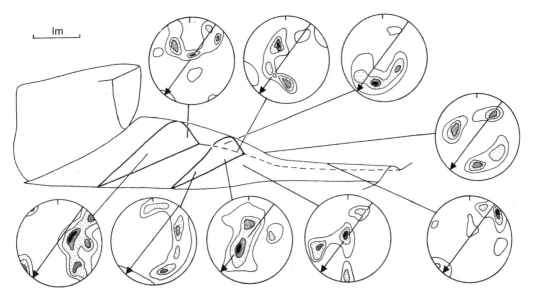

Figure 10.4 The macrofabrics of a till wedge or prow on the down-ice side of a ploughed boulder and an incipient fluting developed down flow from the prow. The fabrics show clast dips towards the boulder on the proximal face and down ice on the distal side, with ice-flow-parallel alignments developing on the fluting (from Boulton, 1976). Arrows on stereonets show ice flow direction.

flutings and ploughed boulders provides us with a clear indication that coarse-grained tills, unlike the finer-grained materials of MSGL with strong fabric signatures (e.g. Spagnolo *et al.*, 2016; Figure 8.4), do not develop strongly ice-flow-aligned fabrics indicative of steady-state strain even though fluting construction is a feature that would not be unusual in the operation of a fault gouge or the base of an erodent layer (Figure 6.6), where shear strains are considerable. Internally, the fabrics of flutings vary according to their location with respect to the fluting summit and flanks (Figure 8.10). Boulton (1976) identified within such patterns clast alignments relating to: (a) pre-flute till deposition, represented by ice-flow-parallel basal samples; (b) flute formation by deformation of sediment inwards towards the flute crest and away from the flanking troughs, represented by the intermediate samples; and (c) flute crest streamlining due to shearing by the overriding ice, recorded in the surface samples. As we discussed in Chapter 8, the bi-modal to spread bi-modal clast macrofabrics and high dip angles in the cores of flutings, as well as other subglacial landforms such as crevasse-squeeze ridges, boulder prows and till eskers, indicate clast mobility due to till squeezing or flowage. This clearly demonstrates that there are field conditions in which steady-state strains are not achieved in tills. Cases have been made above that this likely relates to tills with coarse grain size characteristics within which there are widespread clast collisions and interruptions of matrix deformation patterns by large clasts. The discovery of such tills on recently deglaciated forelands of predominantly active temperate or surging glaciers suggests that the prime location for weak fabric development is under the relatively lower basal shear stresses of sub-marginal settings, where flutings, marginal crevasse-squeeze ridges, elongate limbs of sawtooth push/squeeze moraines and till eskers develop rapidly in response to till liquefaction and flowage into ice cavities and crevasses (cf. Price, 1970; Evans and Rea, 2003; Christoffersen *et al.*, 2005; Evans *et al.*, 2010, 2015a; Eyles *et al.*, 2015). The occurrence of stronger, ice-flow-parallel fabrics from tills underlying flutings are instructive in that they potentially record the higher strains of till deformation beneath thicker ice prior to the development of sub-marginal till-squeezing processes. The variable fabric strengths that have been reported from such sub-marginal settings are therefore likely a product of Iverson and Hooyer's (2002) strain heterogeneities (cf. Jaeger and Nagel, 1992; Beeler *et al.*, 1996) introduced during late-stage till emplacement and typical of A-horizon development.

The deformation of subglacial materials is manifest in a variety of sedimentological structures in addition to the macrofabrics reviewed in Chapter 8 and discussed above. For example, Menzies (2000), using the principles of microtectonics from hard rock sequences (cf. Passchier and Trouw, 1996), highlighted the progressive development of brittle to ductile microscale structures in materials such as tills when they are subject to shear and become increasingly clay-rich due to comminution and hence develop higher porewater pressures (Figure 10.5). The early stages of subglacial till development, or more precisely replenishment, by deformation (see Section 6.3) have been related to the disturbance of sub-till materials by folding and/or clast or ice keel ploughing to liberate soft clasts and create 'tectonic/depositional slices' (Boulton *et al.*, 2001; Figures 6.21 and 6.22). A similar effect is created by the liberation of bedrock rafts from uneven glacier beds, as will be reviewed in Chapter 14, so that it is often possible to trace inclusions or rafts back into glacitectonite or undisturbed parent materials. This incorporation and then attenuation and eventual homogenisation of substrate materials into a deforming layer has been referred to as 'cannibalisation' (e.g. Hicock and Dreimanis, 1992a, b; Evans, 2000a, b) and has a long pedigree (Lamplugh, 1911; see Chapter 2 and Section 6.5). The operation of this deforming layer is significant geologically in that it is effective in creating an erosional shear zone that can cut downwards into substrates (cf. Boyce and Eyles, 1991; Eyles *et al.*, 2016) or can accrete and stack deforming layers, in tandem with clast ploughing and lodging, in vertical sequences (Figure 6.6), depending on the location of erosional and depositional regimes beneath an ice sheet or glacier at specific times (Figure 9.1). This concept was encapsulated

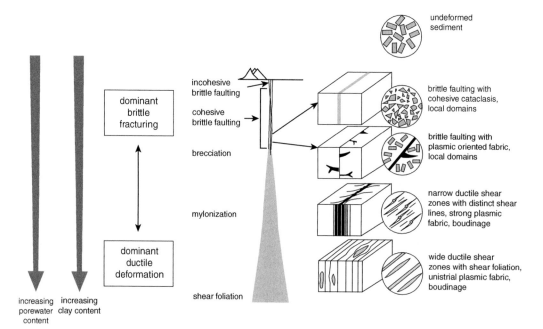

Figure 10.5 Diagram to illustrate the progressive development of brittle to ductile microscale structures in tills when subject to shear (from Menzies, 2000).

in the terms 'excavational' and 'constructional' deformation (Figure 10.6) by Hart *et al.* (1990), Hart and Boulton (1991) and Hart (1995). The excavational scenario has been linked to the operation of what Eyles *et al.* (2016) term an 'erodent layer', after the 'erodent layer hypothesis' from tribology, wherein wearing surfaces are created by 'erodents', or in the present case the ploughing clasts that protrude from the deforming layer base.

A large number of case studies can be used to illustrate the process–form continuum of cannibalisation, homogenisation and till deformation, most of which serve also as excellent examples of glacitectonite production (see Chapter 14). The cannibalisation process was well demonstrated in the study of the Sunnybrook drift on the north shore of Lake Ontario by Hicock and Dreimanis (1989, 1992a, b), in which the basal contact zone of a massive diamicton (Sunnybrook Till) with either the underlying glacilacustrine, interbedded sands and muds (Sylvan Park Member) or older deltaic sands (Scarborough sands) is characterised by a range of erosional, deformation and ingestion features, not necessarily always occurring together. First, the sharp erosional contacts are marked by the tectonic disturbance and down-flow displacement of the stratified sediments, truncated by striated clast pavements and sole casts at the boundary of the till and stratified sediments. Second, a lower ≤0.5 m zone lies within the till that displays features diagnostic of ingestion, at various stages of disaggregation, of the stratified sediments, including silt clasts and lenses, large partially sheared sand/mud intraclasts and a lighter-coloured, highly silty, 5-cm-thick basal zone with few clasts. Additionally, a further

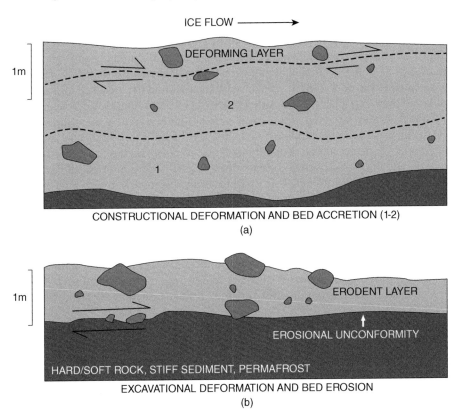

Figure 10.6 The concepts of constructional deformation or till accretion (a) and excavational deformation (b), where deforming till sheets <1 m thick either vertically build composite till sheets or erode and streamline the underlying substrate, respectively (from Eyles *et al.*, 2016, after Hart (1997).

striated clast pavement is developed at the junction between the basal $\leq$0.5 m zone and overlying massive, fissile diamicton up to 4.5 m thick. Structural and fabric data indicate a strong and unidirectional strain signature. These observations prompted Hicock and Dreimanis (1989, 1992a, b) to interpret the lower $\leq$0.5 m of the Sunnybrook Till as a deformation till and its overlying fissile component as lodgement till. Similar basal zones within tills have been identified by Evans *et al.* (2012a) in Alberta, where not only glacilacustrine but also fluvial sands and gravels and local Cretaceous bedrock rafts have been cannibalised and ingested into deforming layers where they are termed 'amalgamation zones' and may also contain penecontemporaneous fluvially infilled scours indicative of localised subglacial meltwater drainage (Figure 10.7; see Chapters 11 and 14). Similarly, Hart and Roberts (1994) summarise a range of intraclast and intrabed features in East Anglian tills in a model of substrate cannibalisation that elaborates on Boulton *et al.*'s (2001) model of 'tectonic/depositional slices' (Figure 6.22), wherein the liberation of pods of underlying sediment into the base of the deforming layer and then their attenuation (glacitectonic lamination) and disintegration down flow (Figure 10.8). This leads ultimately to sediment mixing or homogenisation (cf. boundary mixing; Hooyer and Iverson, 2000; Hoffmann and Piotrowski, 2001).

Amalgamation zone development or cannibalisation followed by diffusive mixing at till bases can be patchy, as demonstrated by Hoffmann and Piotrowski (2001), Larsen *et al.* (2004), Piotrowski *et al.* (2004) and Tylmann *et al.* (2013), using examples of lateral variability in the deformation of glacilacustrine rhythmites and glacifluvial sands and gravels beneath till at a number of sites around the southern margins of the Scandinavian Ice Sheet. They provide evidence that over short lateral distances till may have a sharp contact with underlying undisturbed sediments or a gradational contact or mélange comprising a vertically developed zone of deformed stratified sediments and mixed till and stratified sediment rafts (Figure 10.9). This highly variable appearance of basal till zones and their lower contacts is related to the operation of subglacial till mosaics and the localised disposition of glacifluvial canal fills and their variable cannibalisation by ice–bed recoupling, as is discussed further in Chapter 16. The high strains recorded in this basal zone of till development are indicated by average S_1 eigenvalues as high as 0.81 (e.g. Tylmann *et al.*, 2013), although Evans *et al.* (2012a) have identified that clast macrofabric strengths can drop off rapidly in overlying till layers (Figure 10.10), even though Piotrowski *et al.* (2006) discovered consistently strong fabrics in multiple till layers in their Polish case study illustrated in Figure 8.3. This variable strain signature between multiple till stacks from different field localities reflects complex styles of emplacement, which will be critically reviewed in following chapters.

The cannibalisation model illustrated in Figure 10.8 indicates that often the clearest macroscale evidence for deformation in tills is the presence of deformed inclusions or intraclasts. These may take a variety of forms depending on their stage in the ingestion and homogenisation process (Figure 10.11). Deformation is clear wherever intraclasts are streaked-out or folded pods of sand or soft rock, which can highlight patterns of strain in the surrounding till or exhibit pressure-shadow effects and boudinage (Berthelsen, 1979; Hart and Roberts, 1994; Evans *et al.*, 1995; Benn and Evans, 1996). This has been demonstrated by Piotrowski and Kraus (1997) and Piotrowski *et al.* (2001) with their concept of 'dispersion tails', which extend from soft-sediment clasts and can be used to differentiate between the effects of ploughing and deformation (Figure 10.12). Deformation of till around a soft clast that has been partially disaggregated should develop two dispersion tails extending up- and down-glacier, thereby reflecting the velocity gradients within the till (Hart and Boulton, 1991). Ploughing at the ice–bed interface on the other hand should create only one tail extending down glacier from the upper surface of the soft clast, similar to the liberation of the tops of soft substrates as depicted in Figure 10.8. As discussed above, such deformed inclusions are likely to be particularly abundant near till bases due to them being areas where cumulative strains are lower and the material less

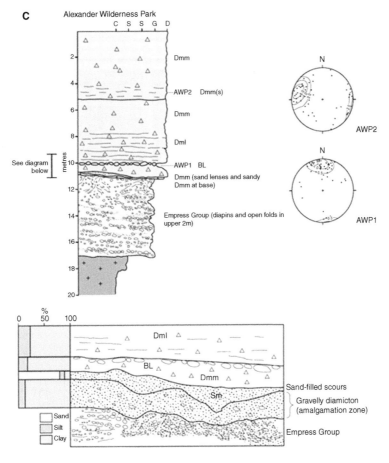

Figure 10.7 Example of an amalgamation zone (gravelly diamicton) at the boundary between preglacial (Empress Group) gravels and sands and till in the Lethbridge area, Alberta, Canada (from Evans *et al.*, 2012a). Note also the subglacial fluvially infilled scours and clast pavement separating lower and upper tills.

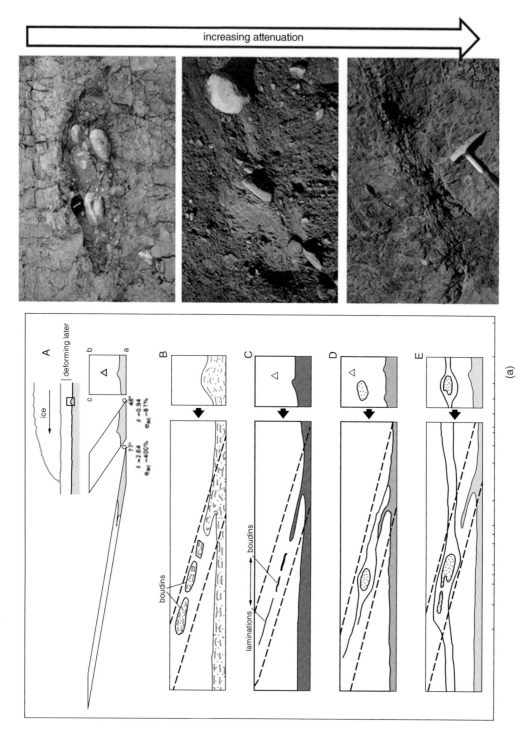

Figure 10.8 The origins of shear zones and intraclasts by substrate cannibalisation or the creation of 'tectonic/depositional slices': (a) left panel shows the various ways in which simple shear can liberate rafts of substrate into a subglacial deforming layer (from Hart and Roberts, 1994). Right panel shows examples of intraclasts in diamictons at increasing levels of attenuation during deformation, including a gravelly sand pod (upper), a sand boudin (middle) and a highly attenuated shale raft (lower); (b) an example from the island of Rügen, north Germany, of folding and attenuation of sub-till stratified sands and gravels and their ingestion and attenuation into the base of the deforming layer to produce glacitectonic lamination. The box in the right hand photograph shows the area of the enlarged view of glacitectonic lamination at bottom left.

(a)

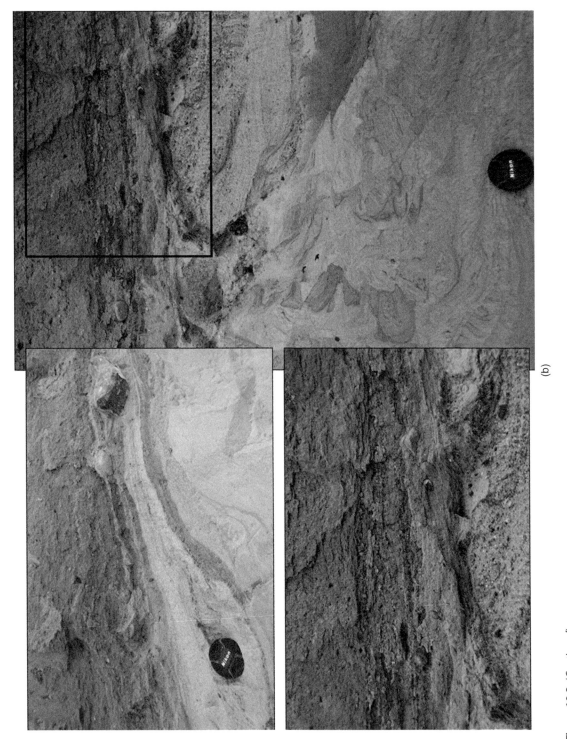

Figure 10.8 (Continued)

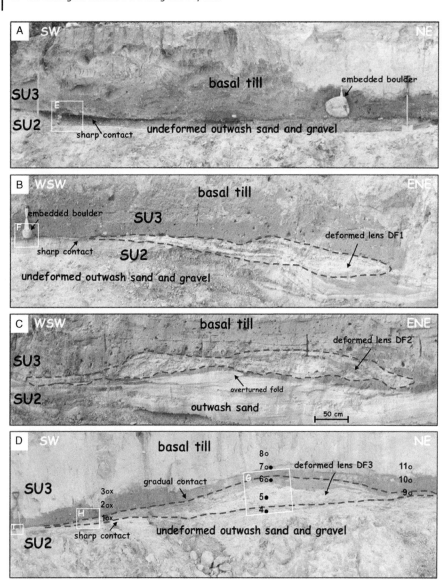

(a)

Figure 10.9 Examples of the patchy nature of amalgamation zone development at till–substrate contacts: (a) details of the contact between till (SU3) and stratified sediment (SU2) in the former sub-marginal zone of the Scandinavian Ice Sheet in northern Poland (from Tylmann *et al.*, 2013) showing: (A) sharp, erosive contact between undeformed outwash deposits and massive basal till, with no evidence of deformation; (B) deformed lens (DF1) of gravel and sand mixed to varying degrees with diamicton; (C) deformed lens (DF2) of partly homogenised gravel, sand and diamicton; (D) upward-convex and flat-based, deformed lens (DF3) of heavily deformed sand, gravel and till with admixture of till in the form of tongues, lenses and stringers being most prominent at the top of DF3. Open circles indicate macrofabric samples (1–3 and 6–11), crosses are AMS fabrics (1–3), and black dots mark samples for grain size and petrographic analysis (4–7).

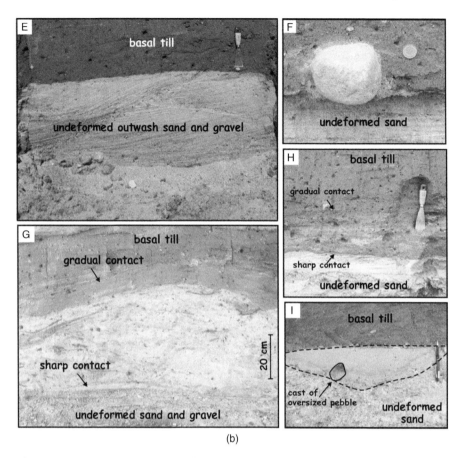

(b)

Figure 10.9 (b) detailed images identified in (a) and showing: (E) sharp contact between the massive basal till and undeformed and truncated trough crossbedded outwash deposits; (F) base of massive till with embedded pebble overlying undeformed, sub-horizontally-bedded outwash sand; (G) middle part of lens DF3 showing heavily deformed gravel, sand and diamicton with contrasting top (gradual) and bottom (sharp) boundaries; (H) flank of lens DF3 consisting of 30-cm-thick diamicton mixed with outwash deposits; (I) cut-and-fill structure interpreted as a subglacial N-type channel at the contact between outwash sand and basal till.

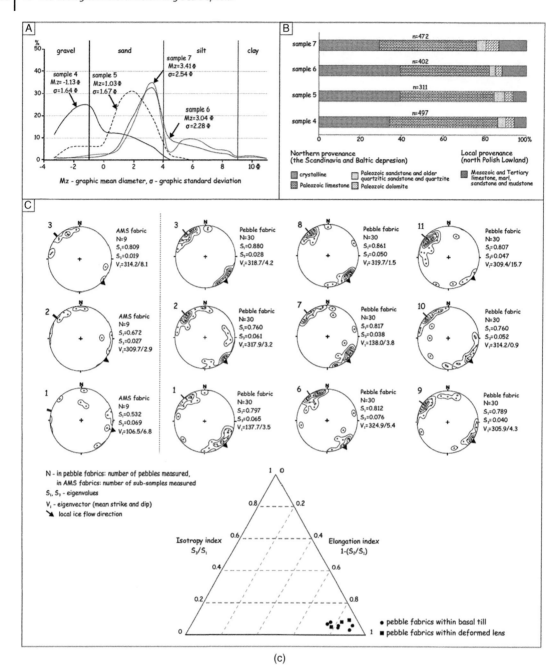

(c)

Figure 10.9 (c) clast macrofabric, AMS fabric, grain size and petrographic data from located sampling points (from Tylmann *et al.*, 2013).

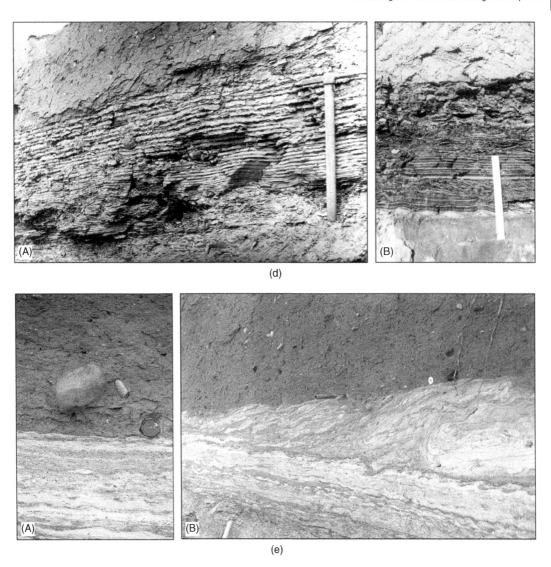

Figure 10.9 (d) glacilacustrine sediment capped by till in the Leipzig area of eastern Germany, showing: (A) largely undisturbed varved clay with sharp upper contact (photo by Prof. L. Eissmann); and (B) the same varved clay with heavily deformed upper half, grading into till through a zone of intensive mixing; (e) glacilacustrine sediment capped by till at Knud Strand, northwest Jutland, Denmark, showing (A) largely intact stratified sand, silt and clay with sharp upper contact and no apparent diffusive mixing with the till; (B) the same outcrop but located ca. 150 m distant to (A) is an area of heavily disturbed glacilacustrine sediment below the till contact (from Piotrowski *et al.*, 2004).

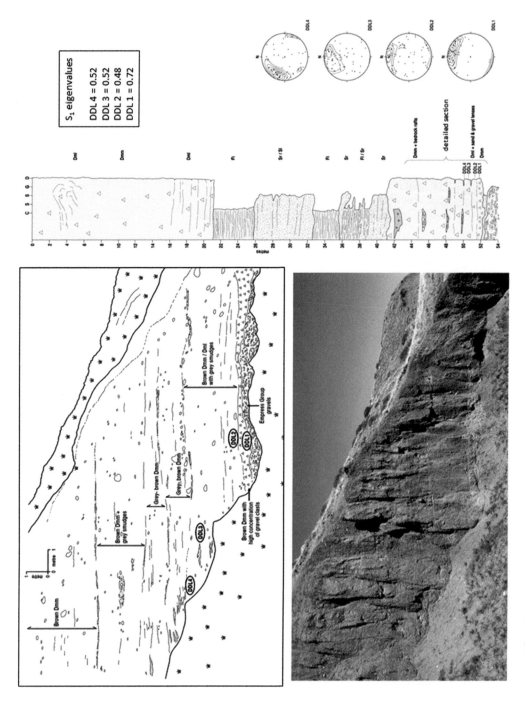

Figure 10.10 Till and intervening stratified sediments in the Lethbridge area of Alberta, Canada, showing details of basal sequence of diamictons with deformed bedrock rafts or smudges separated by discontinuous sand and gravel lenses and stringers or distinct partings (from Evans et al., 2012a). An amalgamation zone occurs at the contact between the till and underlying Empress Group sands and gravels. The clast macrofabric strengths in the basal till sequence weaken immediately above the amalgamation zone.

Figure 10.11 Examples of deformed inclusions or intraclasts: (a) stratified sands and gravels folded and attenuated into the base of a matrix-rich till, Filey Bay, eastern England; (b) small rafts of rippled sand in the base of a matrix-rich till, liberated from underlying climbing ripple drift deposits, Whitburn, northeast England; (c) preglacial sands rafted into basal shear zone of till sequence at Drayton Valley, Alberta, Canada. Sand intraclasts vary in shape from elongate blocks to boudins to attenuated lenses/laminae; (d) attenuated chalk rafts (glacitectonic laminae) in the Skipsea Till of eastern England; (e) deformed sand pods in clay-rich diamicton, Elk Point, Alberta, Canada.

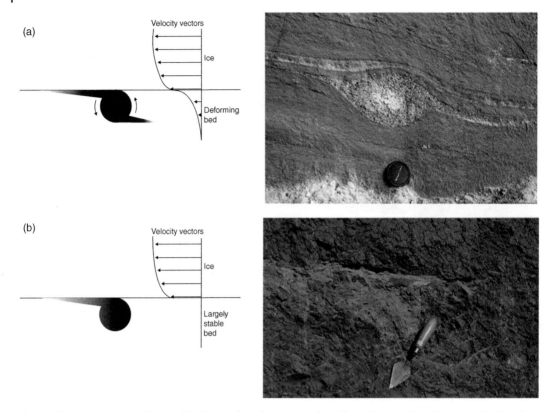

Figure 10.12 Dispersion tails extending from soft-sediment intraclasts (from Piotrowski and Kraus, 1997). Sketch shows two possibilities: (a) a deforming bed where the vertical distribution of velocity causes the clast to rotate and dispersion tails develop on its up-ice lower side and down-ice upper side (photograph shows example from Funen, Denmark); (b) a stable bed subject to ice sliding where the dispersion tail is at the sliding interface only (photograph shows sand lens with truncated upper contact with overlying till, Elk Point, Alberta, Canada).

homogenised, forming a gradational contact with underlying glacitectonite (Figures 10.7–10.10; e.g. Hicock and Dreimanis, 1989, 1992a, b; Hart, 1995; Benn and Evans, 1996; Evans, 2000b; Evans and Twigg, 2002).

At their more advanced stages of attenuation, intraclasts take on the form of glacitectonic lamination and hence can be difficult to differentiate from highly deformed stratified sediments or glacitectonites, especially if deformation takes place in pre-existing diamictons that have accumulated subaqueously (i.e. ice-rafted debris/dropstone muds) or by gravity mass flows (see Chapter 14). This introduces a particularly difficult problem for glacial sedimentologists in the form of laminated or banded diamictons (Figure 10.13). Strictly speaking, where intraclasts are deformed to produce the horizontal banding it should be called 'pseudo-lamination', because it has not been created by primary subaqueous sedimentation processes. However, highly attenuated glacitectonites that originated from stratified sediments could indeed have inherited true lamination. This inheritance process has been demonstrated using East Anglian till types by Hart and Roberts (1994) and Roberts and Hart (2005), who provide a set of criteria to attempt differentiation between their 'Type 1 laminae', produced by intergranular pervasive shear (glacitectonic lamination), and 'Type 2 laminae', created by the deformation of subaqueous deposits (glacitectonite). Laminae that originate

Figure 10.13 Examples of banded or (pseudo) laminated diamictons: (a) crude banding in the Skipsea Till at Barmston, Yorkshire, England. This deposit appears massive when first exposed but develops a banded appearance after wave erosion; (b) strong lamination likely derived from stratified (glacilacustrine) deposits in the Bacton Green Member of East Anglia, England; (c) folded laminations in the base of the Filey Bay diamicton ('till'), Yorkshire, England; (d) strongly laminated Bacton Green Member, East Anglia, England, with boudin created by deformation of coherent (initially frozen?) sand body.

from subaqueous deposits should still display some form of grading inherited from their primary sedimentation process, whereas glacitectonic laminae are typically non-graded. Discontinuous units such as pods, especially if they display boudinage or tails, are more typical of glacitectonically created (pseudo-laminated) sequences. Often difficult to interpret are dropstone-like clasts, which diagnostically in subaqueous deposits will bend underlying bedding and be draped by on lapping sediments, but when in association with glacitectonic laminae will have developed fold structures in the laminae as they act like augens and create pressure shadows. Needless to say, inheritance of mass flow diamictons in deformation sequences will be very difficult to decipher and poses one of the greatest challenges to glacial sedimentologists at present (see Chapter 14). This is not a trivial problem, as the origins of thick 'tills' may be fashioned initially by the accumulation of mass flow diamictons, as we shall investigate in Chapter 16.

Structures indicative of small-scale brittle deformation include discrete microshears (particularly in clay-rich sediments), brecciation, boudins or augen, and crushed quartz grains (e.g. Owen and Derbyshire, 1988; Menzies and Maltman, 1992; van der Meer, 1993; Menzies *et al.*, 2006). Microshears can be observed in macroscale patterns in the form of discrete fissility, although such structures could be created by unloading and hence are diagnostic of shear only if individual partings are adorned with slickensides (Figure 10.14). Under the microscope, thin sections sampled from tills display a range of

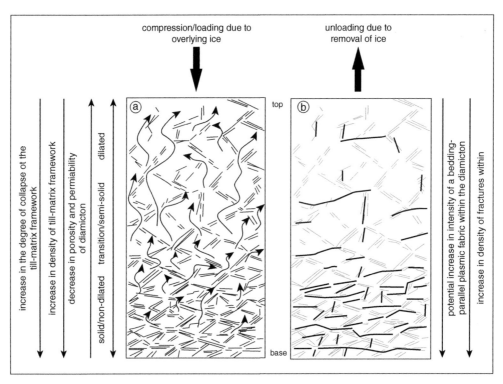

Figure 10.14 Fissility development in diamictons: Upper photographs show strongly developed, densely spaced fissility with slickensides in compact diamicton (left) and crudely developed, widely spaced fissile structures in clast-rich diamicton (right). Lower panel shows conceptual diagrams to explain how dilatancy impacts on framework development and hence potential fissility in tills (from Evans *et al.*, 2006b): (a) compression/loading leads to an increase in density of the till-matrix framework with depth and hence stronger density of fractures; (b) unloading also leads to the increase in intensity of bedding-parallel structures with depth.

microscale structures that both replicate and augment the macroscale features discussed above (e.g. Menzies and Maltman, 1992; Seret, 1993; van der Meer, 1993; Menzies, 2000; Menzies *et al.*, 2006; Neudorf *et al.*, 2013; Figures 4.9 and 7.1). These have been related by van der Meer (1993) to processes identified in the vertical deformation profile compiled by Alley (1991) based upon subglacial observations and acknowledging A and B horizon development and ploughing and sliding at the ice–bed interface (Figure 10.15). Most fundamentally, the clay-rich parts of deformed tills may show strong birefringence in cross-polarised light, a feature referred to as plasmic fabric, a direct result of clay platelets being aligned parallel to each other under the influence of shear stresses (for overviews, see van der Meer, 1993, Carr, 2004 and Menzies *et al.*, 2006). The partitioning of deformation into the clay-rich parts of tills results in strain hardening and brittle failure or faulting (see Section 4.3) and is a non-uniform response by heterogeneous subglacial materials to deformation. Also prevalent in such materials are till pebbles or rounded, soft-sediment inclusions/intraclasts (van der Meer, 1993), which form wherever small-scale variations in composition, grain size and/or water content lead to the solidification of patches or fragments in the till matrix, which otherwise is in a dilated or liquefied state (Figure 10.1). The isolation of till pebbles creates parallel-walled fissures which reflect the plane of shear and develop a typical marble bed structure (Figure 10.16; van der Meer, 1996, 1997; Hiemstra and van der Meer, 1997). This is part of a process of pore space modification in shearing till (Figure 10.17; Kilfeather and van der Meer, 2008), whereby initial shearing forces particles to rotate and fill void spaces to create turbate structures and thereby increases the sediment density. Continued shearing then develops lineations or fissile partings, once again increasing till porosity and encouraging water flow through interconnected shear planes (van der Meer *et al.*, 2003). The development of fissility is thought to relate to shearing under high porewater conditions, whereas marble bed structures appear to develop in situations where water pressures are low (cf. van der Meer *et al.*, 2003; Kilfeather and van der Meer, 2008).

The polydeformed nature of subglacial materials, particularly tills, has prompted Phillips *et al.* (2011) to construct a protocol for identifying clast microfabric domains that record multiple phases of deformation in thin sections. Displayed on standard stereonets, the clast microfabric data (Figure 10.18; cf. Evenson, 1970, 1971; Johnson, 1983; Carr *et al.*, 2000; Carr, 2001; Carr and Rose, 2003; Thomason and Iverson, 2006) are then combined with the microstructures identified in Figures 10.15–10.17 to create a microstructural map of thin sections employing the well-established terminology used in metamorphic petrology (Figures 10.19 and 10.20a). An idealised summary sketch of the typical features identified by Phillips *et al.* (2011) in subglacial tills is presented in Figure 10.20b, which illustrates the use of microfabric and microstructural data to identify five areas within a thin section. To explain the phased development of the microstructures and fabrics, Phillips *et al.* (2011) invoke a changing till rheology, including an early phase of viscous behaviour in highly dilated, water saturated material followed by its development into a much stiffer deposit capable of undergoing solid-state deformation, and thereby taking up folds, fabric and faulting. This switch in behaviour is when clast microfabric starts to develop within the matrix. These changes are recorded in early-phase arcuate grain alignments and turbate structures (Figures 4.9 and 10.17; van der Meer, 1993, 1997), related to rotational deformation in water saturated and dilated till, in which the free rotation of clasts caused them to become variably coated with finer-grained matrix. Till dewatering and falling dilation then increases matrix stiffness and the locking of larger clasts in position, thereby initiating clast microfabric development. This brings about the development and collapse of the 'till-matrix framework' of Evans *et al.* (2006b; Figure 10.1) to generate fabric. In the idealised case study in Figure 10.20, the first (S_1) microfabric dips down-ice and is formed by the passive rotation of sand-grade detritus into the plane of the developing foliation. The till is still subject to relatively elevated porewater content and/or pressure towards the end of this process and hence stresses result

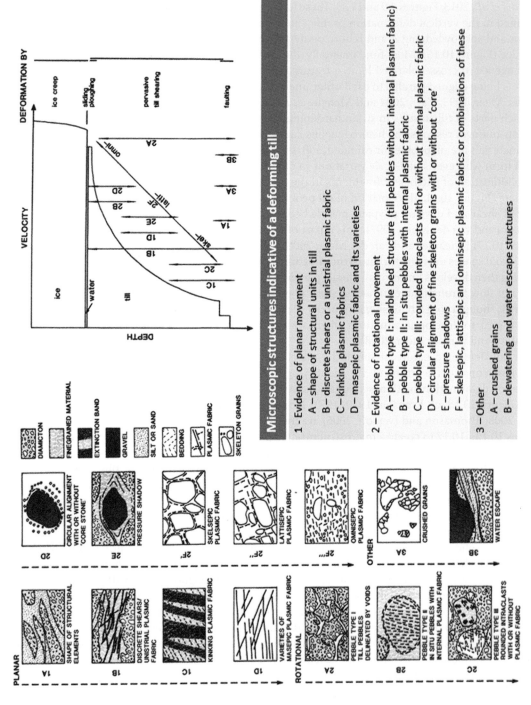

Figure 10.15 Microscale features related by van der Meer (1993) to processes operating in the vertical deformation profile compiled by Alley (1991).

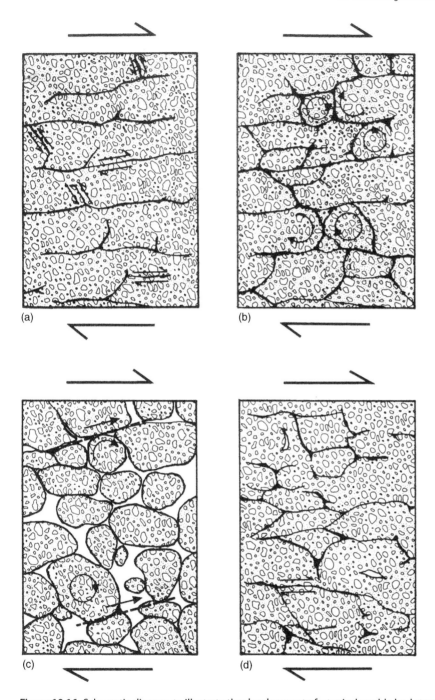

Figure 10.16 Schematic diagram to illustrate the development of a typical marble bed structure by the production of till pebbles or rounded, soft-sediment inclusions/intraclasts during shear (from Hiemstra and van der Meer, 1997): (a) shear zone development after dissipation of water with displacement along discrete planes with strain hardening; (b) brecciation of dry till with aggregation and progressive shear creating marble-bed appearance; (c) additional strain causes till to dilate with water entering voids; (d) collapse of dilated structure due to reduction in strain rate and expulsion of water and stiffening of till. This may trigger further shear stress and a return to marble bed.

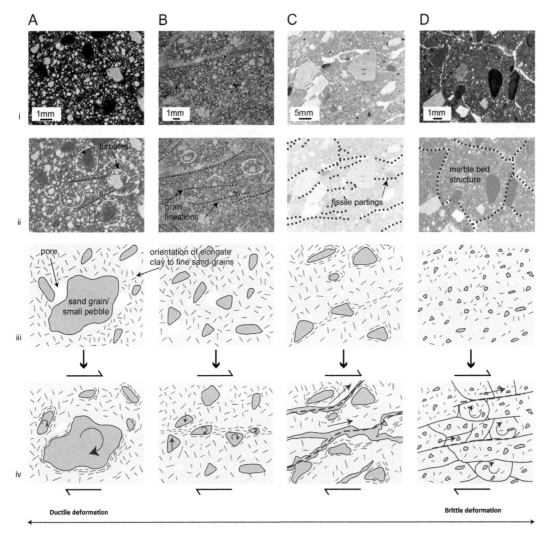

Figure 10.17 Pore space modification in shearing till (from Kilfeather and van der Meer, 2008): (i) and (ii) thin section examples of structures that effect the development of pores; (iii) schematic sketches to show the initial state of sediment prior to subglacial deformation; (iv) schematic sketches to show interpretations of the forms of deformation that result in the development and destruction of pores and other microstructures. This includes: (Aiv) long axes of elongate small grains align along the sides of a rotating pebble. Particles find the paths of least resistance and infill pores; (Biv) till shearing to form grain lineations or plasma. Porosity may decrease along these shear zones as particles migrate towards them and find the paths of least resistance to infill pores; (Civ) brittle break-up during shearing, possibly associated with water-escape, forming fissile partings that increase the porosity; (Div) break-up of a dense and dry till bed during high shear stress and rotation of individual aggregates to form marble-bed with increasing porosity.

clasts included within till

arcuate grain aggregates

linear grain aggregates

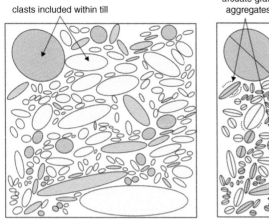

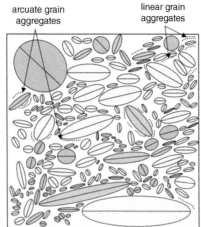

Stage I : high resolution scan of thin section

Stage 2: measure orientation of long axes of clasts

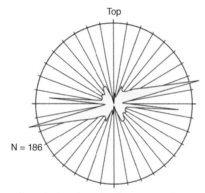

Top

N = 186

Stage 3: plot orientation data on rose diagram

long axis of clasts

microfabric domains

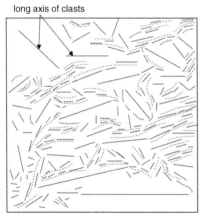

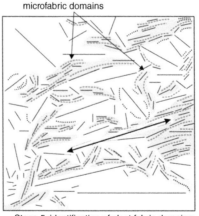

Stage 4: identification of main clast fabrics

Stage 5: identification of clast fabric domains and final interpretation of clast fabrics

.... arcuate and linear grain aggregates ⁄ long axis of clasts

.... microfabric defined by long axes of clasts ◄────► orientation of main fabric(s)

Figure 10.18 The collection and display of clast microfabric data (from Phillips *et al.*, 2011). The five stages are: (1) import high-resolution scans of thin sections into the graphics package; (2) measure the orientation of the clast long axes; (3) plot orientation data on a rose diagram; (4) identify main clast microfabrics; and (5) identify clast microfabric domains and make final interpretation.

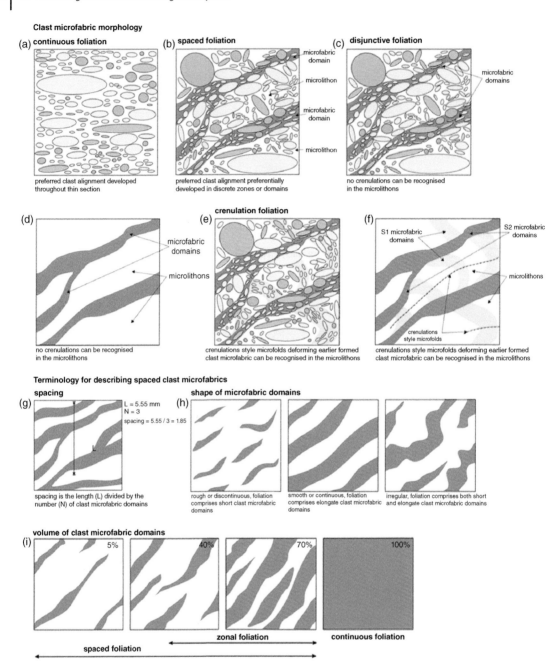

Figure 10.19 Illustration of the proposed non-genetic terminology for the morphological description of clast microfabrics in glacial sediments (from Phillips *et al.*, 2011), based upon the system used for the description and classification of cleavage and/or schistosity in metamorphic rocks (cf. Passchier and Trouw, 1996).

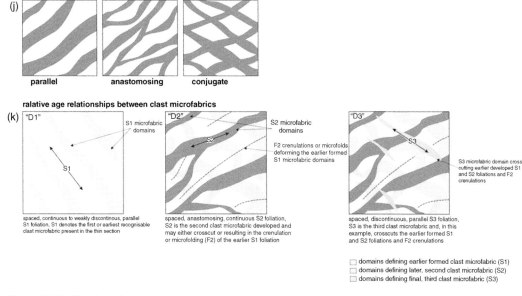

Figure 10.19 (*Continued*)

in the heterogeneous collapse of the till–matrix framework and localised liquefaction and renewed dilation. This could result in the localised loss or partial overprinting of the S_1 microfabric as well as remobilisation of liquefied till so that it is squeezed out of relatively higher-pressure areas where the till–matrix framework is collapsing. Phillips *et al.* (2011) speculate that these changes may equate to till 'collapse', liquefaction and remobilisation (Hiemstra and van der Meer, 1997) which in turn may drive the mobile operation of sticky spots (cf. van der Meer *et al.*, 2003; Piotrowski *et al.*, 2004); this could also relate to the stick–slip motion identified in subglacial observations (Fischer and Clarke, 1997; Boulton *et al.*, 2001; Section 6.3). The continuation of deformation in our idealised case study results in the early fabric becoming progressively folded and the development of a more pervasive second (S_2) microfabric. Its occurrence throughout the till indicates that it has dewatered sufficiently to restrict or retard further deformation induced liquefaction. The last (S_3) phase of deformation is focused into narrow zones of ductile shear due to the fact that the tills have dewatered and stiffened so that deformation begins to 'lock up', bringing about the development of discrete shears and small-scale brittle faults within the matrix. The complete sequence of phased deformation development depicted in Figure 10.20 equates to the gradual change from predominantly ductile to more brittle deformation, also depicted in progressive pore space modification in Figure 10.17 (Kilfeather and van der Meer, 2008).

The analyses of thin sections as outlined above reveals features that replicate at microscale what is observed as representative of till deformation at the macroscale. For example, Phillips *et al.* (2007) have identified the cannibalisation of stratified rafts into the base of an overlying till, in association with water escape conduits that trace the passage of pressurised water (hydrofractures) from the mildly deformed upper layers of underlying glacilacustrine rhythmites (Figure 10.21a). The sample was taken from the till–glacitectonite contact at the Drumbeg ice-marginal till stacks illustrated in Figure 8.12, where the effects of pressurised water in modifying all traces of till shearing are traced throughout the stacked till layers using micromorphology (Evans and Hiemstra, 2005). Similarly, the microscopic details of a heavily glacitectonised, inter-digitated contact between glacitectonite

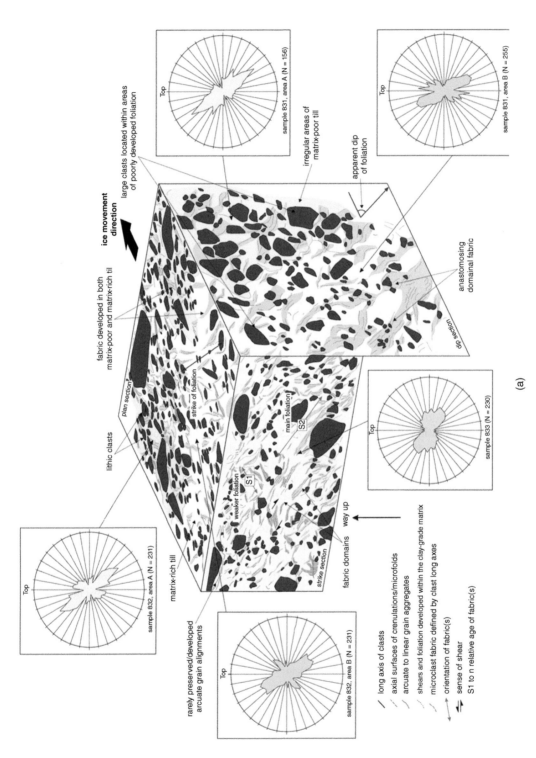

Figure 10.20 Examples of the employment of the microstructural mapping protocol proposed by Phillips *et al.* (2011): (a) schematic 3D block diagram showing the relationships between the various microfabrics developed within a fluting. There is a highly irregular boundary between the two lithologically distinct areas of diamicton, across which the main clast microfabrics cut; (b) schematic diagram to show the possible microfabric relationships developed in response to polyphase deformation associated with the formation of a subglacial till.

(a)

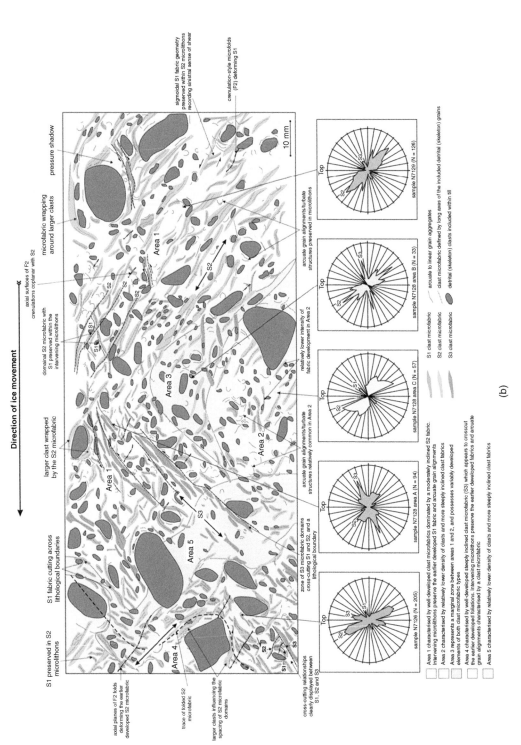

Direction of ice movement

S1 preserved in S2 microliths

S1 fabric cutting across lithological boundaries

larger clast wrapped by the S2 microfabric

domainal S2 microfabric with S1 preserved within the intervening microliths

axial surfaces of F2 crenulations coplanar with S2

microfabric wrapping around larger clasts

pressure shadow

sigmoidal S1 fabric geometry preserved within S2 microliths recording sinistral sense of shear

crenulation-style microfolds (F2) deforming S1

Area 1

S2

arcuate grain alignments/turbate structures preserved in microliths

Area 3

Area 2

relatively lower intensity of fabric development in Area 2

Area 1

S3

arcuate grain alignments/turbate structures relatively common in Area 2

Area 5

zone of S3 microfabric domains cross-cutting S1 and S2, and a lithological boundary

Area 4

axial planes of F2 folds deforming the earlier developed S2 microfabric

trace of folded S2 microfabric

larger clasts influencing the spacing of S2 microfabric domains

cross-cutting relationships clearly displayed between S1, S2 and S3

10 mm

Top

S3

S2 S1

sample N7126 (N = 205)

Top

S3

S2 S1

sample N7128 area A (N = 94)

Top

S1

S2 S3

sample N7128 area C (N = 57)

Top

S3

S2 S1

sample N7128 area B (N = 33)

Top

S3

S2 S1

sample N7729 (N = 126)

arcuate to linear grain aggregates

clast microfabric defined by long axes of the included detrital (skeleton) grains

detrital (skeleton) clasts included within till

S1 clast microfabric

S2 clast microfabric

S3 clast microfabric

Area 1 characterised by well-developed clast microfabrics dominated by a moderately inclined S2 fabric. Intervening microliths preserve the earlier developed S1 fabric and arcuate grain alignments

Area 2 characterised by relatively lower density of clasts and more steeply inclined clast fabrics

Area 3 represents a marginal zone between areas 1 and 2, and possesses variably developed elements of both clast microfabric types

Area 4 characterised by well-developed steeply inclined clast microfabric (S3) which appears to crosscut the earlier developed foliations. Intervening microliths preserve the earlier developed fabrics and arcuate grain alignments characterised by a clast microfabric

Area 5 characterised by relatively lower density of clasts and more steeply inclined clast fabrics

(b)

Figure 10.20 (*Continued*)

developed in rhythmites and an overlying till is illustrated in Figure 10.21b, as reported from Raitts Burn, Strathspey, Scotland, by Phillips and Auton (2000) and Phillips *et al.* (2011). This illustrates a distinct conjugate patterned microfabric within the till, with S_1 dipping down-ice and parallel to an inclined stratification and layer-parallel plasmic fabric. The S_1 orientation data are co-planar to the unistrial plasmic fabric/ductile shears and normal faults developed during deformation within the underlying rhythmites. The pervasively developed S_2 clast microfabric in the till dips up-ice and is

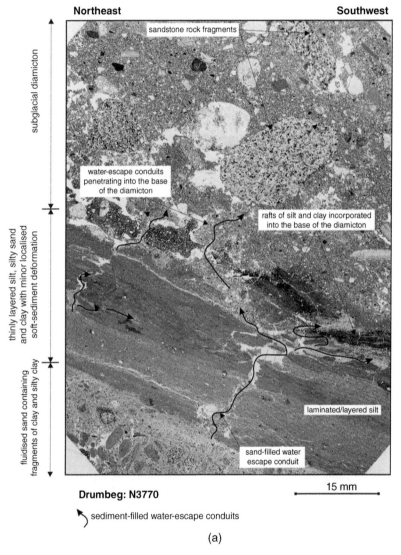

(a)

Figure 10.21 Examples of microstructures associated with till deformation: (a) cannibalisation of stratified rafts and development of water escape conduits at a till–glacitectonite contact (from Phillips *et al.*, 2007); (b) a microstructural map of polydeformed, thinly laminated sand silt and clay overlain by subglacial till from Raitts Burn, Strathspey, Scotland (from Phillips and Auton, 2000; Phillips *et al.*, 2011).

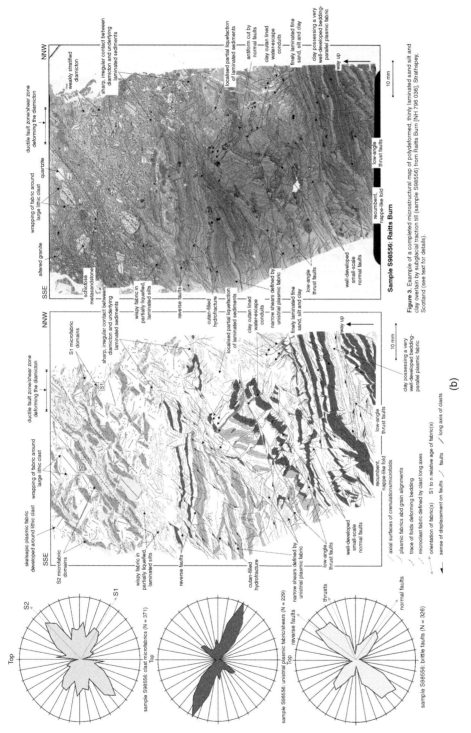

Figure 3. Example of a completed microstructural map of polydeformed, thinly laminated sand silt and clay overlain by subglacial traction till (sample S98556) from Raitts Burn [NH 796 036], Strathspey, Scotland (see text for details).

Sample S98556: Raitts Burn

Figure 10.21 *(Continued)*

(b)

parallel to a heterogeneously developed unistrial plasmic fabric in the matrix, suggesting that they were created simultaneously. This helps define a 10–12-mm-wide zone of enhanced ductile shear which offsets the till–rhythmite contact and can be traced into the underlying laminated sediments where it occurs parallel to a set of well-developed reverse faults. The microstructures developed within both the diamicton and underlying rhythmites relate to formation in response to the same stress regime imposed by ice overriding the site, and the differences in style and apparent intensity of deformation are directly related to the potentially much higher porewater content and/or pressure occurring within the till during subglacial deformation.

Micromorphology has been employed by Roberts and Hart (2005) to verify the details of their 'Type 1 laminae' (glacitectonic lamination) and 'Type 2 laminae' (glacitectonite). An example of Type 1 laminae is illustrated in Figure 10.22a, which is a chalky 'stratified' diamicton comprising a mixture of sub-horizontal, discontinuous, chalky and silty sand stringers with sharp and undulatory contact boundaries in a matrix with low anisotropy. The chalk stringers are composed of reworked chalk with secondary inclusions of quartz and feldspar and appear locally attenuated. Isoclinal, recumbent microfolds are evident in the pressure shadows of chalk clasts, which are both deformed and undeformed and associated with 'dropstone-like structures'. The variability of chalk clast coherence in glacially deformed materials is marked at both macro- and microscale and dictates the extent to which clasts are attenuated. This indicates that the yield strength of the chalk is locally reduced, possibly as a result of bedrock weathering prior to their rafting and isolation as intraclasts. Where relatively incoherent the chalk is attenuated into stringers which often have become detached from their source clasts and now appear isolated or form small folds in the lee-side pressure shadows of clasts. An example of Type 2 laminae is illustrated in Figure 10.22b, which depicts a silty, sandy stratified diamicton with low anisotropy in which laminae tend to be continuous and exhibit reworked soft-sediment clasts, although silty laminae are more discontinuous and attenuated. The contacts between laminae are sharp and conformable and display dropstone structures. The laminae are thought to be derived from suspension rainout and mass flow activity but are then subject to deformation and hence are strictly glacitectonites (see Chapter 14).

The attenuated nature of intraclasts is clearly diagnostic of subglacially deformed diamictons at both macro- and microscale (Figures 4.2, 10.8 and 10.11). Hence, they can be employed in the interpretation of diamictons recovered from cores or borehole samples (e.g. van der Meer and Hiemstra, 1998), although differentiation of Type 1 from Type 2 examples is challenging, especially when they are at their more advanced stages of attenuation and appear as 'pseudo-laminated' or 'pseudo-stratified' diamictons.

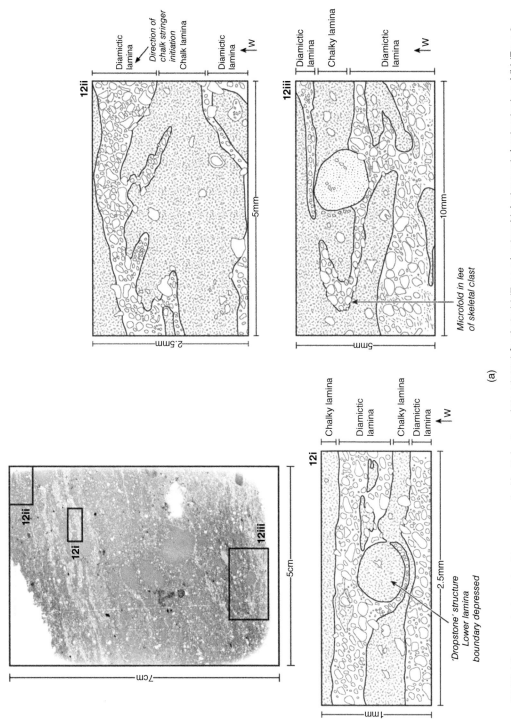

Figure 10.22 Micromorphological evidence compiled by Roberts and Hart (2005) for their: (a) 'Type 1 laminae' (glacitectonic lamination) and (b) 'Type 2 laminae' (glacitectonite).

(a)

12ii

Diamictic lamina

Direction of chalk stringer initiation

Chalk lamina

Diamictic lamina

w

5mm

2.5mm

12iii

Diamictic lamina

Chalky lamina

Diamictic lamina

w

10mm

5mm

Microfold in lee of skeletal clast

7cm

5cm

12ii

12i

12iii

12i

Chalky lamina

Diamictic lamina

Chalky lamina

Diamictic lamina

w

2.5mm

1mm

'Dropstone' structure Lower lamina boundary depressed

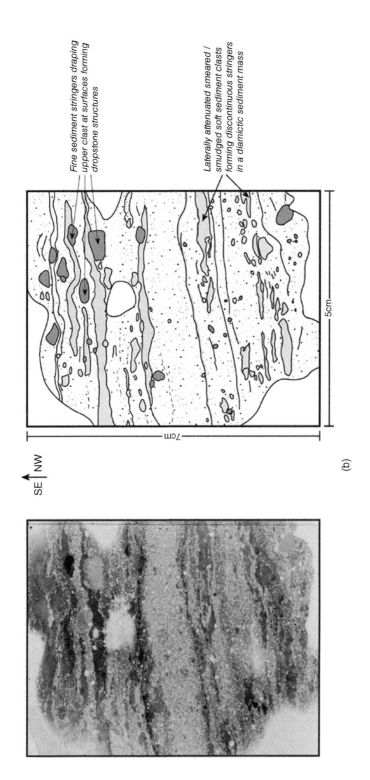

Fine sediment stringers draping upper clast at surfaces forming dropstone structures

Laterally attenuated smeared / smudged soft sediment clasts forming discontinuous stringers in a diamictic sediment mass

NW

SE

7cm

5cm

(b)

Figure 10.22 (*Continued*)

11

The Geological Record: Sliding Bed Deposits

This seam of sand was doubtless quickly deposited by running subglacial water; and it was thenceforward undisturbed while the deposition of the till continued over its whole extent.

Upham (1889, p. 239)

In Section 6.3, the concept of ice–bed decoupling and its implications for subglacial meltwater processes was briefly reviewed and it was stressed that ancient glacigenic deposits have been critical to developing an understanding of till-related meltwater sedimentation. Certainly, a strong glaciological case for glacier sliding and water flow over till beds is well established (cf. Alley, 1989a, b, 1992; Engelhardt and Kamb, 1998; Fischer and Clarke, 2001; Kavanaugh and Clarke, 2006; Truffer and Harrison, 2006; Figure 11.1), but a range of till-related stratified sediment facies and structures also have been interpreted as the products of intermittent water flow at the ice–bed interface (e.g. Eyles *et al.*, 1982; Brown *et al.*, 1987; Clark and Walder, 1994; Evans *et al.*, 1995; Piotrowski and Kraus, 1997; Piotrowski and Tulaczyk, 1999; Piotrowski *et al.*, 2001; Figure 6.13), and although not always uncontroversial, these sedimentologically based conceptual models are central to appreciating the former occurrence of ice sheet and glacier bed decoupling and what has become known as 'soft-bed sliding'.

The concept of 'canal fills', the infills of former braided canal systems that develop at the ice–bed interface during periods of decoupling (6 on Figure 6.13), was introduced by Clark and Walder (1994) based upon the Walder–Fowler theory that meltwater drainage at the ice–soft-bed interface would be organised in anastomosing, wide and shallow channels or canals with flat roofs and low flow velocities (Walder and Fowler, 1994). The sedimentary evidence cited in support of this theory was that of fluvial channel fill deposits reported by Eyles *et al.* (1982) from the stacked subglacial ('lodgement') tills of northeast England (Figure 11.2). This explanation was later applied by Evans *et al.* (1995) to similar stratified sediment bodies in the tills of Holderness, eastern England (Figure 11.3). The proposed canal fills comprise stratified lenses with flat upper surfaces and concave bases (convexo-planar; Figure 4.2), containing primary depositional structures or bedforms, in addition to thin sand layers and stringers. Where preserved they occur as intra-till bodies and may be undeformed but are usually intensely deformed at their upper contacts or throughout their depth (Figure 11.4). Hence, ice–bed recoupling and shearing of the bed following on from a phase of meltwater drainage is usually clearly demonstrated. Stacked sequences of tills and intervening stratified interbeds are reported from the 'Northern Till' in Ontario, Canada, by Boyce and Eyles

Till: A Glacial Process Sedimentology, First Edition. David J A Evans.
© 2018 John Wiley & Sons Ltd. Published 2018 by John Wiley & Sons Ltd.

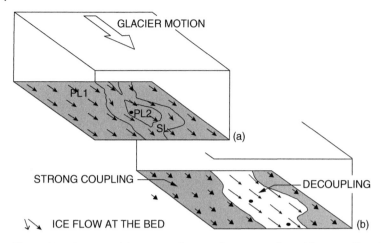

Figure 11.1 Conceptual diagram to illustrate the perceived variable rates of basal sliding due to changing ice–bed coupling (from Fischer and Clarke, 2001). Ice flow direction and magnitude are depicted by arrows during periods of: (a) low subglacial water pressures, and (b) high subglacial water pressures in the connected region of the bed (outlined). The locations of ploughmeters (PL1 and 2) and sliding sensor (SL), upon which this reconstruction is based, are marked.

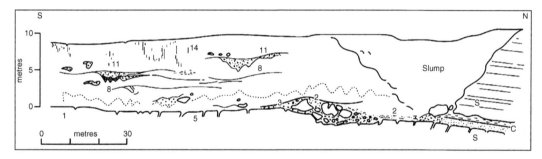

Figure 11.2 The stratigraphy of subglacial tills and associated sediments at Sandy Bay, northeast England (from Eyles *et al.*, 1982). Bedrock cliffline at north end of bay comprises Carboniferous sandstone (S) with a thin coal and shale band (C). Other features are: (1) striated rockhead; (2) bedrock rafts; (3) coarse rubbly till (lee-side cavity fill); (5) intrusion of till into bedrock joints; (8) glacifluvial channel fills or till interbeds (canal fills); (11) glacitectonised upper surfaces of canal fills with rafts of fill material in base of overlying till; (14) vertical jointing in till.

(2000; Figure 11.5), demonstrating that subglacial deforming layers can be laid down in complex sequences that record pulses of alternate deformation and sliding in settings where there is a net vertical till accretion. Clearly, erodent layers are not effective in such settings, which as we discussed in Section 6.3 are most likely to be near glacier and ice sheet margins where shear stresses and meltwater discharges tail off enough to stack the materials that are advected from up ice (cf. Alley *et al.*, 1997; see Chapter 16). In these zones of incremental thickening and marginal till wedges, the net vertical accretion is critical to the preservation of meltwater sediments created during the soft-bed sliding process. Elsewhere, for example, where erodent layers are effectively removing most if not all such sediments, or where subglacial deformation is particularly intense after each phase of ice–bed separation, the remnants of former canal fills might be represented only by attenuated

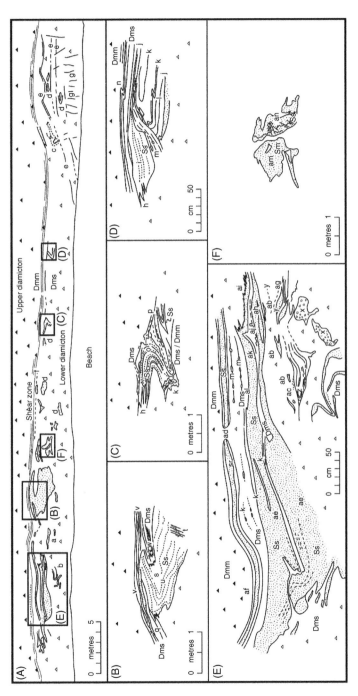

Figure 11.3 Sedimentary structures associated with the Skipsea Till of Yorkshire, eastern England (after Evans *et al.*, 1995), indicative of meltwater canal fill activity between till emplacement events: (A) overview of section face showing Skipsea Till and associated intrabeds of stratified sediments; (B–F) details of structures outlined by boxes in A with labels locating the following structures: (a) deformed stratified, poorly sorted, coarse and pebbly sands; (b) concentrations of chalk clasts; (c) concentrations of rounded pebbles; (d) deformed sand lenses; (e) crude lamination in the till due to subtle grain size variations; (f) major discontinuity in the till, marking the top of the zone of deformed sand lenses; (g) strongly developed vertical joints; (h) interdigitisation of till and sand; (i) crudely planar stratified sands; (j) laminated till comprising red, grey-brown and buff lamina derived from cannibalised soft bedrock; (k) chalk stringers; (l) laminated till; (m) laminated till with chalk stringers; (n) chevron fold; (o) stratified sand with interstratified minor diamicton beds <10 mm thick; (p) till with weak stratification at base and chalk stringers subparallel to lower contact; (q) dark brown till with chalk stringers; (r) sand with stratification parallel to base of overlying till; (s) cross-stratified sands; (t) normal faulted sands; (u) gentle folds; (v) light brown stratified diamicton with sandy intercalations; (w) folded stratified diamicton; (x) concentrations of surrounded chalk pebbles, some of which are deformed into stringers within the surrounding till; (y) sand stringers in till; (z) sandy diamicton interstratified with cm-thick beds of clayey sand folded into a recumbent fold; (aa) deformed stratified clayey sands; (ab) stratified sand lenses (dm wide) with convex tops and flat bases; (ac) shears in till lined with sand; (ad) laminated till grading into massive till; (ae) moderately-to-well-sorted, coarse-to-fine sands; (af) rippled medium sands; (ag) deformed lenses of rippled sands; (ah) fine to very coarse, poorly sorted clayey sands containing pebbles towards the top; (ai) stratified diamicton interdigitated with sandy clay; (aj) chalk stringer deformed by a diapir; (ak) coarse to fine sands with low-angled cross-stratification deformed by small normal faults; (al) smooth base of laminated till parallel to the laminations of the underlying sands; (am) massive to moderately well-sorted coarse sands; (an) faulted, stratified, poor to moderately sorted, medium to fine sands.

Figure 11.4 Details of the subglacial canal fills of eastern England: (a) series of partially connected canal fills with heavily glacitectonised tops in the Skipsea Till at Skipsea, Yorkshire; (b) laminated fines in a series of canal fills in the Horden Till at Whitburn, Durham, where the subglacial tunnel locally widened into a slack water cavity near the glacier grounding line in Glacial Lake Wear (cf. Davies *et al.*, 2009); (c–e) details of the glacitectonic structures developed at the tops and margins of the Skipsea Till canal fills.

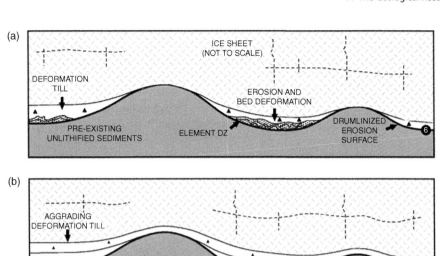

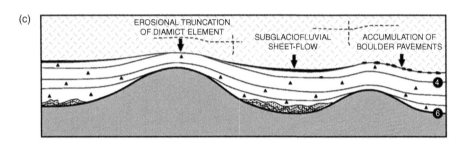

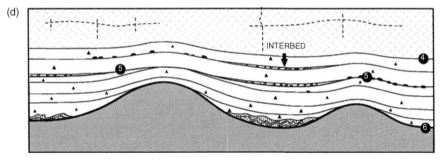

Figure 11.5 A conceptual model for the evolution of tills and their stratified interbeds (from Boyce and Eyles, 2000): (a) erosion and deformation of pre-existing unlithified sediments to form drumlinised surface; (b) conformable aggradation of deformation tills on drumlinised surface; (c) subglacial fluvial reworking of diamicton to form sheet-like interbeds and boulder pavements; (d) continued aggradation of deformation tills and stratified interbeds.

sand and gravel pods and might thereby be potentially indistinguishable from rafts/intraclasts (as discussed above in Chapter 10) or even clast pavements (see Section 9.3). In some circumstances, a subglacial fluvial origin might be recorded by the concentration of calcium carbonate precipitates, which are preferentially deposited in subglacial meltwater systems (cf. Hallet, 1976). Because canal fills are the thinner up-ice extensions of sub-marginal drainage networks, it is not surprising to see them thicken towards eskers and/or ice-contact subaqueous fans or grounding line fans, excellent examples of which occur on the Holderness and Durham coasts of eastern England where glacial lakes

Figure 11.6 Examples of proposed ice–bed separation features (from Piotrowski and Tulaczyk, 1999), showing a lack of significant deformation and with sand layers thought to be the product of thin water films at the ice–bed interface: (a) sub-horizontal, often slickensided fissures, in places filled with syn-depositional sorted sediments; (b) sub-horizontal fissures (upper part) and thin stringers of sorted sediments interbedded with till matrix (lower part); (c) horizontally stratified till consisting of mm-thick sand layers intercalated with till matrix; (d) single horizontal stringer of stratified sand in till matrix.

Holderness and Wear, respectively, were the depo-centres into which the sub-marginal meltwaters drained (Davies *et al.*, 2009; Evans and Thomson, 2010; see Chapter 16). Fine-grained rhythmites that fill the canals excavated in the Horden Till of the Durham coast indicate that Glacial Lake Wear sediments were being laid down in the slack water areas of the wider ends of the subglacial canals near the grounding line (Davies *et al.*, 2009).

A variety of thin and laterally extensive intra-till stratified beds (Figure 11.6) have been interpreted by Piotrowski and Tulaczyk (1999) as examples of ice–bed separation features, because they all lack significant deformation and their sandy grain size characteristics would be typical of thin water films at the ice–bed interface. Specific characteristics, as illustrated in Figure 11.6 include sub-horizontal, often slickensided fissures, in places filled with syndepositional thin stringers of sorted sediments interbedded with the host diamicton matrix. Higher stringer densities can give the host diamicton the appearance of being horizontally stratified. In contrast, single horizontal and discontinuous stringers can occur in otherwise massive diamictons. The similarity of these forms to the Type 2 laminae of Roberts and Hart (2005) is striking and problematic in terms of differentiating such deposits and assigning subglacial sliding bed versus glacitectonite origins to them. This equifinality dilemma is revisited in the following chapters and remains one of the greatest challenges in glacial sedimentology.

Sub-till stratified units, especially on hard or relatively impermeable beds (Figure 11.7), have been interpreted as evidence for water films at sliding till bases and indeed Evans *et al.* (2006) speculated that Hindmarsh's (1996) theoretical concept of 'till sliding' could give rise to the generation of thin stratified beds by sub-till water films. A modern example of this has been reported by Kjær *et al.* (2006) from the margin of the Icelandic surging glacier Bruarjokull (Figure 11.8) where water escape structures beneath glacitectonite near the outer zone of a fluted till sheet are interpreted as evidence for a dual-coupled model where the glacier is coupled to its deforming bed and the substrate is decoupled from the bedrock, thereby leading to fast ice flow due to over-pressurised water being forced along a near-impermeable bedrock surface, at least near the ice margin.

Finally, during till deformation meltwater may drain through the till by pipe flow (3 in Figure 6.13), a process that has been observed at the base of the Rutford Ice Stream, Antarctica (King *et al.*, 2004).

Figure 11.7 Example of a sub-till stratified lens lying beneath a thin ice stream bed till and overlying relatively impermeable substrate of shale in Alberta, Canada.

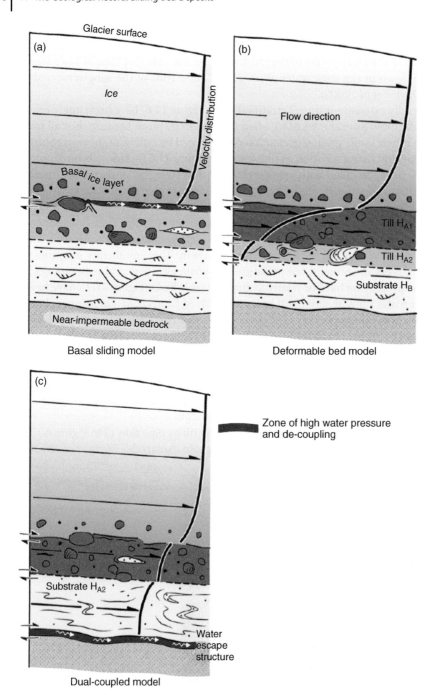

Figure 11.8 Schematic diagrams summarising the various modes of basal motion associated with ice streams and surging glaciers (from Kjær *et al.*, 2006): (a) decoupling sustained by enhanced basal sliding across the glacier–till interface with limited or no subglacial deformation; (b) glacier–bed coupling and fast ice flow sustained through subglacial deformation of water-saturated sediment and the development of a fast-deforming sediment (H_{A1}) over a more slowly deforming sediment (H_{A2}) that in turn is superimposed on a stable horizon (H_B); (c) a dual-coupled model where the glacier is coupled to its bed to create slow subglacial deformation, while the substrate is decoupled from the bedrock by a water film leading to fast ice flow.

The potential sedimentary signatures of this process could be Alley's (1991) 'mini eskers' but sedimentologically such features would be very difficult to differentiate from rafts cannibalised from older sediments or canal fills, all of which would be folded and/or attenuated after their emplacement in a mobile deforming layer. Evidence to affirm the prediction by Alley (1992) that till could squirt into channels in and on soft beds could be manifest in the interdigitisation of till and stratified sands around the margins of canal fills (Figure 11.9).

Figure 11.9 Sedimentological evidence for till squirting into canal fills, demonstrated by interdigitisation of till and stratified sands around the margins of the Skipsea Till canal fills, Yorkshire, England.

12

The Geological Record: Impacts of Pressurised Water (Clastic Dykes)

> *The nature of the basal hydraulic system is not only a key determinant of glacier dynamics but must also play a vital role in determining till processes.*
>
> Boulton *et al.* (2001, p. 26)

Clastic dykes are sediment-filled fissures that cut across, and are therefore post-depositional infills of, fractures in pre-existing, host materials (Figure 12.1). Genetically, they are described as hydrofracture fills, water escape structures and burst-out structures. They are associated with the escape of water and liquefied sediment under pressure, which is a common process in glacitectonically and subglacially deformed materials and closely linked with the operation of subglacial meltwater systems and their interactions with groundwater (Mandl and Harkness, 1987; Boulton and Caban, 1995). In Chapter 11, an example of water escape structures from an Icelandic surging glacier (Figure 11.8) demonstrated how clastic dykes may be created in sub-till settings by interacting with a rapidly pressurised subglacial drainage system. More commonly, the pressurised water and fluidised sediment are not confined entirely to such horizontally aligned beds but instead branch out from the source aquifer along hydrofractures or tensional cracks which cross-cut neighbouring, predominantly overlying, deposits (Lowe, 1975; Nichols *et al.*, 1994).

The idea that clastic dyke production could be initiated by glacier imposed stress has a long history (e.g. Hansen, 1930; Lundqvist, 1967; Berthelsen, 1974). Early work on clastic dykes attracted significant critical discussion (e.g. Mörner, 1972, 1973a, b, 1974; Dreimanis, 1973, 1992; Worsley, 1973; Dionne and Shilts, 1974; Elson, 1975; Humlum, 1978; van der Meer, 1980; Åmark, 1986) and tended to highlight diamicton-filled features and hence often used the term 'till wedges' to refer to features created by downward injection into the glacier bed; mixed sediment dykes were reported by Åmark (1986). In a more recent review of clastic dykes, van der Meer *et al.* (2009) tended to highlight only finer-grained examples, specifically because they associated the dominant influence of water through flow, predominantly upwards through sediment towards the glacier sole, as diagnostic of clastic dyke classification. However, they did also briefly present examples of much coarser-grained dykes in which cobble gravels and diamictic sediments dominate. Clastic dykes are now widely recognised in glacitectonites and tills at both macro- and microscale and are regarded as a product of both water-dominated fracture and infill as well as sediment injection.

A range of clastic dyke infills, created by downward injection of remobilised subglacial till into underlying proglacial outwash during a readvance of the snout of Sólheimajökull, Iceland, are reported by Le Heron and Etienne (2005). The clastic dykes form a three-dimensional, reticulate

Till: A Glacial Process Sedimentology, First Edition. David J A Evans.
© 2018 John Wiley & Sons Ltd. Published 2018 by John Wiley & Sons Ltd.

Figure 12.1 Examples of clastic dykes: (a) burst-out structure of gravel dyke and its branches in the Horden Till at Whitburn, Durham, North East England; (b) vertical gravel-filled dykes in crudely stratified diamictons, Clifden, Connemara, western Ireland; (c) sand and silt-filled dykes in locally liquefied and subglacially sheared rhythmites beneath till at Swarthy Hill, northwest Cumbria, England; (d) and (e) hydrofracture infills containing stratified clay, silt, sand and granule gravel and cross-cutting granule gravel glacifluvial outwash capped by till, Slettjökull foreland, Iceland.

swarm and bifurcate and dip predominantly away from the former hydraulic head source, which was the glacier terminus readvance position. The laminated nature of the sediment infills is explained by Le Heron and Etienne (2005) as a product of dyke injection of sediment in a non-fluidised state (Figure 12.2), either by: Model 1, sustained, continuous intrusion, whereby the dyke material underwent a rheological change from viscous to plastic during injection to form deformation bands (cf. Jonk *et al.*, 2003); or Model 2, in which repeated fracture and expansion allowed sediment squeezed in to the fracture to smear on to the fracture walls (cf. Hayashi, 1966). Lamination in finer-grained clastic dykes is interpreted in a different way by van der Meer *et al.* (1992, 2009) based upon the typical characteristics of clay, silt and sand laminae, occasional intraclasts and fine gravel, various directions of internal grading, cross-cutting relationships, 'drip-shaped patterns' and parallel alignments to fracture walls (Figure 12.3). Similar to Le Heron and Etienne's (2005) Model 2, the hydrofracturing and infilling of the space is thought by van der Meer *et al.* (2009) to be an ongoing process, thereby explaining the incremental and continuous infilling of fractures by cross-cutting laminae; the changing and variable directions of grading are a product of changing hydraulic pressures and the 'drip-like patterns' interpreted as a record of pulsating movement of sediment.

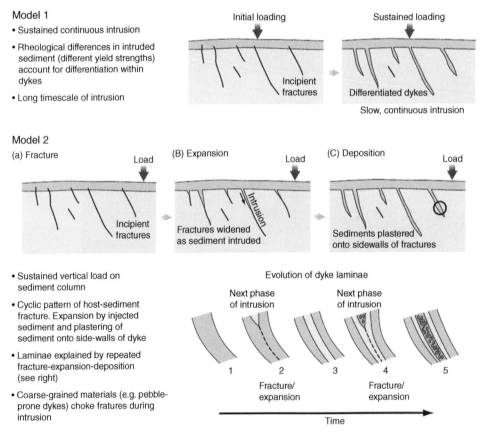

Figure 12.2 Models of clastic dyke infills due to downward injection of remobilised subglacial till into sand and gravel (from Le Heron and Etienne, 2005).

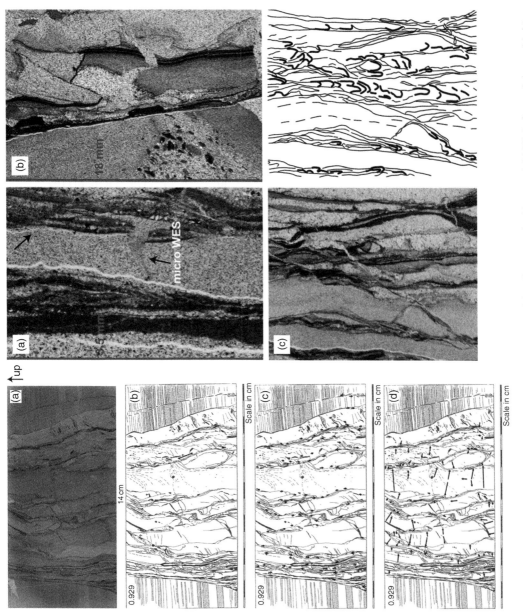

Figure 12.3 Microscale characteristics of the lamination in finer-grained clastic dykes (from van der Meer *et al.*, 2009). Left panel shows: (a) thin section of constricted part of clastic dyke from San Martin de los Andes, Argentina; (b) sketch of laminae in the thin section with single barb arrows pointing to displacement along microfaults and normal arrows pointing to related intraclasts; (c) sketch of the distribution of drip-like structures in the thin section; (d) grading (sense indicated by arrows) in individual laminae. Right panel shows finer details of thin section: (a) small fault (arrowed) in top right corner and micro-WES in centre; (b) broken up intraclasts with small intraclasts in left hand lower corner; (c) image and sketch (right) of downwards oriented drip-like structures along boundaries of laminae.

The interaction of water escape pathways with subglacial till deformation has been reported from a contemporary setting by van der Meer *et al.* (1999) from the sub-marginal environment at Slettjökull, Iceland (Figure 12.4). In this location, a marginal belt of permafrost is thought to have been instrumental in forcing increased subglacial meltwater discharges into sub-till outwash deposits to produce down-glacier dipping water escape structures (WES). A subsequent phase of fracture filling (black WES) was then initiated after the termination of the meltwater drainage event and the till dried out, involving upward fracturing in the till due to water and liquefied sand being driven upwards from the underlying outwash. Subglacial till deformation then partially distorted the structures. The role of permafrost here is a localised one and is not critical to the impedance of groundwater flow and artesian pressure build up that is required to inject sediment bodies with materials bursting out from confined aquifers (Nichols *et al.*, 1994; Boulton and Caban, 1995; Le Heron and Etienne, 2005). The creation of large pressure gradients, regardless of the trigger, can lead to hydrofracturing of pre-existing sediments and clastic dyke production, and many examples of coarse-grained dykes cross-cutting tills and associated deposits have been reported that relate to glacitectonic and subglacial deformation (e.g. McCabe and Dardis, 1994; Dreimanis and Rappol, 1997; Meehan *et al.*, 1997; Evans and Ó Cofaigh, 2003; Davies *et al.*, 2009). In the case study of Dreimanis and Rappol (1997), a range of large (in places >16 m long), down-glacier dipping clastic dykes, some with truncated or displaced tops due to contemporaneous till shearing, have been injected through clays and silts and into overlying stratified and massive diamictons of the Catfish Creek 'till' (Formation). It is thought that they were emplaced by vertically escaping pressurised water moving along pre-existing subglacial tension fractures.

A conceptual model to explain the occurrence of coarse-grained clastic dykes in tills was developed by Rijsdijk *et al.* (1999) based upon a number of impressive examples exposed at Killiney Bay, Ireland (Figure 12.5a). The dykes are regarded as the products of hydrofracture filling that was sourced from a sub-till aquifer, because they are rooted in an underlying gravel layer and extend sub-vertically into the overlying till, where they often terminate in plumes or burst-out structures (Figure 12.5b–e). The characteristics of the dyke infills are that they comprise poorly sorted coarse gravels with clast a-axes and a–b planes aligned parallel to the dyke walls, include till intraclasts which are commonly streamlined and also aligned parallel with dyke walls, and the dyke tops and branches often end in funnel-shaped clusters of clasts. Hydrofracturing of the till aquitard occurred when water pressures were forced to rise in the underlying gravel aquifer, causing fluidised gravel to be injected up to 7 m vertically into the till and eventually terminating as burst-out structures and fan-shaped gravel clusters (Figure 12.5d and e). This process requires that the till must have been initially supersaturated but the post-injection shearing of the tops of some dykes and emplacement of a capping till layer (Figure 12.5f) indicates that aquifer pressurisation was caused by glacier margin readvance over the site. Similar features have been reported from within drumlin till cores by McCabe and Dardis (1994). The intrusion of clastic dykes downwards as depicted in Figure 12.5a is based upon prediction by Boulton and Caban (1995) but has been verified by numerous field based observations of features like 'till wedges' from the early literature (see Figure 8.13) as well as the case study of van der Meer *et al.* (1999; Figure 12.4). Examples of till being intruded into fractures in bedrock by squeezing have been reported from a variety of settings (e.g. Meehan *et al.*, 1997; Evans *et al.*, 1998), where it has been implicated in the process of bedrock rafting into subglacial deforming till layers (see Chapter 14).

Clastic dykes, and more especially water escape structures, are well developed in glacitectonites, particularly at the boundaries with overlying tills, where they can be influential in the liberation of soft-sediment rafts (e.g. Phillips *et al.*, 2007; Figure 10.21a). The complexity of structures associated with water escape in such materials, particularly in stratified fine-grained sediments like glacilacustrine or glacimarine deposits (e.g. Phillips and Merritt, 2008), are particularly well illustrated

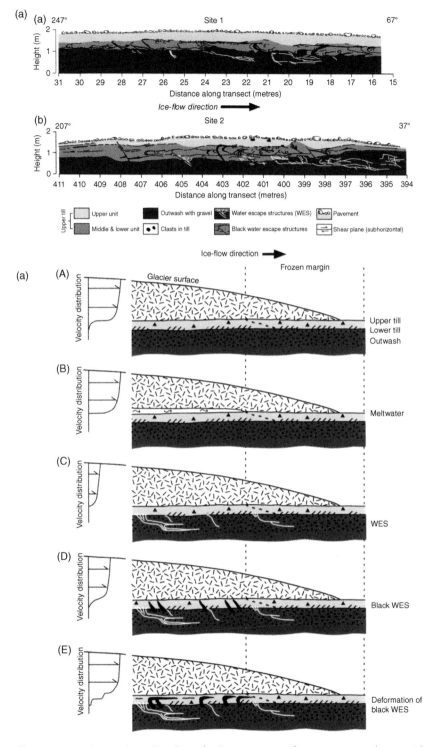

Figure 12.4 Sedimentological evidence for the interaction of water escape pathways with subglacial till deformation at Slettjökull, Iceland (from van der Meer *et al.*, 1999): (a) black WES and till structure at sites 1 and 2; (b) reconstruction of the events responsible for the development of the water-escape structures and till structures.

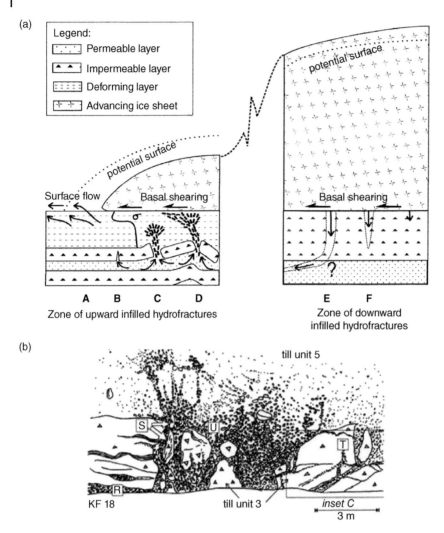

Figure 12.5 Clastic dykes developed in tills based upon examples from Killiney Bay, Ireland (from Rijsdijk *et al.*, 1999): (a) conceptual model for dyke development; (b) and (c) sketched examples of plumes or burst-out structures; (d) proposed stages of the hydrofracturing process at Killiney (for the *P/t* graph, *P* = water pressure, *t* = time, *τ* = maximal tensile shear strength of capping aquifer, *Fls* = fluidisation velocity of sand, *Flg* = fluidisation velocity of gravels). (A) Stage 1 – water pressures within confined aquifer increase when meltwater supply exceeds capacity of the aquifer. (B) Stage 2 – water pressures exceed the overburden pressures and a water blister may form and lifts the till layer. (C) Stage 3 – water pressures exceed the tensile strength (*τ*) of capping till layer and a component of the total normal stress and a hydrofracture forms. A high-pressure potential is generated across the fracture, leading to upward water flow. Convergence of flow at crack base leads to flow velocities exceeding the minimal fluidisation velocity of coarse gravels (*Flg*), causing them and anything smaller to fluidise. The ejected fluid supersaturates the overlying diamict, allowing ejected materials to settle through the matrix to form a crudely stratified diamict. When flow through the fractures ceases, fluidised sediments settle; (e) schematic diagrams showing stages of branching of clastic dykes. Left (a) shows that during burst-out through a vertical hydrofracture, hydraulic pressure builds up locally along the dyke walls, where pre-existing joints (broken lines) may dilate and begin to fill. Right (b) shows that during filling of the joints, the tensile strength of the capping till is locally exceeded, small-scale hydrofractures may form and water bursts out to form smaller-scale vertical offshoots; (f) broad stratigraphic architecture of the Killiney Bay site showing the post-injection shearing of the dyke tops by a capping till layer.

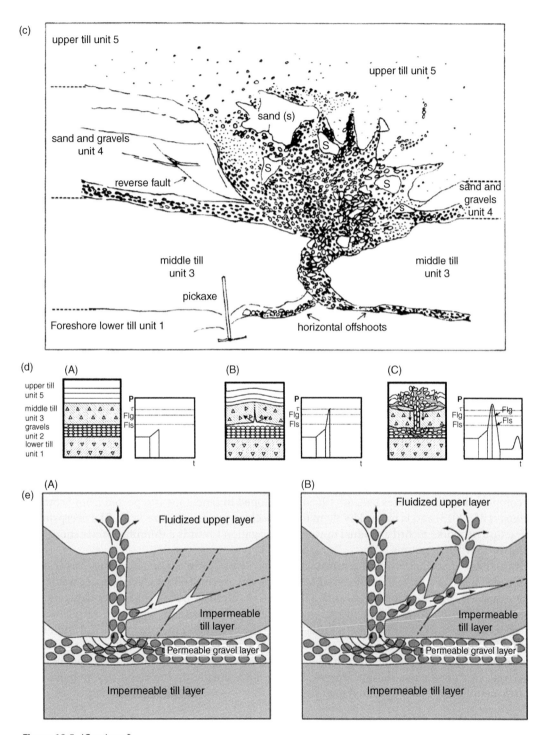

Figure 12.5 (*Continued*)

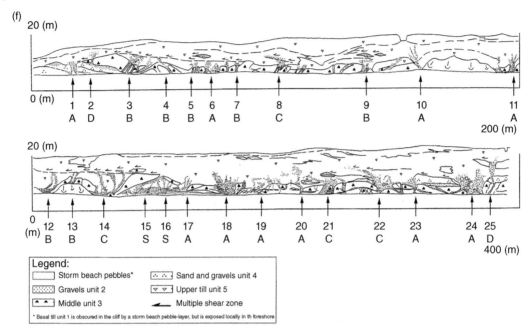

Figure 12.5 (*Continued*)

at microscale (Figure 12.6). In addition to the graded laminae described by van der Meer *et al.* (2009), hydrofracture zones may display microfaults and thrusts, anastomosing thrusts, brecciation, flame-like disturbance structures and patches of dispersed sediment in pipes where bedding has been totally destroyed by liquefaction.

Some bedrock types can also be heavily disrupted by hydrofracture development driven by elevated subglacial groundwater pressures and this is critical to the liberation of bedrock rafts (see Chapter 14) as well as the development of subglacial meltwater systems. An excellent example of a multi-phase complex of hydrofractures beneath till and developed in sandstone is reported from Scotland by Phillips *et al.* (2013) and is likened by them to the fracking process (Figure 12.7). The overprinted hydrofracture networks record a gradual upwards propagation towards a thinning glacier snout due to falling overburden pressures.

Finally, another important intrusive feature, which represents a form continuum from a clastic dyke to an intruded intraclast driven into host sediments by over-pressurised subglacial groundwater, are clay/silt intraclast concentrations or swarms. These are not derived from primary resedimentation of pre-existing deposits by fluvial or subaqueous reworking (Figure 12.8). They are conspicuous in that they occur as blocks of various sizes, cross-cutting the bedding of their host material and often traceable back through clouds of increasing density to branching, tentacle-like dykes and then back to their source strata, the upper contact of which possesses a ragged and/or interleaved contact with the overlying host materials. These features are most likely created in the same way as burst-out structures identified in coarser-grained materials.

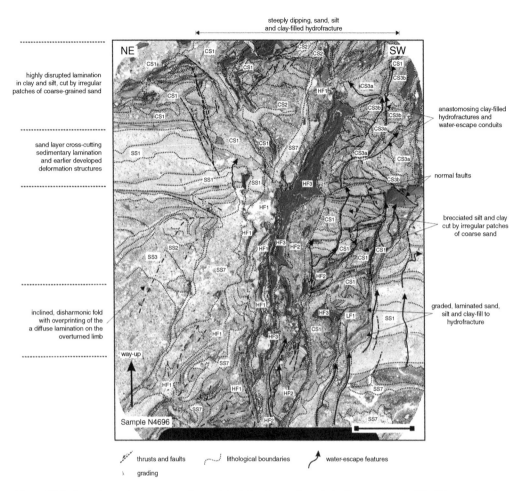

Figure 12.6 Complex water escape and associated structures viewed at microscale in stratified fine-grained glacimarine sediments occurring as thrust bound rafts at Clava, North East Scotland (from Phillips and Merritt, 2008).

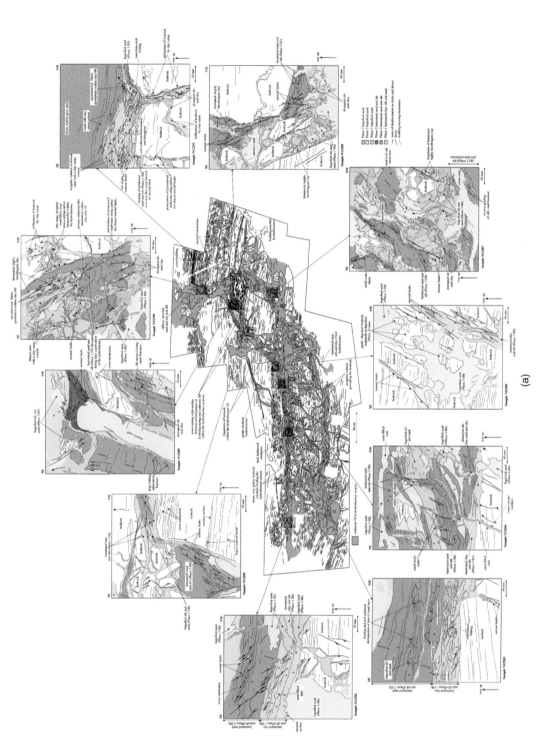

Figure 12.7 Case study of multiphase hydrofracture development in sandstone beneath till at Meads of St John, Inverness, North East Scotland (from Phillips *et al.*, 2013): (a) compilation of numerous thin section images and their locations in the Meads of St John hydrofracture system developed in sandstone bedrock; (b) set of schematic diagrams to show the evolutionary stages of the hydrofracture complex; (c) interpretive sequence of events linking the recession of the local Findhorn Glacier with the development of the hydrofracture complex.

(a)

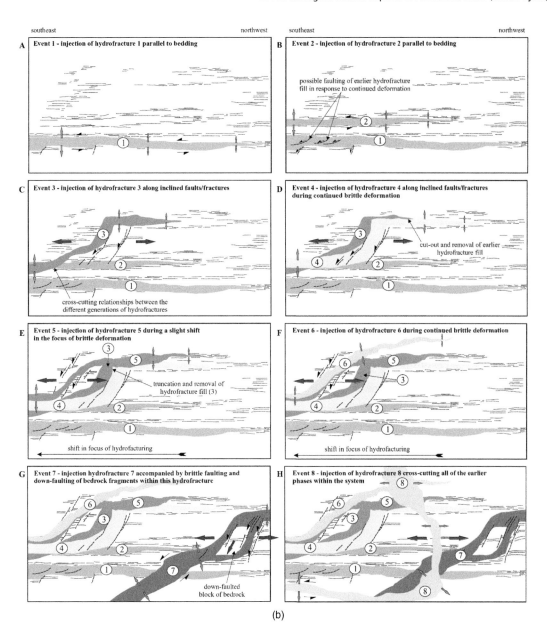

(b)

Figure 12.7 (*Continued*)

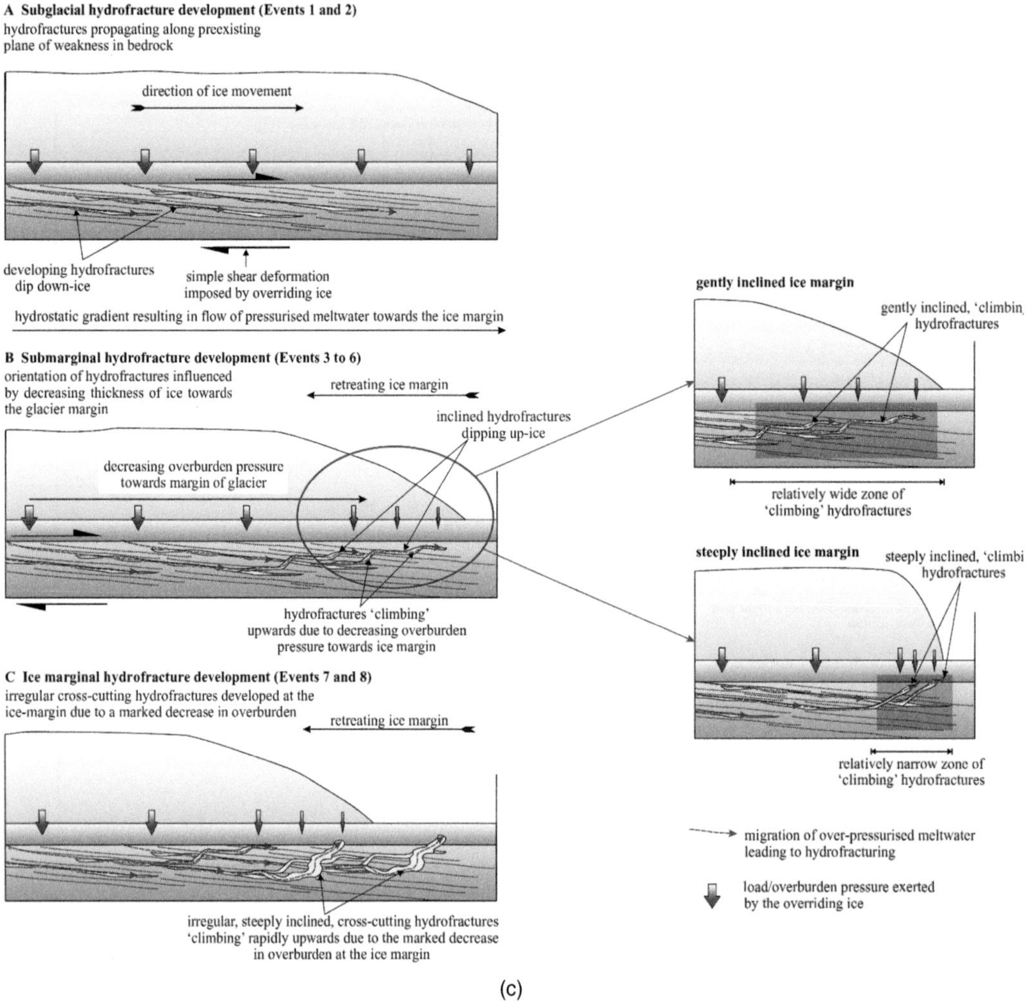

A Subglacial hydrofracture development (Events 1 and 2)
hydrofractures propagating along preexisting
plane of weakness in bedrock

direction of ice movement

developing hydrofractures
dip down-ice

simple shear deformation
imposed by overriding ice

hydrostatic gradient resulting in flow of pressurised meltwater towards the ice margin

B Submarginal hydrofracture development (Events 3 to 6)
orientation of hydrofractures influenced
by decreasing thickness of ice towards
the glacier margin

retreating ice margin

inclined hydrofractures
dipping up-ice

decreasing overburden pressure
towards margin of glacier

hydrofractures 'climbing'
upwards due to decreasing overburden
pressure towards ice margin

C Ice marginal hydrofracture development (Events 7 and 8)
irregular cross-cutting hydrofractures developed at the
ice-margin due to a marked decrease in overburden

retreating ice margin

irregular, steeply inclined, cross-cutting hydrofractures
'climbing' rapidly upwards due to the marked decrease
in overburden at the ice margin

gently inclined ice margin

gently inclined, 'climbing'
hydrofractures

relatively wide zone of
'climbing' hydrofractures

steeply inclined ice margin

steeply inclined, 'climbing'
hydrofractures

relatively narrow zone of
'climbing' hydrofractures

migration of over-pressurised meltwater
leading to hydrofracturing

load/overburden pressure exerted
by the overriding ice

(c)

Figure 12.7 (*Continued*)

Figure 12.8 Example of clay/silt intraclast concentrations or swarms. The clay/silt bodies have been injected into deltaic sands in a glacially overridden ice-contact delta at south Loch Lomond, Scotland, and connect up to form a vein-like network.

13

The Geological Record: Melt-out Till

It is … the 'stratified drifts' … the root cause of the trouble. One and all they point to subglacial accumulations due to the undermelt of a simple ice-sheet, stagnant and decaying. Once that is grasped our troubles vanish.

Carruthers (1953, p. 1).

The long-established process-based definition of melt-out till and some of the difficulties surrounding its preservation and differentiation from other till types were reviewed in Section 6.4, where it was established that subglacial melt-out till strictly defined should be a sediment released by the melting of stagnant or slowly moving debris-rich glacier ice, and directly deposited without subsequent transport or deformation. So interwoven are the sedimentary characteristics of proposed ancient melt-out tills and the envisaged process of melt-out till production that we need to consider both process and form in tandem in this chapter. A case has been made by Carlson (2004) that an accumulating stack of embryonic melt-out till, if it has a low hydraulic conductivity, can accommodate the transport of up to three orders of magnitude more water than is typically produced at the base of a glacier. The most significant of the controls on melt-out till production are those associated with thaw consolidation or volume reduction as ice is melted and removed from the sediment stack. This becomes more problematic in terms of sediment disruption wherever the original debris content of the ice is low, because thaw consolidation will be significant. On the other hand, high debris contents may conceivably result in a final sediment volume that is only slightly less than that of the parent ice, and hence any delicate englacial structures may be preserved. This is determined by the origin of the basal debris-rich ice facies, which at the classic Matanuska Glacier type site (Figures 6.15–6.18) is by supercooling and hence debris concentrations are very high. Even higher debris concentrations may result from downward migration of the freezing front from the base of the glacier into sub-marginal water saturated sediments, the specific origins of which will dictate the final stratification of the melt-out product. For example, downward freezing into stratified sand and gravel (Harris and Bothamley, 1984) or apron incorporation by overriding anything from proglacial outwash to finer-grained raised marine sediments (Shaw, 1977; Evans, 1989; Fitzsimons, 1990) and even permafrost/buried glacier ice (Astakhov and Isayeva, 1988; Ingólfsson and Lokrantz, 2003; Murton *et al.*, 2005) will create debris-rich basal ice facies in which the ice acts merely as a cement to hold grains together. Significant debris can also be frozen on in ice sheet interiors or at the base of ice streams during periods of quiescence, as illustrated by Christoffersen and Tulaczyk (2003) and Christoffersen *et al.* (2010), as well as beneath soft-bedded glaciers, as predicted by Iverson (2000) and observed

Till: A Glacial Process Sedimentology, First Edition. David J A Evans.
© 2018 John Wiley & Sons Ltd. Published 2018 by John Wiley & Sons Ltd.

in sub-marginal settings by Matthews *et al.* (1995), Krüger (1996) and Evans and Hiemstra (2005). Hence, the sedimentology and extent of melt-out tills can be predicted to be as variable as the range of parent sediment types frozen on to the ice base. Additionally, Ham and Mickelson (1994) report a situation in which lodgement processes appear to have been superimposed on debris-rich ice as the glacier continued to move over its stagnant basal ice facies.

Clearly, debris concentrations are critical to melt-out till preservation, but other local site characteristics also need to be recognised when assessing the efficacy of thaw consolidation. The behaviour of the debris once it is released during thaw consolidation is dictated by the balance between meltwater production and drainage away from the site (Figure 13.1; Paul and Eyles, 1990). Meltwater draining away freely during melt out at a rate equal to or higher than the rate of its production creates little or no disturbance other than that due to consolidation. This potentially creates a consolidation fabric in a sediment stack that already inherits the commonly strong clast orientations of the basal ice zone (e.g. Lawson, 1979a, b; Fitzsimons, 1990; Ham and Mickelson, 1994; Hart, 1995; Larson *et al.*, 2016). In contrast, any impedance of meltwater drainage will result in elevated porewater pressures as melt-out progresses, decreasing the frictional strength of the freshly released debris and increasing the likelihood of failure and remobilisation, especially where the sediment is fine-grained (Paul and Eyles, 1990). Another important local site characteristic is underlying bed slope, because remobilisation of saturated sediment can occur on slopes as low as 8° (Paul and Eyles, 1990). This is compounded on sloping glacier surfaces where melt-out gradually thickens over the debris-rich basal ice, which continues to deliver meltwater to the accumulating sediment body; at the Matanuska Glacier, where melt-out tills appear to be developing, Lawson (1979a, b, 1981a, b, 1982) identified significant supraglacial remobilisation of melted-out debris in a variety of gravity mass flow types (see Chapter 15).

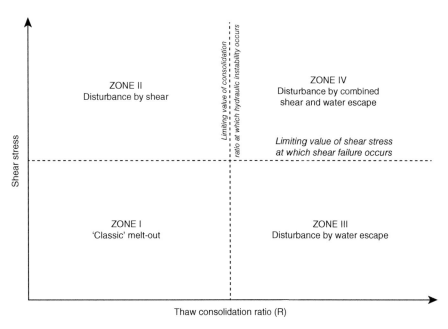

Figure 13.1 The classification scheme proposed by Paul and Eyles (1990) for soft-sediment deformation zones defined by the occurrence of shear instability and/or hydraulic instability during the melt-out process.

The ubiquitous nature of such sediment remobilisation in the debris cover on debris-rich glacier snouts (Figure 13.2) has been highlighted by Paul and Eyles (1990), who isolate excess porewater pressures and strong pressure gradients in the accumulating melt-out sequences as critical to the extent of their remobilisation. They identify two important processes stemming from these pressures, specifically shear instability and hydraulic instability, which lead to deformation of the accumulating sediment pile by shear failure and water escape. Shear instability arises when the available strength of the sediment is exceeded by the applied shear stress, which can be in response to elevated pore-water pressures in the accumulating sediment pile, and the loss of strength also increases rapidly with increasing thaw consolidation ratio. Any slope or bed protuberances in the substrate will initiate shear stresses in the melt-out sequence. Hydraulic instability arises when the forces created by meltwater seepage overcome the self-weight and frictional contact forces acting upon sedimentary particles, which are determined by the pore pressure gradient. The seepage forces are at a maximum where the pore pressure gradients are greatest, so this tends to take place at the base of a subglacial melt-out sequence; the magnitude of the gradient also increases rapidly with thaw consolidation ratio. Hydraulic instability should also be produced where effective stresses are low, because interparticle forces are reduced in such cases. Once interparticle movement starts to take place, water escape structures will be generated in the accumulating sediment so that a situation arises that is similar to quicksand development or fluidisation (Lowe, 1975; Postma, 1983).

A body of debris-rich ice facies which has been isolated from a receding glacier snout and largely covered by an accumulating gravity mass flow-fed supraglacial debris cover, nourished by melt-out on the ice surface, is subject to the predominantly downwards migration of meltwater. Paul and Eyles (1990) propose that at this stage interparticle movement due to thaw consolidation is governed by the lateral restraint on sediment movement and the internal friction of the sediment. The processes of shear and hydraulic instability then operate in combination and bring about four deformational environments or zones, which are defined by the level of shear stress relative to the sediment strength and the level of the pore pressure gradient relative to the critical hydraulic gradient necessary for

Figure 13.2 Supraglacial debris undergoing failure and debris flow mobilisation on the surface of Longyearbreen, Svalbard.

fluidisation (Figure 13.1). The zones are defined by two sets of conditions, which create the boundary lines on Figure 13.1. First, the shear failure boundary is the failure envelope for the sediment as determined by the material properties and the porewater pressure, which in turn is determined by the thaw consolidation ratio. With respect to the controlling environmental conditions, this can be defined as the maximum shear stress, which in turn is normally related to ice surface or substrate slope. Second, the hydraulic disruption boundary is defined by the value of the thaw consolidation ratio at which the level of effective stress and hydraulic gradient permit local grain separation and thereby the development of internal structures. This model prompted Paul and Eyles (1990) to predict specific depositional facies, predominantly diamictic, that should evolve as a response to the controlling conditions created in each environment depicted in Figure 13.1. In Zone I, the shear stress and porewater pressure on the thawing debris is always below that required to produce deformation and fluidisation, respectively, so that only vertical settlement takes place and 'classic' melt-out till is created through the grain-by-grain aggregation of englacial debris. In Zone II, shear stresses are high enough to produce failure in fresh melt-out materials but porewater pressure and seepage conditions are insufficient to generate fluidisation. The occurrence of the highest porewater pressures at the thaw interface brings about repeated failure and therefore the sediment can display evidence of shearing and hence resemble deformed subglacial till. In Zone III, porewater pressure and seepage conditions are sufficient to generate fluidisation but there is no shear failure. Porewater escape causes a loss of fines and the disruption of clast and matrix fabrics, which could be pervasive but is more likely restricted to zones of water escape structures or piping. In a heterogeneous sediment, this will generate load structures due to the creation of reverse-density gradients. Finally, in Zone IV, the operation of both shear failure and fluidisation can create a deposit with a wide variety of deformation structures as well as textural sorting, which can generate crude layering or shear-banding.

By acknowledging the changing nature of the deformation environment as controlled by the thaw consolidation ratio and the level of applied shear stress, Paul and Eyles (1990) proposed that typical field conditions would constrain the formation and preservation of melt-out till. Moreover, the model of 'classic' melt-out till genesis is applicable only under well-defined and very restricted boundary conditions and beyond those conditions the melting out debris will inevitably undergo extensive syn- and post-depositional deformation. Each deformational zone depicted in Figure 13.1 predicts a specific end-member deposit, but it is likely that the nature of the drainage and consolidation conditions would change over time, giving rise to a final depositional facies that displays any combination of the characteristics typical of zones I–IV. Therefore, 'classic' melt-out till is likely to be part of a more complex stratigraphic sequence, especially if capped by supraglacial mass flow diamictons.

So, what is 'classic' melt-out till and can we be confident in our genetic classification of potential melt-out deposits? With all till types we need to be constrained as much as possible in our sedimentological definitions by safe or demonstrable process–form relationships, such as those that can be observed in modern glacial environments. In Section 6.4, we saw that the most compelling modern analogue for 'classic' melt-out till was that of the Alaskan glacier snouts, and that of the Matanuska Glacier in particular (Figures 6.15–6.18; Lawson, 1979a, b, 1981a, b; Larson *et al.*, 2016). The characteristics of the melt-out tills at this location were summarised by Lawson (1981a, b; Figure 6.15) as strongly aligned clast orientations parallel to ice flow and the presence of discontinuous lenses and poorly sorted sediment inherited from pre-existing stratified englacial debris, which may all be distinct from the surrounding, low-porosity diamicton in terms of texture, composition or colour. These lenses are draped over larger clasts as a result of differential compaction. The clast macrofabrics are similar to those of the overlying englacial debris but display flatter dips and more girdle-like patterns due to the compaction process (Table 8.1). Larson *et al.* (2016) further define the deposits as materials that display discontinuous and contorted and sheet-like lamina, layers and

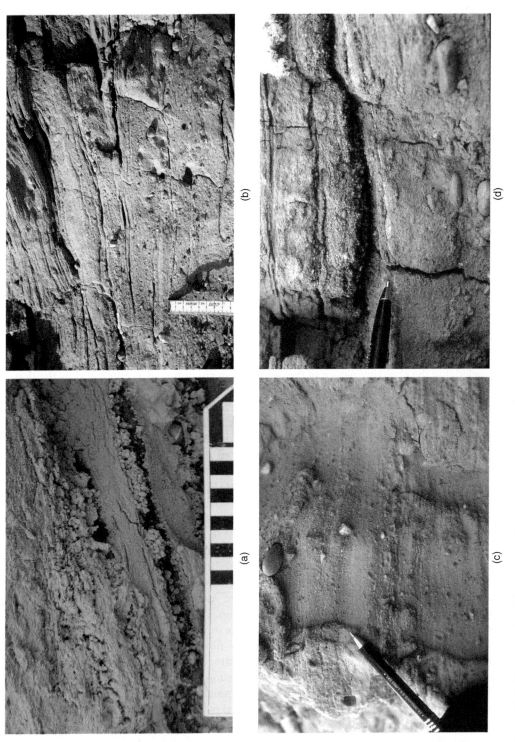

Figure 13.3 Characteristics of the melt-out till sequence at the margin of the Matanuska Glacier, Alaska (from Larson *et al.*, 2016): (a) detail of the transition zone between debris-rich ice and overlying melt-out till sequence, showing accreted silt aggregates at the base of the diamicton; (b) the pseudo-stratified diamicton, showing laminae, bedding and scattered clasts; (c) details of laminations, showing coarse sand bed at tip of pen with pebbly silt laminae above and below; (d) bed of coarse sand between pebbly silt beds and aligned clasts at base of image.

lenses, textural and compositional banding with indistinct contacts and internal flow structures and undeformed, intra-till soft-sediment or soft-bedrock rafts (Figure 6.16), an appearance that is best described as 'pseudo-stratification'. Problematic in relying upon the Matanuska Glacier case study is the inheritance at that location of supercooled ice characteristics in particular, so that signatures of other types of englacial debris are under-represented in our modern analogues. Hence, supercooled melt-out till can be well constrained sedimentologically as a deposit that we described in Section 6.4 as displaying silt or sandy-silt lamina, *en echelon* silt lamina, irregular lenses of coarse to fine sand that offset or cross bedding planes, short lenses of gravel, sand and silt, thin lenses of silt and clay – silt aggregates, dispersed sub-rounded, but non-striated pebbles and cobbles around which lamina bend (Larson *et al.*, 2016; Figures 6.16 and 13.3). The appearance at microscale of porewater pathways and compressed and lightly deformed clay – silt clasts, together with an absence of shear indicators, provide additional diagnostic criteria (Figure 6.17). Certainly, this modern analogue can be utilised in other settings prone to supercooled meltwater, for example, in the southern Vatnajökull outlet glaciers (Roberts *et al.*, 2002; Cook *et al.*, 2010), where Cook *et al.* (2011b) have proposed that the sedimentary characteristics of supercooled ice facies may be preserved in the grain size signatures of marginal tills. In terms of direct facies signatures, recently exposed subglacial till sequences on the foreland of Skaftafellsjökull (Figure 13.4) display discontinuous and relatively thin units of pseudo-stratified to fissile-structured diamictons containing irregular lenses of sand and fine gravel, typical of supercooled melt-out till (compare Figures 6.16, 13.3 and 13.4b, c).

A small number of examples of melt-out sedimentation have been reported from the hyper-arid environments of Antarctica, where Shaw (1977, 1989) proposed that sublimation was capable of creating the purist type of melt-out till or sublimation till (Figure 6.14). Typical deposits are reported by Lundqvist (1989) and Fitzsimons (1990), who record excellent preservation of stratified ice facies, closely spaced jointing or fissility, 'pellet structures' or a small-scale blocky appearance defined by anastomosing fissility, and 'let-down structures', where bedding drapes clasts (Figure 13.5). Importantly, Fitzsimons (1990) regards the horizontal fissility as potentially indicative of dewatering and therefore not particularly diagnostic of melt-out. Additionally, the arid polar melt-out tills do not display particularly strong clast macrofabrics when compared to other classic melt-out localities (Table 8.1), which Fitzsimons (1990) regards as indicative of the dispersed nature of the debris and hence its tendency to be reorganised during melt-out.

Beyond supercooled and arid polar melt-out products, the diagnostic criteria that have been proposed for melt-out tills are derived from the geotechnical predictions of Paul and Eyles (1990; Figure 13.1) and interpretations of specific sedimentary structures in ancient tills (e.g. Haldorsen, 1982; Haldorsen and Shaw, 1982; Shaw, 1982, 1983; Piotrowski, 1994; Munro-Stasiuk, 2000). To date, very few attempts have been made to draw comparisons between the predictions of Paul and Eyles (1990) and potential melt-out till facies, especially in relation to zones II – IV in Figure 13.1. Paul and Eyles (1990) cite the descriptions of Bouchard *et al.* (1984) of coarse-grained, stratified basal tills, displaying differential compaction around clasts, water escape structures and diapiric deformation structures, which potentially formed under zone III conditions. The dewatering structures that are predicted for zones III and IV by Paul and Eyles (1990) have been further qualified by Carlson (2004) and proposed by him to be diagnostic of poor drainage or debris-poor ice facies. Debris contents >40% by volume with a well-developed drainage network on the other hand could drain effectively and dewater without significant disruption. Indeed, some melt-out tills are structureless and massive, and lack any visible foliation (Lawson, 1979a). The penetration of some till sequences by vertical pipes, often filled with sorted sediments (i.e. clastic dykes), has been related to dewatering during melt-out, but a hydrofracturing origin, as discussed in Chapter 12, is also likely, especially if the dykes can be linked to intra- or sub-till aquifers.

Early work on the development of melt-out till genesis based on ancient glacial landforms was directed at the Rogen moraine and associated Sveg tills (Figure 13.6; Lundqvist, 1969a, b; Aario, 1977; Shaw, 1979), located in the Sveg region of central Sweden. Lundqvist (1969a) described the deposit as 'a mixture of thin lenses of sorted sediments and compact basal till' and Shaw (1979) provided a representative stratigraphic sequence (Figure 13.6a), comprising a lower massive till, an intermediate

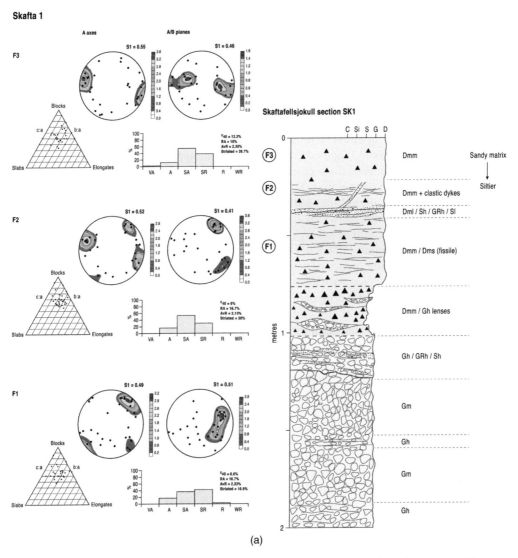

(a)

Figure 13.4 Likely melt-out till at the location of a previously reported exposure through supercooled ice at the margin of Skaftafellsjökull, Iceland: (a) vertical profile log of tills overlying outwash on the foreland, together with clast macrofabric and form data. The potential melt-out till is classified as Dml/Sh/GRh/Sl and is only 10–15 cm thick and lies between subglacial tills; (b) photograph of section top, showing upper diamictons containing thin lens of potential melt-out till; (c) detailed photograph of the likely melt-out till, showing pseudo-stratified diamictic and sandy to fine gravelly laminations and capped by brecciated to fissile till.

(b)

(c)

Figure 13.4 *(Continued)*

stoney or clast-rich till and an upper 'stratified till'. This upper stratified diamicton (Sveg Till; Figure 2.9) is critical to the development of melt-out till theory. Although Lundqvist (1969a, b) and Aario (1977) regarded the deposit as an 'ablation till' or supraglacial debris assemblage, Shaw (1979) highlighted the horizontally interbedded nature of the stratified and diamicton layers, their lateral continuity of ≥10 m, and the consistent clast macrofabric orientations between all three diamicton

Figure 13.5 Sublimation tills from Antarctica. Upper photograph shows debris-rich stratified ice passing upwards abruptly into sublimation till, the junction being marked where bedding changes angle (photo by S. Rubulis from Vega Island). Lower photograph shows stratified diamicton interpreted as sublimation till at Sørsdal Glacier, Vestfold Hills (from Lundqvist, 1989).

(till) units in the sequence to propose a melt-out origin for the Sveg Till in particular. Shaw (1979, 1982) envisaged meltwater drainage from melt-out stacks along debris-poor folia and between the debris-rich folia that produced the diamicton, resulting in intra-diamict stratified lenses and beds (Figure 13.6b). Clasts that bridged ice folia were then draped by the accumulating interbeds. Associated with the resultant interbedded diamicton and stratified sand and gravel sequences in this model, Shaw (1982, 1983) later identified scour fills underneath individual clasts (Figure 13.7), which he proposed to be diagnostic of subglacial meltwater scouring beneath clasts while they were held in the overlying debris-rich ice; after melt-out the clasts and scour fills would penetrate from the diamicton bed into underlying stratified sediments. Although Shaw (1979) reported on undeformed stratigraphic sequences of Sveg Till, extensive deformation was noted in other

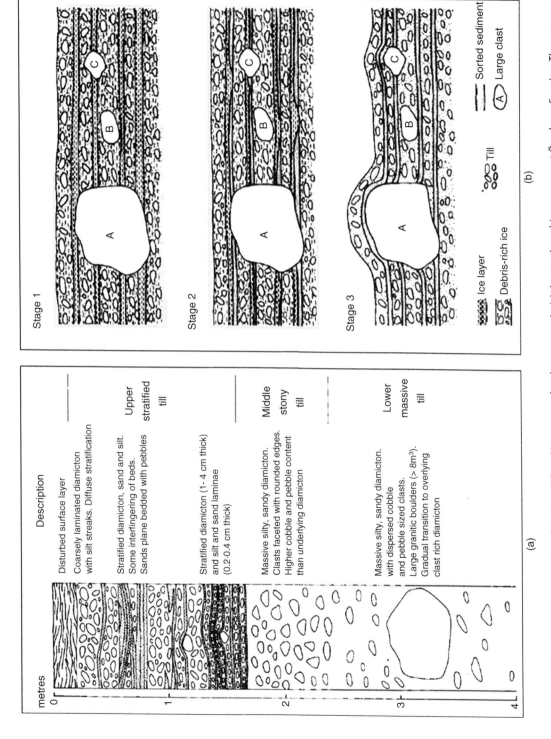

Figure 13.6 Shaws' (1979) presentation of the Sveg Till and its proposed melt-out genesis; (a) the stratigraphic sequence at Överberg, Sweden. The upper stratified till was critical to the melt-out interpretation; (b) the evolutionary sequence of stratified layers and enclosed clasts proposed by Shaw to be related to melt-out of debris-rich ice.

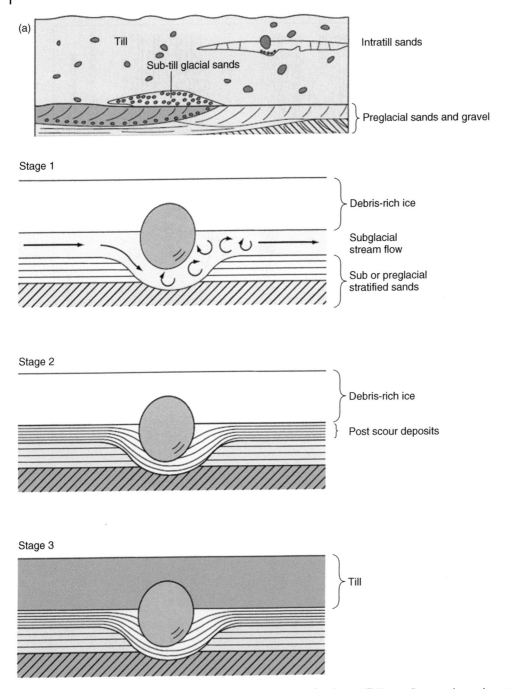

Figure 13.7 Features proposed by Shaw (1983) to be diagnostic of melt-out till. Upper diagram shows the nature of stratified lenses in sub-till and intra-till positions, with a clast lying across the upper boundary of the intra-till lens and associated with a scour infill. Such a feature is explained in the lower diagrams, which show three stages in the development of a scour fill beneath a clast held in basal ice during melt-out. The upper till layer is created by the passive melt-out of debris-rich basal ice.

exposures by Lundqvist (1969a), which Shaw (1979) proposed to be related to later reactivation of the overlying ice after Sveg till deposition.

This model of till production in Rogen (ribbed) moraine has been applied by Möller (2010) and Möller and Dowling (2015) to landforms in southern Sweden, where large tracts of debris-rich ice at the receding southern margin of the Fennoscandinavian ice sheet are proposed to explain the extensive outcrops of the till type. Critical to their interpretations of tills in the area as melt-out in origin are the exposures at Åbogen and Horgeboda in particular (Figure 13.8). At Horgeboda, an exposure reveals three sediment facies (Units A–C) containing what is regarded as melt-out diagnostic criteria. Above Unit A, which is a massive diamict with a strong preferred clast axis orientation, there is a gradual upwards coarsening of the diamictons of Units B and C and an increasing frequency in the occurrence of undeformed sorted sediment intrabeds within which sand beds thicken below and bend over and below boulders and there are gradational contacts between sorted sediments and overlying diamicts. Each of these observations in the melt-out theory would be interpreted as evidence for meltwater flowing through alternating debris-rich and debris poor ice foliae. Möller (2010) also provides a variant on Shaw's (1982, 1983) sub-boulder scour infills where he documents the restriction of the infills to the vertical boundaries of boulders, interpreted as 'debris melt-out shadows' below clasts held in cavity roofs. At an outcrop scale, banding and crude stratification occurs in the Unit C diamicton, which dips at 20–25° towards the proximal side of the moraine and is cross-cut by the younger sub-horizontal sorted sediment beds and hence regarded as inherited from parent glacier ice, especially as it dips northwards and perpendicular to the moraine ridge crest-line and hence could be related to compressive shear stacking and/or folding of debris-rich as proposed by Shaw (1979) for the landform–sediment associations. Although Möller (2010) and Möller and Dowling (2015) confidently conclude that the sediments described in their study of southern Swedish ribbed moraine can only be interpreted as melt-out till based upon the combined evidence, most importantly because they display no direct indications of the shear stresses typical of subglacial till production, we will see later in Chapters 14 and 16 that the melt-out indicative (diagnostic) criteria can be interpreted as the products of quite different depositional scenarios. Significantly, in this respect, it is the changing basal thermal regimes invoked by Möller (2010) and Möller and Dowling (2015) to create the debris-rich ice required to form melt-out tills that are also potentially influential in the superimposition of till sedimentary characteristics. It is important that we view such case studies as hypotheses for further testing in relation to the melt-out process–form regime and its differentiation from other till and glacitectonite forming processes, as we will discuss below.

The criteria of Shaw (1982, 1983) were employed also by Munro-Stasiuk (1999, 2000, 2003) in an interpretation of a sequence of stratified sediments and diamictons thought to have been deposited in a large subglacial cavity beneath the southwest Laurentide Ice Sheet in southern Alberta, Canada (Figure 13.9a). A re-interpretation of this sedimentary sequence by Evans *et al.* (2006a) is an example of how the diagnostic melt-out criteria for some researchers are open to distinctly different interpretations by others. Munro-Stasiuk (2000) interpreted her Facies 3 in the area as a melt-out till based upon the characteristics of interbeds of laterally extensive massive, fine-grained diamicton, containing striated and facetted clasts, and thin, continuous beds of sand and/or silt with occasional channel scour fills with gravel bases. The sand and silt beds specifically were interpreted as the product of subglacial sheet flows that periodically broke down into channels to produce scour-fills, which because they often occur below clasts were attributed to the melt-out process, based upon Shaw's (1983) interpretation of such features. An alternative origin for these clasts and scours was proposed by Evans *et al.* (2006a), who invoked the role of anchor ice in glacilacustrine sedimentation and interpreted the whole Facies 3 sequence as proximal glacilacustrine deposits. Scouring beneath clasts protruding from the base of lake anchor ice has been documented by Reimnitz *et al.* (1987), and shallow

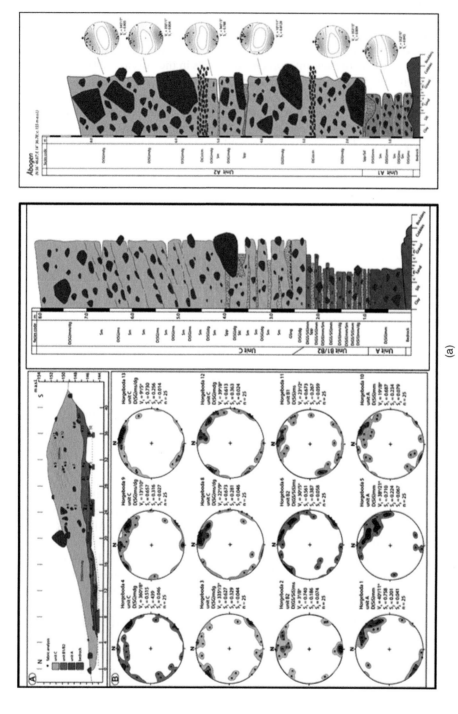

Figure 13.8 Stratigraphy, sediment characteristics and clast macrofabric data from Åbogen and Horgeboda, Southern Sweden, interpreted by Möller (2010) and Möller and Dowling (2015) as melt-out till sequences: a) the stratigraphy and clast macrofabric data from the two sites; b) the characteristics shows the characteristics of the diamictons of the study area in which (A) – (G) are characteristics of the unit C diamict at Horgeboda; (A) lateral and vertical grading between indistinct lenses of gravelly and sandy diamictons, and lenses of higher or lower degree of sorting; (B) indistinct, small lenses of well-sorted sand (two marked by arrows); (C) weak stratification in diamicton inclined towards the proximal side of the moraine ridge; (D) thin sub-horizontal sand intrabed, cutting inclined stratification in diamicton; (E) sand bed, conformably bent under a boulder; (F) anticlinal sand bed that conforms to the underlying boulder; (G) trough cross-laminated sand underneath a large boulder; (H) thick pocket of interbedded silt and fine sand in diamicton underneath and between the vertical projection of a large boulder in the Åbogen section; (I) massive sand bed underneath a boulder but which does not continue outside the vertical projection of the boulder at Hälsegylet section.

(a)

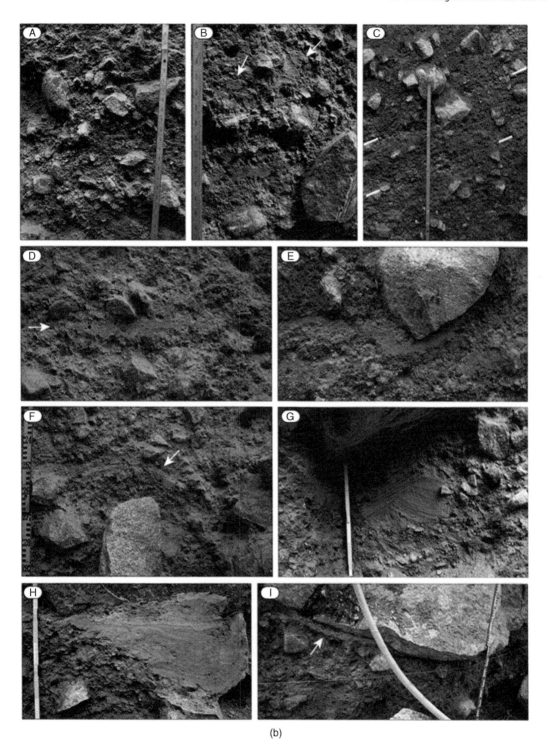

(b)

Figure 13.8 (*Continued*)

and/or fluctuating lake water depths are widely invoked as the driving mechanism in the adfreezing of bottom sediment and clasts and their later release after buoyant lifting and drift of the lake ice fragments during spring melt (e.g. Heron and Woo, 1994; Smith, 2000). Scouring and filling beneath clasts protruding from the base of the lake ice is undertaken by currents that are initiated by meltwater that plunges down through ice surface cracks and then flows between the lake ice sole and the lake bed after grounding (e.g. Squyres *et al.*, 1991). Diamictons in such environments can also be related to lake ice through shallow lake bed turbation and rafting of adfrozen sediment (e.g. Dionne, 1979). The clast macrofabrics in the Facies 3 and underlying Facies 2 subaqueous deposits were also interpreted differently in that Munro-Stasiuk (2000) regarded them as typical of melt-out and Evans *et al.* (2006a) as indicative of glacitectonically disturbed glacilacustrine sediments which had been subject to deformation partitioning in a vertically increasing shear signature (Figure 13.9b). Hence, the sediments have been alternatively interpreted as either melt-out tills or glacitectonites developed in glacilacustrine deposits (see Chapter 14).

In a study of 'supraglacial melt-out tills' melting out from the surface of buried glacier ice on the Tuktoyaktuk Coastlands, Canada (Figure 13.10), Murton *et al.* (2005) provide a clear illustration of the preservation potential of such sequences. Although they conclude that the prediction of Paul and Eyles (1990) that melt-out till has a very low preservation potential 'is sometimes unduly pessimistic', the very fact that their melt-out till overlies at least 13 m of buried glacier ice or intra-sedimental ground ice does indeed demonstrate that final preservation is unlikely; the Murton *et al.* (2005) melt-out till is in fact not a till *per se*, as it has not yet been deposited and requires the passive removal of at least 13 m of underlying ice before it is. So, where exactly should we expect melt-out tills to be preserved and what is their likely sedimentology, architecture and extent?

Fundamental to melt-out till production is the creation of thick sequences of debris-charged basal ice facies from which such till can be derived. Hence, melt-out till will only be preserved where significant debris is incorporated englacially, for example, in small pockets around Alaskan and Icelandic glacier snouts, where supercooling locally operates on the adverse slopes of overdeepenings. Other prime settings are in polar environments where apron entrainment by advancing glacier margins of buried glacier ice, entombed within the permafrost, instantaneously creates thick sequences of debris-charged basal ice. However, deglaciation is likely retarded in such environments and melt-out till production does not operate until enough supraglacial mass flow debris or melt-out material (e.g. Murton *et al.*, 2005) has accumulated over the debris-rich ice facies, to not only insulate it and thereby slow down the ablation rate but also to entomb it in the permafrost; at that point ice can only be removed by sublimation and hence delicate englacial structures are preserved (Fitzsimons, 1990).

Figure 13.9 Sequence of, locally heavily contorted, stratified sediments and diamictons at McGregor Lake, Alberta, Canada, employed in two alternative interpretations of glacial depositional environments in the area by Munro-Stasiuk (melt-out till/subglacial lake undermelt; 1999, 2000, 2003) and Evans *et al.* (glacitectonised proglacial lake deposits; 2006): (a) selected outcrop sketches, photographs and clast macrofabric data, showing the heavily deformed and stratified nature of the diamictons and associated sediments; (b) reconstruction of depositional history of the deposits proposed by Evans *et al.* (2006a), including: (1) early glacier advance recorded by grey till sandwiched between glacilacustrine sediments and overlying preglacial Empress Group gravels; (2) glacitectonic disturbance folds and stacks the sedimentary sequence after more glacilacustrine sediment is deposited in front of the advancing ice; (3) overriding ice erodes the glacitectonised sequence and a thick glacilacustrine sediment pile is deposited, which continues to accumulate and is glacitectonised during ice readvances. Shearing in the sediment pile is recorded by macrofabrics that developed through distributed shear/deformation partitioning. Variability in total relative strain with depth is represented by decreasing fabric strengths through 'Facies 3 and 2'; (4) spillway incision results in the dissection of the sedimentary sequences; (5) postglacial incision results in the further dissection of the sedimentary sequence.

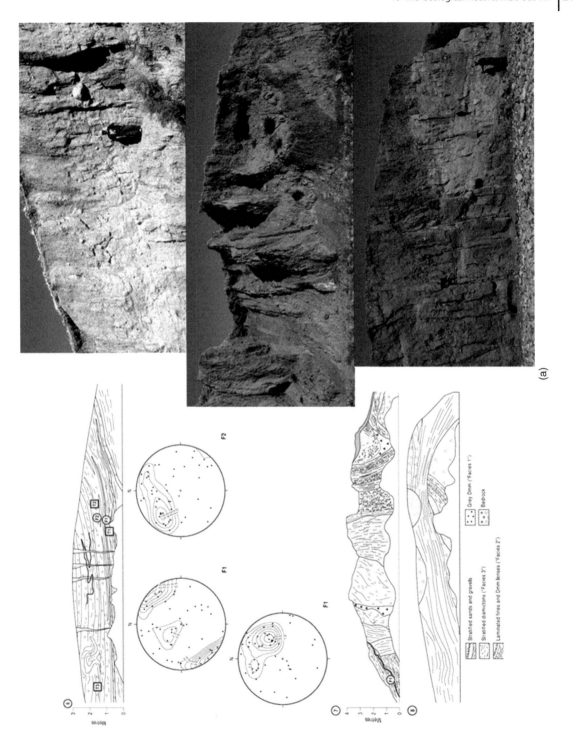

(a)

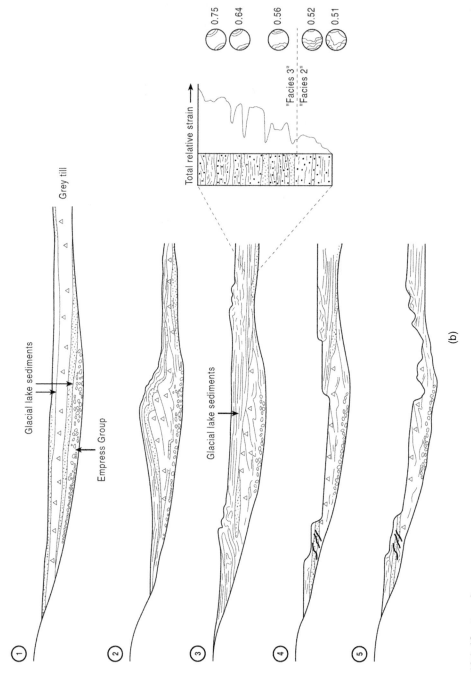

Figure 13.9 (Continued)

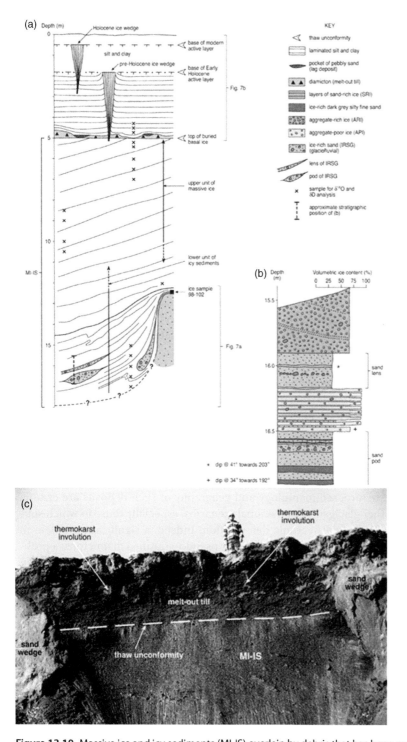

Figure 13.10 Massive ice and icy sediments (MI-IS) overlain by debris that has been melted out from them and therefore termed 'supraglacial melt-out tills' on the Tuktoyaktuk Coastlands, Canada (from Murton *et al.*, 2005): (a) schematic log showing stratigraphic context of MI–IS; (b) detailed log through the lower unit of icy sediments; (c) photograph of MI–IS overlain by diamicton (interpreted as melt-out till) above the thaw unconformity. The diamicton contains irregular shaped bodies of sand or thermokarst involutions.

However, complete de-icing of such terrains is required before melt-out till is released over large areas of former ice sheet beds, a process that, since the last ice sheet deglaciation, is incomplete. Zones of vigorous freeze-on in polythermal glaciers may vary spatially and temporally, as observed for example in Svalbard glaciers (e.g. Lovell *et al.*, 2015) giving rise to arcuate zones of controlled moraine (*sensu* Evans, 2009) fed by thick debris-rich ice facies and hence serving as an appropriate analogue for the regional landform–sediment associations linked to derivations from melt-out by Möller (2010) and Möller and Dowling (2015; cf. Shaw, 1979).

Nevertheless, a critical incompatibility arises in the process–form regime envisaged for melt-out till production based upon ancient till sequences, specifically in the coincident operation of englacial foliation preservation (requiring passive melt-out or sublimation) and extensive meltwater drainage and stratified sediment deposition between debris-charged foliae (requiring significant *in situ* ice melting). Moreover, it is unclear as to why meltwater streams would be restricted to horizontal ice facies over large distances when englacial drainage networks are prone to do quite the opposite, by exploiting moulins and crevasse networks and linking to rapidly expanding englacial and subglacial tunnel networks during deglaciation. Hence, we are obliged to return to those few case studies where melt-out till process and form can be unequivocally demonstrated in order to derive predictions of its sedimentology, architecture and extent. Predominantly such case studies indicate that melt-out tills are stratified or pseudo-stratified; the production of massive diamictons by melt-out has been explained by Lawson (1979a) as the product of slow drainage of meltwater from highly debris-charged ice. The lateral and vertical extent of melt-out tills that might accumulate in such circumstances is restricted by various factors. First, the volume of debris-charged glacier ice might be unrealistic, at least over very large distances. For example, Boulton (1996b) calculated that in order to produce a melt-out till sequence in excess of 10 m would require more than 100 m of debris-rich ice of average debris concentration, which is certainly uncommon in present day basal ice sequences at least. Second, as highlighted above, the operation of debris entrainment processes that conceivably could create such debris-rich ice facies are localised. We also need to contemplate a not entirely purely semantic issue of whether or not we can call apron-incorporated permafrost or frozen-on lake sediments a 'till' after they melt-out from glacier ice? Third, as illustrated by the alternative interpretations of proposed melt-out tills in Alberta, Canada, by Munro-Stasiuk (2000) and Evans *et al.* (2006a), the significant lateral extent, structure, sedimentology and geography of such deposits are criteria that can be accommodated in other glacigenic depositional scenarios, especially those in which subaqueous sedimentation is coeval with glacitectonic deformation. Indeed, a significant problem in till sedimentology surrounds the differentiation of stratified or pseudo-stratified diamictons, a problem compounded wherever primary melt-out or subaqueous deposits are subject to deformation and glacitectonite production (see Chapter 14). Nevertheless, some specific differences between pure melt-out till and other 'bedded glacigenic sediments' have been highlighted by Larson *et al.* (2016). These include differences between melt-out till and subaqueous deposits, such as the tapering, overlapping conformable silt beds <3 m long, irregularly shaped and internally massive sand and gravel lenses and lenses of openwork silt aggregates found in melt-out tills and not subaqueous deposits, as well as the lack of current bedforms. Also identified are differences between melt-out tills and sediment gravity flows (see Chapter 15), which include the lack of grading, and thin granule and clast lag horizons in melt-out bedding. The irregular shaped lenses of openwork gravel that occur in melt-out tills are also absent from sediment gravity flows. At microscale, sediment gravity flows display deformation structures but melt-out tills do not. In fact, deformation structures are absent at all scales in the Matanuska melt-out tills, contrasting them with the wide range of glacitectonites and deformed subglacial tills.

In conclusion, we can summarise exactly where we might expect to find melt-out till and what it should look like as informed by the modern analogue case studies (see Table 6.2). Where permafrost has been temporarily coupled with advancing glacier ice and hence converted into basal ice facies, it is effectively glacitectonised permafrost and even though it has acted as the deformed basal ice of an ice sheet or glacier it is more than a trivial semantic issue whether or not we classify it as melt-out till once it melts, because it is effectively glacitectonite (see Chapter 14). Fitzsimons (1990) and Larson *et al.* (2016) propose that preservation of melt-out till is enhanced by the burying of a debris-rich basal ice layer within a detached, stagnant part of a receding glacier snout. Insulation of the ice mass by accumulating supraglacial debris is crucial as it slows the melt rate and thereby the release the meltwater by groundwater seepage. Because preservation of expansive areas of the snout is unlikely due to the distribution of crevasse and meltwater tunnel networks that will focus enhanced melting and collapse of the stagnant ice mass, melt-out till could be preserved in small outcrops underlain by subglacial tills and draped by collapsed supraglacial mass flow and outwash deposits (Figures 6.16, 13.3 and 13.4), ancient examples of which have been proposed (e.g. Johnson and Gillam, 1995; Johnson *et al.*, 1995). Regardless of the production process for englacial debris-rich ice facies, 'pseudo-stratification' or pseudo-lamination is likely to be prominent and will feature discontinuous and contorted and sheet-like lamina, layers and lenses, textural and compositional banding with indistinct contacts and internal flow structures and undeformed, intra-till soft-sediment or soft-bedrock rafts (Figures 6.15–6.17 and 13.3–13.5). Microscale investigations should reveal similar features, in addition to porewater pathways, compressed and lightly deformed clay – silt clasts and an absence of shear indicators. Clast macrofabrics are often reported as being very strong, but, in reality, this is not a consistent pattern (Table 8.1).

14

The Geological Record: Glacitectonite

Confusion baffles classification and reduces the observer to despair.

Sollas and Praeger (1894, p. 14)

In Section 6.5, 'glacitectonite' was defined as rock or sediment that has been deformed by subglacial shearing (deformation) but retains some of the structural characteristics of the parent material, which may consist of igneous, metamorphic or sedimentary rock or unlithified sediments (Benn and Evans, 1996). It is therefore a mélange, as defined in Section 4.2. The term 'glacitectonite' was introduced by Banham (1977) as it was analogous to the metamorphic petrology term 'tectonite', which referred to the tectonic imprint in a rock unit as recorded by structures and fabrics. He further identified 'exodiamict glacitectonite' or sheared materials that retain some primary structures, and 'endiamict glacitectonite' or diamictons in which all primary structures have been destroyed by shear (Figure 14.1). The use of the term 'glacitectonite' herein relates specifically to Banham's (1977) exodiamict glacitectonite and defines materials in which pre-existing structures are still recognisable and hence can be used as strain markers. Endiamict glacitectonite refers to fully homogenised subglacial materials that originated as glacitectonites (see Chapter 10).

An alternative subdivision of glacitectonites according to deformation intensity was proposed by Benn and Evans (1996) and is adopted herein. This classification scheme recognises 'Type A or penetrative glacitectonite', which displays evidence of penetrative deformation, and Type B glacitectonite, which has undergone non-penetrative deformation, so pre-deformational sedimentary structures are merely folded and faulted (Figure 14.1). It is important to stress that the classification scheme encompasses a continuum and hence there is a gradation of structural signatures from undeformed substrate through Type B to Type A to laminated and then homogenised diamicton that reflects glacitectonite maturity. The arrangement in Figure 14.1 as a vertical sequence reflects a pattern of increasing deformation up-section, as the result of patterns of cumulative strain in the substrate (Banham, 1977; Pedersen, 1989; Hart et al., 1990; Hart and Boulton, 1991; Benn and Evans, 1996; Hiemstra et al., 2007). In addition to reflecting measurements of deformation at the base of modern glaciers (Sections 6.2 and 6.5; Figures 6.9–6.11), where overburden pressures and hence sediment frictional strength decrease upwards so that the most intense deformation and the highest cumulative strains will be found at the top of the sequence, this vertical pattern of deformation intensity also reflects the cannibalisation, rafting and attenuation of substrates by simple shear (Figures 4.6 and 6.19–6.22). This represents therefore the shear zone horizons: (i) lower (S_r) zone of minimal strain with rooted structures such as drag folds; (ii) intermediate (S_b) zone of strongly deformed material with transposed foliation of boudins, rootless folds and highly attenuated bands; and (iii) upper (S_h) zone of maximum strain of complete homogenization into diamicton (Figure 6.21).

Till: A Glacial Process Sedimentology, First Edition. David J A Evans.
© 2018 John Wiley & Sons Ltd. Published 2018 by John Wiley & Sons Ltd.

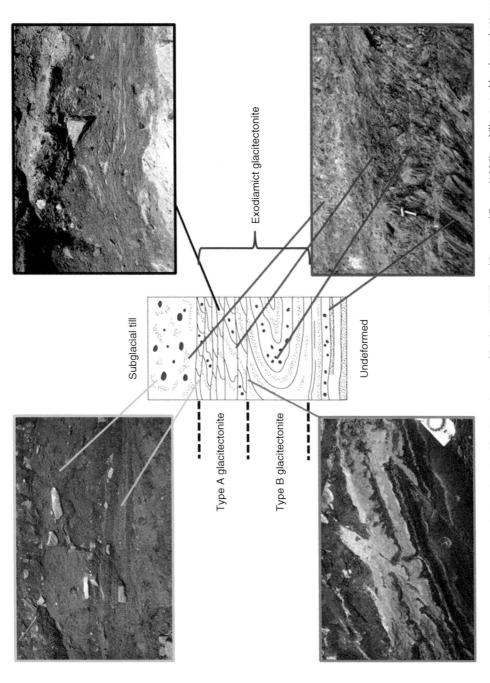

Figure 14.1 Styles of glacitectonite, combining the classifcation schemes of Banham (1977) and Benn and Evans (1996) and illustrated by: (green) attenuated periglacial slope deposits and overlying till, Whiting Bay, southern Ireland; (black) glacitectonised glacilacustrine deposits, Loch Quoich, Scottish Highlands; (blue) thrust-faulted and locally soft-sediment deformed fluvial–estuarine deposits and peats in a thrust moraine, Brennhola-alda, Breiðamerkurjökull foreland, Iceland; and (red) bedrock glacitectonite and subglacial till continuum from Bannow, South East Ireland.

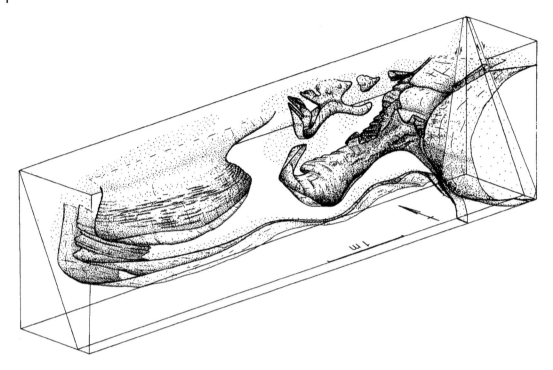

Figure 14.2 Three-dimensional sketch of complex glacitectonic deformation structures logged by van der Wateren (1995b).

Glacitectonites are the most varied of all glacial deposits and mélange types, because they can be derived from any type of pre-existing material, both bedrock and sediment. Evident within the structures imposed on such pre-existing materials by glacial shear stresses are indicators of brittle or ductile deformation, or a combination of the two (Figure 14.1). Brittle styles of deformation are evident in the brecciated nature of some glacitectonites, where materials have been fractured and displaced along fault planes. Ductile deformation is evident in various styles of folding all the way up to pseudo-laminated diamictons which are the products of the advanced stages of pervasive shearing (Figure 14.1). The greatest complexity in deformation structures is created in inhomogeneous materials (Figure 14.2), because they have widely varying rheologies and strengths, common in glaciated basins where pockets or strata of diamictons, gravels, sands, silts and clays are juxtaposed in infinite variety, each material type displaying a different behaviour in response to stress, primarily due to contrasts in sediment frictional strength and permeability. It is common therefore to see highly deformed fine-grained sediment bodies with complex elongated folds adjacent to coarser-grained pods and lenses that have resisted deformation under the same strain regime (e.g. Berthelsen, 1979; Hart and Boulton, 1991; Hart and Roberts, 1994; Benn and Evans, 1996; Evans *et al.*, 1999a; Roberts and Hart, 2005; Benn and Prave, 2006; Hart, 2007; Lee and Phillips, 2008; Figure 14.3). As discussed in Section 6.5, the pods of relatively stiff sediment within glacitectonites will develop streamlined forms (augens or boudins) during deformation, as a result of the gradual erosion and deformation of their edges during simple shear within more rapidly deforming material (Hart and Boulton, 1991; Hart, 2007; Figure 14.4).

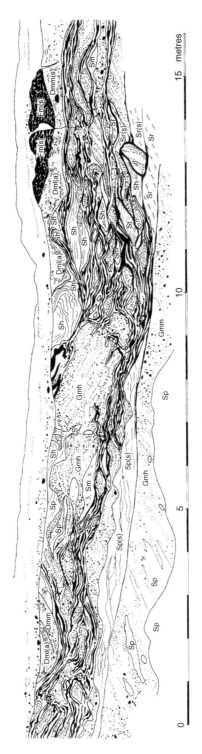

Figure 14.3 Scaled section drawing of glacitectonite developed from the top of an ice-contact delta overrun by glacier ice, south Loch Lomond, Scotland (from Benn and Evans, 1996). Highly attenuated laminated silts and clays are depicted as black.

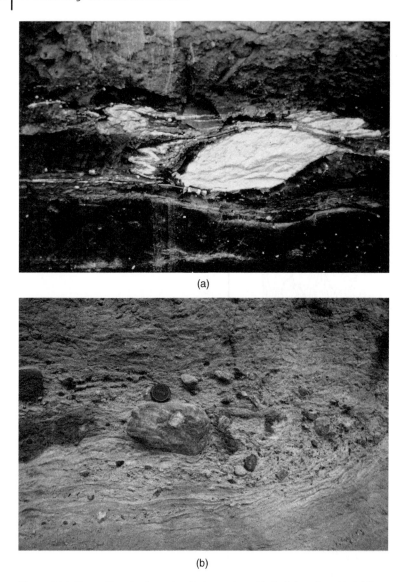

(a)

(b)

Figure 14.4 Typical boudin forms in glacitectonites: (a) chalk boudin with folding in pressure shadow, West Runton, Norfolk, England (from Hart and Roberts, 1994); (b) attenuated glacilacustrine deposits with boudin of stratified, clast-rich diamicton, Loch Quoich, Scottish Highlands.

Even where deformation patterns are complex, glacitectonites are relatively easy to identify at the lower end of the strain signatures, for example, in Type B glacitectonites. However, at increasing levels of cumulative strain, where Type A glacitectonites develop and evolve into pseudo-laminated and then macroscopically homogenous diamictons, differentiation from other types of banded or stratified deposits (i.e. melt-out till, glacilacustrine/glacimarine sequences, gravity mass flows) has proven controversial. If we consider that glacitectonite can be also derived from such banded or stratified deposits our problems are compounded (see below). Type A glacitectonites, because they have undergone high cumulative strains can exhibit a distinctive sub-horizontal banding, leading ultimately to

Figure 14.5 Highly attenuated glacilacustrine deposits (glacitectonite) intruded into crevices on rockhead surface, Loch Quoich, Scottish Highlands.

banded or laminated tills. This appearance is strictly pseudo-lamination because it has not necessarily been driven by primary subaqueous sedimentation and even if this is the case, in its present form the lamination is a highly attenuated remnant of primary bedding. Individual bands are commonly lithologically distinct, and represent different rock and sediment types that have been highly attenuated but not mixed during strain. As discussed in Chapter 10, Roberts and Hart (2005) recognised Type 1 laminae, produced by intergranular shear, and Type 2 laminae, which are the attenuated remnants of pre-existing sedimentary bedding.

The beginning of the glacitectonite forming process was reviewed in Section 6.5 and Chapter 10, where the disruption and liberation of substrate material was linked to rafting and cannibalisation (Figures 6.19–6.22 and 10.11–10.13). Soft or well-jointed bedrock is particularly prone to the cannibalisation process, which has been identified in ancient outcrops as taking place either by soft-bed plucking through till/glacitectonite squeezing into crevices (Broster *et al.*, 1979; Harris, 1991; Evans *et al.*, 1998; Figure 14.5) or the downward migration of subglacial shear zones through the exploitation of sub-horizontal jointing/bedding (Knill, 1968; Money, 1983; Harris, 1991; Hiemstra *et al.*, 2007; Evans and Ó Cofaigh, 2008) or the drag folding and tectonic/depositional slice development in soft materials (Hart *et al.*, 1990; Hart and Boulton, 1991; Boulton *et al.*, 2001; Figures 6.22 and 14.6). The liberation of bedrock blocks is likely partly facilitated through the weakening of surface strata, by not only preglacial weathering (e.g. Broster and Seaman, 1991) but also the growth of segregation ice in joints, where they lie parallel to the ice–bed interface. Chalk, shale and well-bedded limestones are excellent examples of the susceptibility of bedrock to such processes, as their sub-horizontal jointing is particularly prone to exploitation by ground ice and cryoturbation (Figure 14.7; cf. French *et al.*, 1986; Murton, 1996; Muron *et al.*, 2001), which when overrun by glacier ice can couple to significant depths with either the ice and/or a deforming layer and facilitate plucking and/or the development of bedrock glacitectonite (Phillips *et al.*, 2013a). The patchy nature of preglacial weathering in materials like chalk also give rise to a range of raft types, disaggregation stages and attenuation intensities, often resulting in glacitectonites with chalk lamination (highly attenuated rafts), chalk boudins and relatively unmodified chalk blocks in close proximity (Roberts and Hart, 2005; Phillips *et al.*, 2008). The

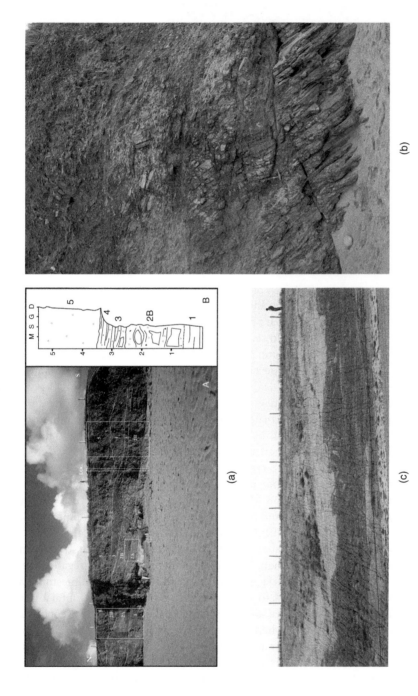

(a)

(b)

(c)

Figure 14.6 Examples of bedrock glacitectonite sequences: (a) Clogher Bay, western Ireland (from Hiemstra *et al.*, 2007), showing a vertical continuum from: Unit 1 – intact, bedded siltstone; Unit 2 (A and B) – deformed (fractured) siltstone with rafts of intact siltstone; Unit 3 – deformed (brecciated) siltstone with occasional rafts; Unit 4 – mix of heavily deformed siltstone and diamicton; Unit 5 – massive diamicton with silty matrix and far-travelled constituents (till). Photograph log in white box at right is used later in Figure 17.5; (b) photograph log located by white box at right-hand side of (a), showing the vertical change from bedrock to deformed and fractured siltstone and capped by till; (c) predominantly ductile glacitectonic deformation of Cretaceous mudrocks and siltstones capped by a thin till veneer (<2 m thick) that displays a prominent fluted surface, Smith Coulee, southern Alberta (Evans *et al.*, 2008).

(a)

(b)

(c)

Figure 14.7 Examples of the liberation of bedrock blocks by periglacial and permafrost processes prior to glacier overriding and the development of bedrock glacitectonites in joint systems lying sub-parallel to the land surface: (a) segregation ice in shale bedrock joints, Melville Island, Arctic Canada (French *et al.,* 1986); (b) cryoturbation structures in Corallian limestone, overlain by diamicton with predominantly local lithologies and an inherited clast macrofabric in its lower 1 m, Filey Brigg, Yorkshire, England; (c) well-developed involutions/pendant structures in the Corallian limestone at Filey Brigg.

largest of such bedrock rafts are spectacular in size (Stalker, 1973, 1976; Sauer, 1978) and therefore exceed the thickness of the deepest subglacial deforming layers as we understand them from modern analogues (i.e. <1 m Table 6.1), although Hiemstra *et al.* (2007) propose raft liberation down to 3 m in siltstone bedrock. Larger rafts are either frozen to the ice base and then displaced along a deeper decollement zone (Figure 6.22c), which is more easily facilitated with thin slabs, or they are prepared in thrust block moraines and then partially deformed as components of a subglacial glacitectonite by overriding ice (e.g. Pedersen, 1996; Figures 6.19 and 14.8).

Vertical sequences that display a range of glacitectonite development stages from undisturbed bedrock through to crushed and homogenised diamicton have been reported from a number of sites. A particularly clear example is that of Clogher Bay, Western Ireland, where Hiemstra *et al.* (2007) describe a vertical transition from undisturbed siltstone bedrock, to deformed bedrock characterised by large-scale anastomosing shear faults, to crushed and brecciated bedrock, to massive matrix-supported till with far-travelled erratics (Figure 14.6a, b). The initial stages of disruption involve the liberation of bedrock fragments and rafts up to 3 m thick, which are then brecciated and

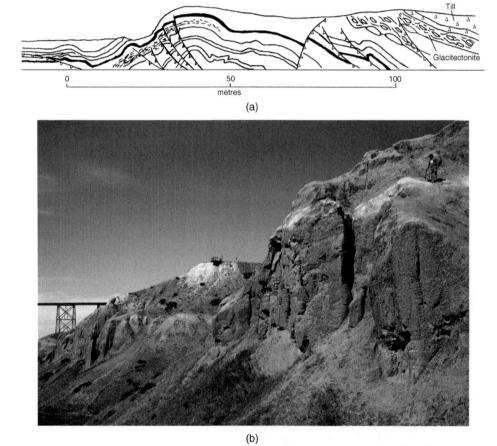

(a)

(b)

Figure 14.8 Examples of bedrock raft liberation in subglacial shear zones/glacitectonites: (a) proglacial to sub-marginal glacitectonic deformation structures in a soft-bedrock substrate at Feggeklit, Denmark (from Pedersen, 1996), showing the liberation of rafts in the sub-marginal shear zone at right; (b) prominent horizon of Cretaceous bedrock megarafts (light grey horizon below figure) in multiple diamicton sequence, Lethbridge, Alberta, Canada.

crushed to produce the matrix of a till. The brecciation and crushing process has traditionally been associated with the creation of 'comminution tills' (Elson, 1961; see Chapter 5) and has been identified as an important till forming process in the same area of Ireland by Croot and Sims (1996) who refer to 'glaciotectonic breccia'. A similar vertically graded sequence is described from Anglesey by Harris (1991), where well-foliated and densely sub-horizontally-jointed metasedimantrey rock strata display weak to penetrative deformation grading upwards to subglacial till (Figure 14.9). Slabs of bedrock have been thrust upwards into a zone of crushed bedrock and display the crushed and sheared edges typical of glacitectonic rafts. A similar sequence has been reported from Anglesey, North Wales, by Phillips *et al.* (2013a). A comminution origin is apparent also in the shale-rich matrixes that characterise the tills of the northeast English coast (Figure 14.10), where Liassic rafts have been rafted onshore by the North Sea lobe of the British – Irish Ice Sheet (Hemingway and Riddler, 1980; Roberts *et al.*, 2013). The raft tops are visible at Upgang, North Yorkshire and display areas of undisturbed, sub-horizontally-bedded to folded Liassic shale as well as zones of reworked brecciated shale or a monolithological clast-supported diamict (Figure 14.10a). This has a sharp contact with overlying red/brown diamicton, but the latter is locally interdigitated with, or injected into the rafts and associated with recumbent folds. Hence, the rafts tops are interpreted as a partly brecciated Liassic shale and mudstone or bedrock glaciotectonite, and the folded and interdigitated/injected contact is a thin and patchy amalgamation zone documenting the plucking of shale fragments and their mixing with overlying diamicton. A similar but less patchy amalgamation of brecciated shale and overlying locally derived till occurs further north on the Northumberland coast at Fenwick, characterised by a <1 m gradational boundary between an unoxidised brecciated and pulverised shale mélange and an overlying oxidised, shale-rich diamicton (Figure 14.10b). Once incorporated into overlying deforming materials, these smaller bedrock rafts, brecciated fragments or plucked blocks can be responsible for an inherited clast macrofabric at all scales, as demonstrated by the case study of Phillips and Auton (2008; Figure 14.10c, d), especially in monolithological diamictons (i.e. Type A bedrock glacitectonite; Figure 14.10e) but also in subglacial deforming tills where far-travelled erratics dilute the local materials.

Once rafts are incorporated into subglacial shear zones, they are subject to significant modification by deformation due to simple shear, which at macroscale is evident in boudinage and pressure-shadow development and tectonic lamination. When combining macro- and microscale observations, the impact of this deformation and the evolution of the raft in the deforming layer can be analysed quantitatively. This is demonstrated by Vaughan-Hirsch *et al.* (2013) for a composite chalk raft in East Anglia, United Kingdom, that has been detached proglacially (cf. Burke *et al.*, 2009) and then moved through a package of younger, Middle Pleistocene deposits called the Happisburgh Till Member (HTM), which is itself a glacitectonite and is discussed in more detail later. Thin section samples (WRA315 – 318 and WRB319) from the underlying HTM, the margins of the rafts and HTM inclusions within the chalk rafts (Figure 14.11a) reveal a range of deformation responses reflecting the total relative strain (Figure 14.11b), specifically confirming that deformation was focussed on the raft margins and the adjacent, relatively weaker but porewater pressurised HTM at low shear strains. It was also possible to summarise the raft emplacement history in four stages as: (1) main transport phase by ductile shearing; (2) continued shear and propagation of narrow ductile shear zone upwards through the raft base, leading to detachment of an elongate chalk block from the main raft; (3) detached block impinges on the deformation at the base of the raft leading to up-ice (reverse) directed shear; and (4) locking of the basal deformation zone during final stages of raft emplacement.

The role of pressurised water in till sedimentology has been well documented in the identification of clastic dykes (Chapter 12), and its influence on glacitectonite development is significant. This has been demonstrated at microscale especially by Phillips and Auton (2000) and Phillips *et al.* (2007),

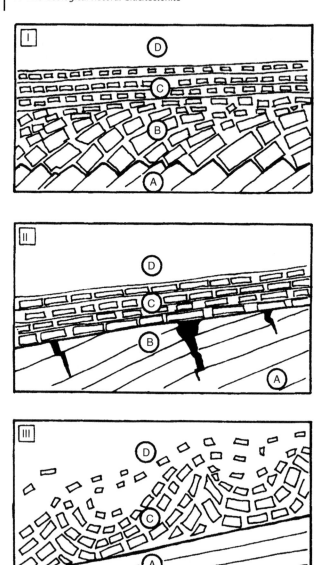

Figure 14.9 Different styles of bedrock glacitectonics based upon the case study of Harris (1991) in the Early Cambrian metasediments of Anglesey, North Wales: Zones are: (A) undisturbed bedrock; (B) weakly deformed bedrock; (C) penetratively deformed bedrock; (D) till. The styles are: (I) zones A, B, C and D form a gradational continuum with bedrock strata being overturned and liberated and contributing to an inherited fabric in zone C and possibly also D (the till); (II) plane of decollement separates zones B and C and the disruption in zone B is restricted to the opening of transverse joints and potential intrusion of till wedges; (III) compressional shortening by ice flow upslope creates folds in zone C, which are separated from zone A by a decollement plane. Note that such structures might be mistaken for, or constitute modifications of, preglacial cryoturbation structures such as those illustrated in Figure 14.7b and c.

Figure 14.10 Examples of bedrock inheritance in glacitectonites/tills: (a) shale-rich matrix of till at Upgang, North Yorkshire, England, where Liassic rafts have been rafted onshore by the North Sea lobe of the British–Irish Ice Sheet (Roberts *et al.*, 2013); (b) amalgamation zone of brecciated shale and overlying local shale-rich till, Fenwick, Northumberland, England; (c) thin-section image showing the dominance of bedrock fragments in the Langholm Till, Dumfries, southwest Scotland, with the orientation of the long and short axes of coarse sand to pebble-sized clasts identified (from Phillips and Auton, 2008); (d) schematic diagram to illustrate how bedrock rafts, brecciated bedrock fragments or plucked blocks can be responsible for an inherited clast macrofabric (from Phillips and Auton, 2008); (e) transition zone between sandstone bedrock and an overlying sandstone-rich till, Prince Edward Island, Canada. The upper photograph is located by the black box in the lower photograph and shows the abrupt contact between till and underlying brecciated bedrock.

(a)

(b)

sample N2850

10 mm

− ve ← → + ve
datum
orientation of clast long axis

/ orientation of clast long axis
.·' moderately inclined clast fabric
‒· gently inclined clast fabric
\ steeply inclined to subvertical clast fabric
\ inclined (-ve slope) clast fabric

(c)

ice movement direction

decrease in preservation potential of 'inherited' class fabric

rock fragments at base of diamicton preserve geometry of bedrock bedding and/or cleavage

Section C   Section B   Section A

diamicton

bedrock composed of interbedded sandstone and mudstone with subvertical joints segmenting the sandstone beds

ridge of bedrock

bedrock   bedrock

rock fragments at base of diamicton preserve geometry of bedrock bedding and/or cleavage

mudstone   sandstone

(d)

Figure 14.10 (*Continued*)

(e)

Figure 14.10 (*Continued*)

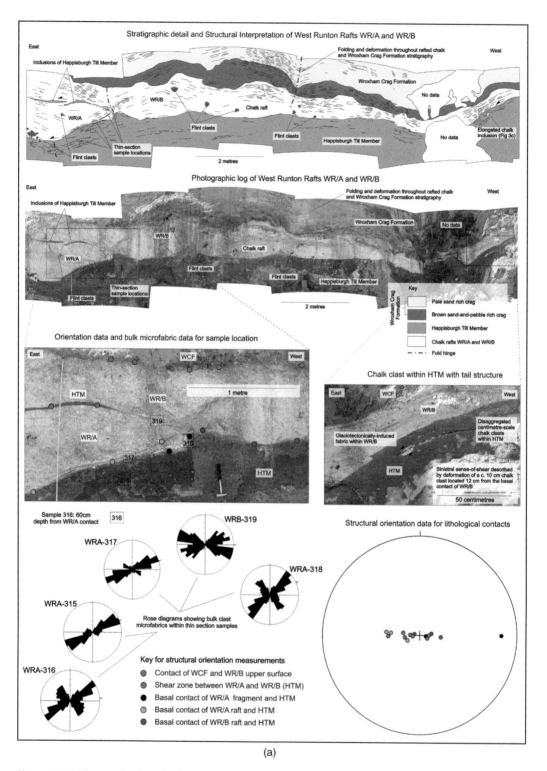

Figure 14.11 Case study of combining macro and microscale observations from an exposure through a composite chalk raft at West Runton, East Anglia, United Kingdom (from Vaughan-Hirsch *et al.*, 2013): (a) annotated photo montages showing the stratigraphy and principal deformation structures associated with the composite raft, together with lower-hemisphere stereographic projection of orientation data obtained for the main lithological contacts (plotted as poles to planes) and a photograph of the raft base showing sample locations (WRA-315-319) used for micromorphological and microstructural analysis.

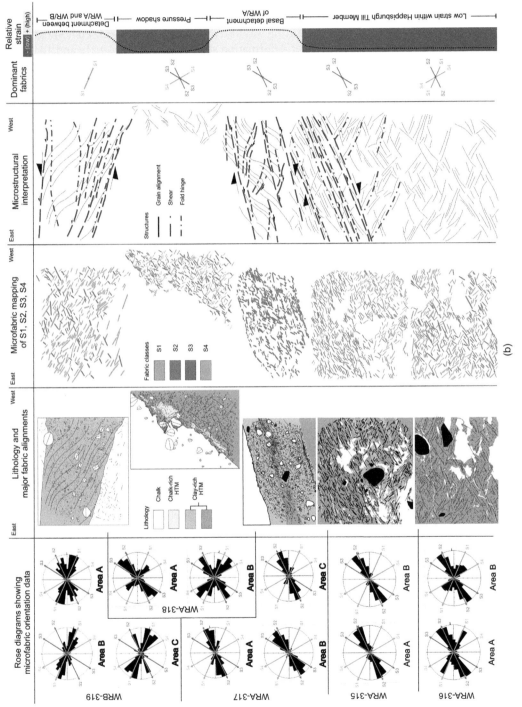

Figure 14.11 (b) schematic interpretation diagram derived from the microstructural analysis, showing the variation in pattern and relative intensity of clast microfabric development within the high-strain zone associated with raft emplacement.

(b)

who identified that most, if not all, of the strain in the subglacially sheared stratified sandy sediments in a range of sub-till settings was accommodated by strain in narrow, water-lubricated detachment zones. Significant fluidisation and soft-sediment deformation is clearly associated with faulting in glacitectonites, and liquefaction and hydrofracturing was also implicated in the localised homogenisation of sediments and the destruction of older deformation structures (Figures 10.21a and 14.12). This overprinting of deformation events at microscale is widely identified in other complex glacitectonite sequences at a range of scales, as we shall investigate at the end of this chapter.

Depending on the extent and depth of deformation beneath tills (Figure 10.9) glacitectonites developed in stratified deposits may be restricted to narrow sub-till shear zones and/or mixing zones in the till base, the latter being classified by Ó Cofaigh and Evans (2001a, b), Evans and Twigg (2002) and Evans *et al.* (2012a) as 'amalgamation zones' or the precursors of homogenised ($S_b - S_h$) zones (cf. Boulton *et al.*, 2001; see Figure 6.21, 6.22 and Chapter c10). The tripartite classification of this sub-till zone by van der Wateren (1995a, b) into lower (S_r), intermediate (S_b) and upper (S_h) zones of vertically increasing deformation intensity or cumulative strain can be identified in structural signatures in finer-grained glacitectonites such as those originating from distal subaqueous deposits or sandy fluvial sequences. For example, the rhythmites of the canal fills in the Horden Till on the Durham coast of North East England (Davies *et al.*, 2009; Figure 11.4) display a complete vertical sequence (Figure 14.13) of rooted drag folds (S_r), grading upwards sharply into a strongly deformed zone of boudins, rootless folds and highly attenuated bands (S_b), and capped by a strongly mixed amalgamation to homogenised zone (S_h) which in turn is truncated by massive matrix-supported diamicton (Horden Till). Often well developed in S_b zones are areas of attenuated sandy to silty bedforms, such as ripples, cross lamination and cross-bedding, which are recognisable as stratified deposits but due to their deformation have been 'smudged' so that the details of the bedform structure are not discernible (Figure 14.13c). Such 'smudged bedforms' (Figure 14.14) are created by elevated porewater pressures, which act to dilate sediment grains enough to displace them relative to one another under simple shear. Relatively greater attenuation in clay-rich laminae in rhythmically or inter-bedded sequences gives rise to boudinage development in the stiffer sandy/silty interbeds around which the clay-rich materials are attenuated into glacitectonic lamination due to them acting as a narrow zone of enhanced shear. An excellent example of this variable response to simple shear and the concomitant development of boudins and tectonic lamination and juxtaposed brittle and ductile deformation in a single shear zone was reported from the Drumbeg glacially overridden ice-contact subaqueous outwash fan and delta sequence at south Loch Lomond, Scotland by Benn and Evans (1996; cf. Phillips *et al.*, 2002; Benn *et al.*, 2004; Figures 8.12b and 14.3). The top of this sequence (Figure 14.15) displays a vertical succession from undisturbed fan/delta sands and gravels, separated from overlying Type A glacitectonites by a planar decollement surface; these are in turn capped by a pseudo-laminated diamicton and stacked sub-marginal till wedges. The Type A glacitectonite comprises detached sand and gravel lenses and pods enclosed within laminated diamicton, sand and silt. Sandy lenses and pods often contain primary depositional structures such as bedforms but these are less common in gravelly lenses and pods, which also display diffuse boundaries indicative of partial disaggregation. The laminated material is highly attenuated silt and clay containing streaks and small augen-like pods of sand and diamicton, with deformation being evident in attenuated overfolds on the down-flow sides of pods. The range of grain sizes in the glacitectonite was inherited from the underlying fan/deltaic deposits and the relatively coarse-grained pods acted as stiff or rigid areas that were rafted along within more rapidly deforming fine-grained sediment. Hence, the glacitectonite

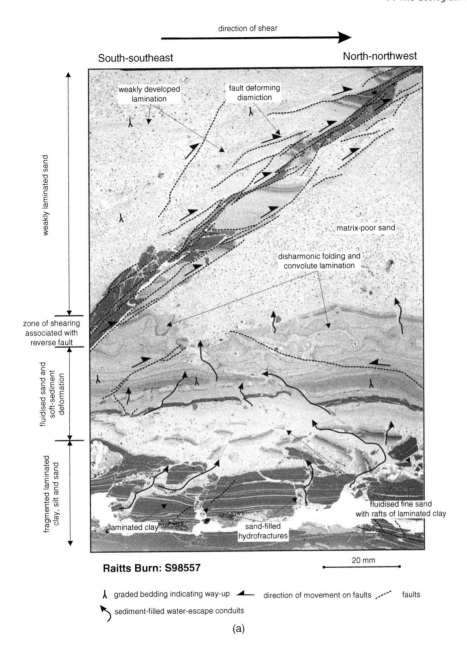

Figure 14.12 Microscale evidence for water-lubricated detachment zones (from Phillips *et al.*, 2007): (a) thin section from Raitts Burn, northeast Scotland, showing deformed laminated silt and clay overlain by sand and containing sand-filled hydrofractures and water escape conduits, soft-sediment deformation structures (disharmonic folds and convolute lamination) and a prominent reverse fault marked by lenses of sheared clay; (b) deformed laminated clay, silt and fine-grained sand exposed at Coire Mhic-sith, Scottish Highlands, showing: (left) low-angle reverse faults and thrusts with minor folding and faulting in the hanging walls; (right) layer-parallel thrusting resulting in localised imbrication and repetition of fine-grained sediments, as well as minor injection of fluidised sand along the reverse faults that deform the laminated sand, silt and clay in the upper part of the sample.

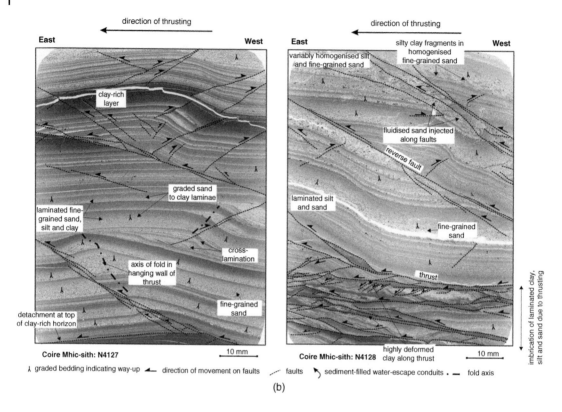

Figure 14.12 (*Continued*)

records differential deformation of heterogeneous sediments exposed to simple shear. The laminated diamicton that separates the Type A glacitectonite from the capping till at the Drumbeg site is an intermediate facies between Type A penetrative glacitectonite and fully homogenised till, the whole deformation horizon being < 3 m thick (Figures 8.12b, 14.3 and 14.15).

The origins of deeper deformation signatures in various types of melange, because they can extend below the depths normally attributed to subglacial shearing (i.e. >3 m), can originate by folding and stacking and/or the overriding of thrust and folded moraine masses (Figure 6.19) and hence can still be the product of glacitectonite genesis. However, in the absence of such thickening evidence, these thick glacitectonites must be viewed as of questionable subglacial origin until unequivocal evidence of deeper subglacial deformation has been proven (see Chapter 16). The spectacularly glacially deformed Quaternary deposits of East Anglia, England, provide us with excellent examples of thick glacitectonites in which, as we have seen above, bedrock rafts are incorporated in various states of disintegration. The precise origins of the glacial deposits in this area vis-à-vis subglacial, subaqueous and subaerial processes have been the subject of much debate (e.g. Perrin *et al.*, 1979; Eyles *et al.*, 1989; Hart, 1990; Lunkka, 1994; Lee, 2001) but their subsequent glacial deformation signature has been central to the wider, long-term development of glacitectonic concepts (e.g. Slater, 1926, 1927a, b; Banham, 1975, 1988; Hart, 1990; Hart and Boulton, 1991; Hart and Roberts, 1994; Roberts

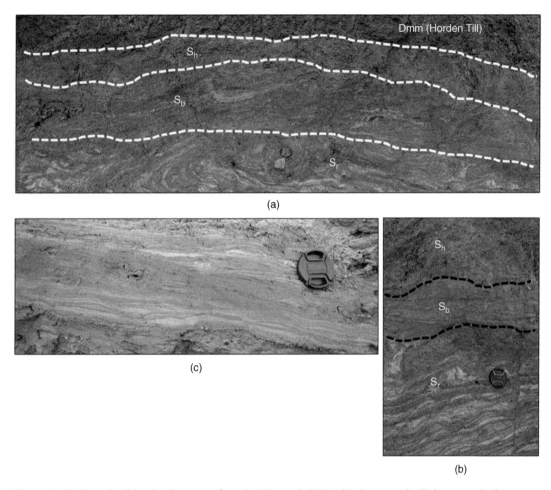

Figure 14.13 Example of the development of van der Wateren's (1995a, b) tripartite sub-till shear zone in the rhythmites of the canal fills in the Horden Till, Durham, (cf. Davies *et al.*, 2009); (a) shows the complete vertical sequence; (b) shows localised details of the three zones; (c) stratified deposits in the S_b zone that display a 'smudged' appearance.

and Hart, 2005). Two sedimentary units are pertinent to the present discussion, specifically the Bacton Green and Happisburgh Diamicton members (Lee *et al.* 2017), which have been previously variously termed 'tills' and diamictons in stratigraphic nomenclature (Figure 14.16a). Also significant in the wider depositional context is the overlying Runton Sand and Gravel Member. In a study of the deformation patterns in these three members, Phillips *et al.* (2008) and Lee and Phillips (2008) have reconstructed a model of onshore ice sheet marginal flow into an ice-dammed glacilacustrine basin, where first subaqueous and then subaerial sedimentation was contemporaneous with glacitectonic deformation (Figure 14.16b). The Bacton Green Member, and likely also the underlying Happisburgh Member (Figure 14.17), were initially deposited subaqueously by sediment gravity flows and potentially also iceberg rafting. The sedimentary sequence was then first proglacially and

Figure 14.14 Examples of smudged bedforms; upper) formally horizontally laminated to ripple-bedded, fine to medium sands whose bedforms have been smudged so that individual features are difficult to discern, Bacton Green Member, West Runton, Norfolk, England; lower) interstratified, horizontally bedded units of medium gravelly sand and coarse to medium sands within a partitioned shear zone lying below a thin till layer, Gutterby, West Cumbria, England. Note that bedforms are indiscernible and that the units have been attenuated to the right.

subsequently progressively subglacially deformed, the proglacial and sub-marginal thrusting and folding being responsible for the thickening of the consequent glacitectonite (Figure 14.16b; cf. H.M. Evans *et al.*, 2011). This is manifest in three clear domains identified by Phillips *et al.* (2008): Domain 1 or proglacial deformation at the eastern end of the coastal sections; Domain 2 or ice-marginal deformation within the central part of the section; and Domain 3 or intense subglacial deformation at the western end of the exposures, all associated with ice advancing from the west. The signature of the subglacial shear zone is contained within the Bacton Green Member (Lee and Phillips, 2008),

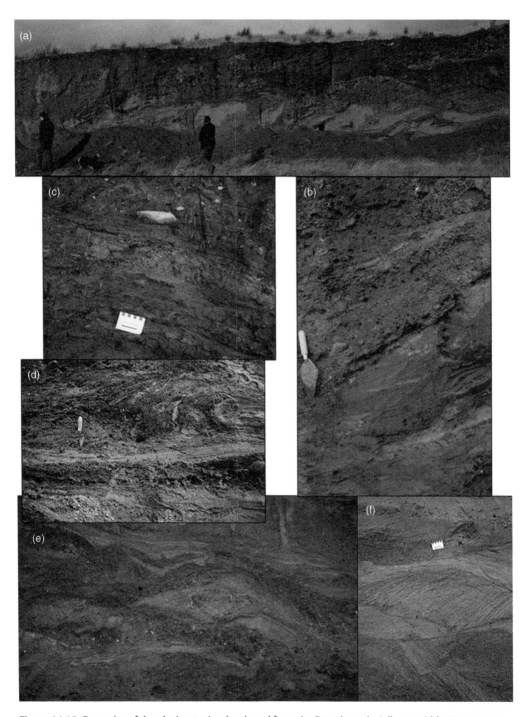

Figure 14.15 Examples of the glacitectonite developed from the Drumbeg glacially overridden ice-contact subaqueous outwash fan and delta sequence at south Loch Lomond, Scotland (after Benn and Evans, 1996; Phillips *et al.*, 2002; Benn *et al.*, 2004); (a) the top of the sequence, showing deformed fan/delta sands and gravels overlain by glacitectonite and sub-marginal till wedges; (b) contact between the crudely pseudo-laminated till carapace and underlying glacitectonite; (c) lower boundary of the pseudo-laminated diamicton and Type A glaciteconite; (d) attenuated folds overlying weakly developed tectonic laminae that had started to form within the stratified sediments beneath the glacitectonite shear zone; (e) and (f) examples of sub-horizontal dislocation of delta deposits to form Type B glacitectonite and augen type rafts of sand and gravel surrounded by attenuated clay–silt rhythmites.

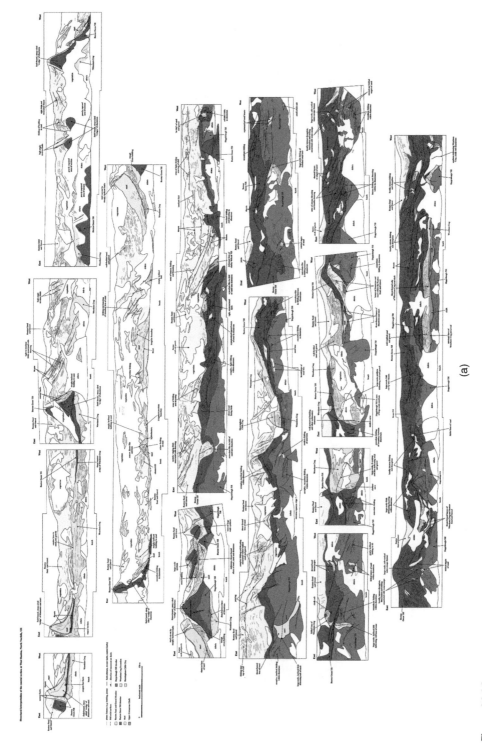

Figure 14.16

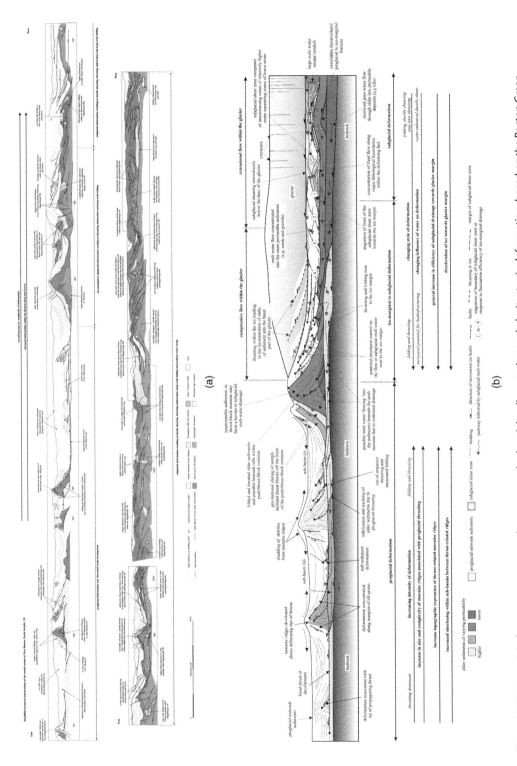

Figure 14.16 A model of contemporaneous subaqueous and subaerial sedimentation and glacitectonic deformation, based on the Bacton Green, Happisburgh and Runton Sand and Gravel members of East Anglia, England (from Phillips *et al.*, 2008); (a) stratigraphic and structural details of the extensive exposures between West Runton and Sheringham. The lower panel is the complete composite structural section, showing the variation in the style and intensity of deformation from east to west; (b) schematic diagram showing the proposed generalised model of progressive proglacial to subglacial deformation observed at the West Runton to Sheringham cliffs.

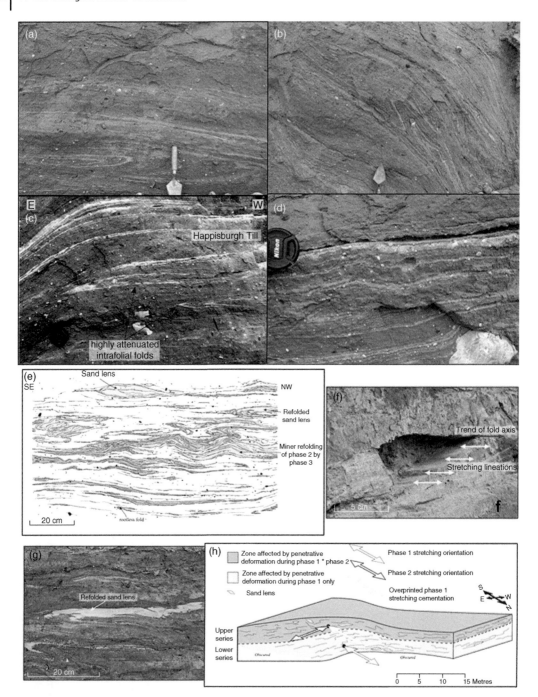

Figure 14.17 Details of the Bacton Green and Happisburgh ('Till') members, East Anglia, England: (a)–(d) pseudo-stratified appearance of the deposits (glacitectonites), showing folded and attenuated bedding, with annotations of features on (c) from Phillips *et al.* (2008); (e) sketch of typical structures in Bacton Green Member, showing the styles of folding in the diamicton (unshaded) and sand (shaded; from Fleming *et al.,* 2013); (f) example of stretching lineation, showing a pronounced stepped nature orientated parallel to the axis of a sheath fold (from Fleming *et al.,* 2013); (g) example of refolding of a sand lens, interpreted by Fleming *et al.* (2013) as indicative of progression of the strain field from extension to compression during overprinting of deformation events; (h) summary model of two-phase evolution of stretching directions based on the deformation signature in the Bacton Green Member and its subdivision into two series by Fleming *et al.* (2013).

which displays a vertical sequence of increasing deformation intensity reflective of Type B to Type A glacitectonites but developed within overprinted thrust slices (Figure 14.18a). Multiple phases of deformation within the Bacton Green Member are identified by Fleming *et al.* (2013) using the detailed complexities of the structural signature in combination with AMS analysis. This facilitated the reconstruction of two main phases of overprinted subglacial shear, recorded in refolded lenses, stretching lineations, sheath folds and magnetic lineations (Figure 14.17g). The overall increase in deformation intensity allows the subdivision of the Bacton Green Member into a lower facies (Type B–Type A glacitectonite) and an upper melange facies (named West Runton Melange Member by Lee *et al.*, 2017; Figure 14.18b). The subaqueous origins of the glacitectonite are apparent in thin sections from the lower facies (Figure 14.18b; cf. Hart and Roberts, 1994; Roberts and Hart, 2005).

Towards the top of the stratigraphic sequences reported by Phillips *et al.* (2008), the syntectonically deposited Runton Sand and Gravel Member is locally juxtaposed with Happisburgh, Bacton Green and West Runton Melange member deposits but, in some outcrops, the deformation styles are indicative of low strains, and large sand intraclasts display well-preserved sandy bedforms (Figure 14.19). Moreover, this low-strain-deformed sequence is up to 40 m thick, and hence Waller *et al.* (2009, 2011) question its subglacial shear zone origin, proposing instead that it represents an area of subglacially deformed, formerly sporadic or patchy permafrost (see Figure 6.23) in which frozen blocks of stratified sand retained their internal integrity (e.g. Menzies, 1990a, b). Although a patchy permafrost environment is entirely consistent with the depositional setting and is corroborated by well-developed ice wedge pseudomorphs in underlying deposits, the occurrence of these areas of low-strain (mostly soft sediment) deformation appear to occur in pockets, around the upper surfaces of what Phillips *et al.* (2008) regard as thrust ridges, and hence could simply be slump-generated folds deforming around coherent (frozen?) blocks of Bacton Green and Happisburgh members and preglacial sand and marl which failed during thrust ridge construction and were then locally overprinted by glacitectonite. Internally coherent soft-sediment intraclasts in melange-type deposits have been interpreted also as diagnostic criteria for melt-out till production (e.g. Hoffmann and Piotrowski, 2001). Whatever its origin, the thickness of the sequence in East Anglia has been explained by Phillips *et al.* (2008) and Waller *et al.* (2011) as a product of sub-marginal thrust stacking of the melange and its subsequent over-ridding and subglacial shearing and boudinage of the constituent beds and the intraclasts (Figure 14.20). Freezing of the intraclasts would have ensured their preservation, as a pore-ice cement (Andersland and Alnouri, 1970) would create rigidity in contrast to the more highly deformed, finer-grained surrounding diamictons and their significantly greater liquid water content; yield stresses of such fine-grained sediments are reduced by the retention of liquid water being adsorbed onto particle surfaces and high interfacial pressures, leading to freezing point depression (Christoffersen and Tulaczyk, 2003). Continued deformation and strain heating over time likely resulted in a breakdown in rheological contrasts and hence the assimilation of intraclasts and diamicton matrix.

The Bacton Green and Happisburgh members are good examples of not just the subaqueous origins of some glacitectonites but also the commonly significant variation in the intensity of deformation through the full vertical depth of heterogeneous, stratified materials, due to the process of deformation partitioning. This variable pattern of deformation has been credited with being the initiator of pseudo-stratified diamictons, some of which have been interpreted as having alternative glacigenic origins (see Figure 13.9b for a melt-out till example). The role of stratification in controlling locally variable porewater pressures has been demonstrated by Roberts and Hart (2005). They highlight the variability in grain size between glacitectonic laminae as a direct determinant of the changing porosity

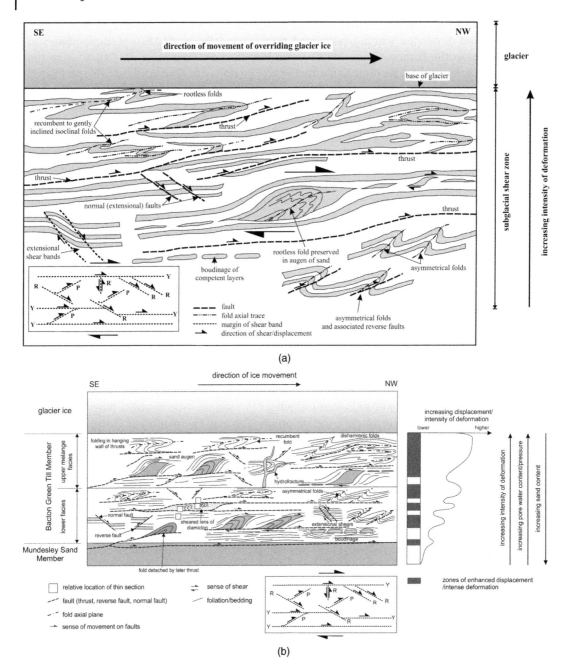

(a)

(b)

Figure 14.18 Subglacial shear zone imprints in the Bacton Green Member (from Lee and Phillips, 2008), showing: (a) a vertical increase in deformation intensity within overprinted thrust slices and (b) its subdivision into a lower Type B–Type A glacitectonite and an upper melange facies (West Runton Melange Member cf. Lee *et al.*, 2017).

Figure 14.19 Examples of the low-strain deformation exhibited in the melange facies at West Runton.

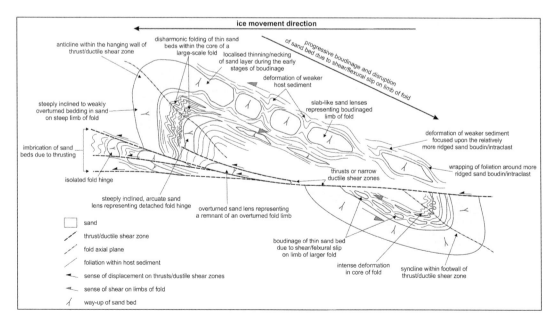

Figure 14.20 Schematic diagram to show how intraclasts likely develop from shear-related thrusting and attenuation of large-scale folds (from Waller *et al.*, 2011).

and hence porewater pressures in a horizontally bedded sequence (Figure 14.21), which drives brittle shear in finer-grained materials and ductile, intergranular shear in coarser materials, allowing them to behave differently and maintain their integrity in a deforming mass. Bedding with similar granulometry, on the other hand, will allow porewater pressure equilibration across laminae and hence potentially promote mixing across boundaries.

Some glacitectonites merely inherit the stratification of their parent materials, recording at the same time proglacial sedimentation followed by ice overriding. In situations where diamictic deposits, particularly stratified diamictons, dominate this will produce a sediment that very closely resembles other subglacial end products, for example, multiple tills and sliding-bed deposits (Chapter 11) or melt-out till (Chapter 13). The conversion of a whole stack of pre-existing subaqueous mass flow deposits into a pseudo-stratified diamicton or glacitectonite has been reported from Feohanagh, western Ireland by Ó Cofaigh *et al.* (2011), where a vertical sequence of sub-horizontally-bedded diamictons and stratified sediments are subdivided into three lithofacies associations (LFAs 1–3) that record different phases of ice-proximal and subglacial deposition and deformation (Figure 14.22). The oldest deposits (LFA 1) comprise stratified and massive diamictons, some clast-rich diamicton, laminated mud and horizontally bedded sand, with the stratification in the diamictons ranging from a fissility, imparted by sub-horizontal, anastomosing partings, to lenses and interbeds of sand with primary stratification. LFA1 is interpreted as the product of

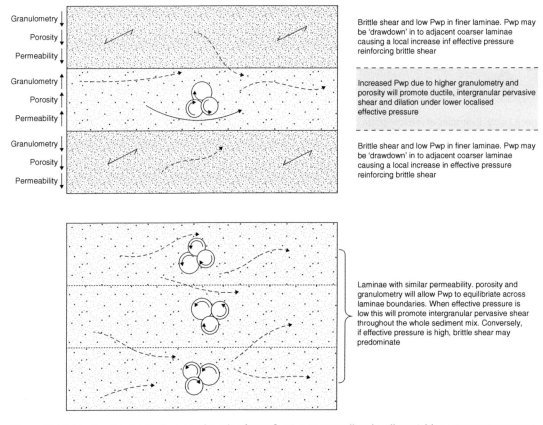

Figure 14.21 Idealised sketch to illustrate the role of stratification in controlling locally variable porewater pressures and concomitant deformation styles (from Roberts and Hart, 2005).

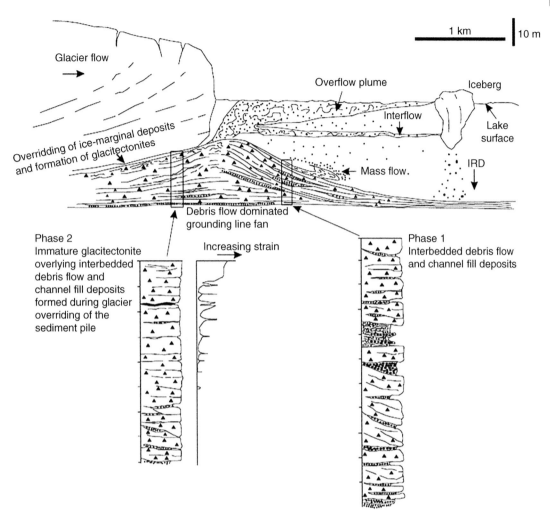

Figure 14.22 Reconstruction of the depositional processes involved in the conversion of a pre-existing subaqueous mass flow depo-centre (phase 1) into a pseudo-stratified diamicton/glacitectonite (phase 2), based upon the stratigraphy at Feohanagh, western Ireland (from Ó Cofaigh *et al.,* 2011).

subaqueous stratified sediment deposition by gravity mass flows and scoured channel fills followed by glacitectonic thrusting and glacier overriding, which initiated shear displacement along the partings or the sorted beds between diamictons (deformation partitioning). This deformation is clearly recorded in the folding and faulting of the mud, sand and gravel units of LFA2, which were thrust *en masse* over LFA1. The youngest LFA 3 comprises a series of laminated, stratified and massive, matrix-supported diamictons, with minor beds of clast-supported diamictons and laminated fines, with lamination/stratification imparted in the diamictons by anastomosing partings or fissility. This is interpreted as subglacially deformed material derived from underlying deposits, especially as it displays an up-sequence transition from heavily sheared (Dml/Dms) to more massive (Dmm/Dcm), consistent with an upward increase in cumulative strain within the sediment. The presence of augen-shaped sandy pods and hook folds within the diamicton are also indicative of subglacial shear, and clast macrofabrics are consistently strongly aligned with former ice flow from the northeast.

The predominantly sub-horizontal shearing and minimal vertical displacement and folding of the pre-existing sediment pile at Feohanagh can be contrasted with other case studies of glacitectonised and cannibalised stratified glacigenic deposits, where proglacial folding and faulting contributed to the displacement of large rafts of substrate in the creation of glacitectonite as ice overrode the site. The Drumbeg case study of Benn and Evans (1996) and Phillips *et al.* (2002; see Figures 8.12b and 14.15) is a clear example of the resulting highly heterogeneous melange created wherever proglacial lakes are overrun and their depositional record displaced in thrust stacks that then undergo significant internal deformation partitioning. Such glacitectonites can be created also in bedrock uplands even though significant subaqueous depo-centres are restricted to localised overdeepenings; in such settings, the lake sediments reworked into glacitectonite are crucial to the derivation and continuity maintenance of otherwise sparse subglacial deforming layers of mountain glaciers. An example from Loch Quoich in the Scottish West Highlands was reported by Evans *et al.* (1998), where glacilacustrine sediment deposited during the final stages of ice sheet deglaciation were glacitectonised by a later readvancing Younger Dryas valley glacier (Figure 14.23; see also

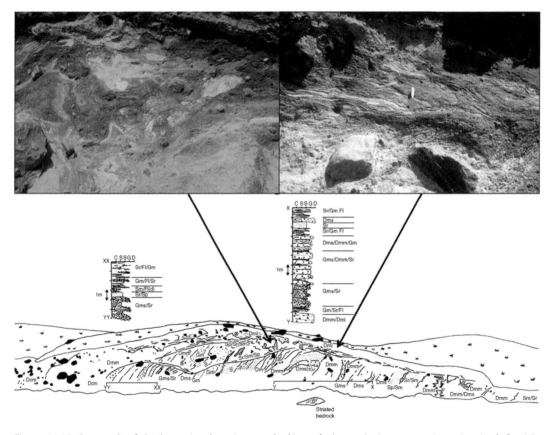

Figure 14.23 Case study of glacilacustrine deposits reworked into glacitectonite in a mountain setting, Loch Quoich, Scottish West Highlands (after Evans *et al.*, 1998). The glacitectonite carapace (photographs) is separated from the underlying folded and hence steeply dipping sequence of glacilacustrine deposits by an abrupt contact representing a thin shearing/decollement zone.

14.1 and 14.4b). The largest exposure shows an anticlinal fold in proximal glacilacustrine deposits, which has been truncated by a shear zone. The glacilacustrine deposits are typified by massive to stratified diamictons, containing micro-laminations and dropstones, interbedded with stratified sands, silts and clays displaying climbing ripples, laminations, mud drapes and dropstone clusters that likely originated as iceberg dump structures; localised stratified diamicton lenses represent subaqueous mass flow deposits. The shear zone represents S_r and S_b type structures (Figure 6.21) developed within the underlying, sub-vertically-inclined glacilacustrine sediments, which have been amalgamated and attenuated. A lower thrust fault that parallels the shear zone was responsible for the sub-horizontal displacement of part of the overturned fold structure as a thrust slice. In and above the shear zone the heavily deformed materials resemble Type III–IV melange and are referred to as a 'glacitectonite Type A carapace' by Evans *et al.* (1998), because they range in appearance from heavily contorted glacilacustrine sediment melanges to laminated diamictons with highly attenuated pods or boudins. The truncated fold beneath the shear zone qualifies as a Type B glacitectonite, because the sediments retain their pre-deformational structures and bedding.

Significantly greater disturbance can be initiated in glacitectonised stratified sediments when pressurised groundwater causes soft-sediment deformation and burst-out structures. The tops of such sediment injections can provide a ready-made melange to the cannibalising deforming layer moving across a site, an excellent example of which occurs at Consort, southern Alberta (Figure 14.24). Here, a sub-horizontally-bedded, stratified sequence of laminated sands, silts, clays and clay-rich diamictons, likely originating as proglacial glacilacustrine deposits, have been injected from below by gravelly diamicton burst-out structures and their tentacle-like dykes or branches. The sequence is then truncated by a largely massive, matrix-supported diamicton with numerous boudin rafts, attenuated lenses and stringers of sands and silts cannibalised from the underlying melange when glacier ice flowed across the site from the north; the clast macrofabric developed in this diamicton is typical of Type A glacitectonites and homogenised but immature subglacial tills (Table 8.1).

Complex multiple shear zones and stacked glacitectonites can be constructed in areas characterised by contrasting but juxtaposed strata, comprising both Quaternary sediment and soft bedrock, as we have seen in the East Anglian examples above. In a similar area of extensively glacitectonically disrupted chalk on the island of Rügen in northern Germany, a site at Glowe displays stacked glacitectonites, of either diamictic or sandy matrix, displaced over tilted chalk bedrock blocks (Figure 14.25). Each glacitectonite contains intraclasts or rafts which display a range of deformation structures, but are most variable in appearance in the thickest dark clay-rich unit (Figure 14.25b i), where the rafts are contorted but not always attenuated. The basal contacts of each glacitectonite comprise shear zones in which underlying materials have been drawn up as drag folds and tectonic slices to form glacitectonic lamination (Figure 14.25b ii and iii). The glacitectonites and chalk are truncated by a horizontal erosional contact which is overlain by a massive matrix-supported diamicton (till), the basal zone of which comprises a shear zone of glacitectonic lamination derived from the underlying glacitectonites (Figure 14.25c i). A large lens of horizontally bedded sandy rhythmites lies between the capping till and the glacitectonites at the western end of the section and at one location displays deformed bedding related to a large boulder that impacted the underlying beds as a dropstone during rhythmite deposition (Figure 14.25c ii). The shear zone that separates these rhythmites from the overlying till is composed of a lower area of smudged rhythmites, cross-cut by anastomosing shears (fissility), grading upwards abruptly into glacitectonic lamination (Figure 14.25c i). Varve-like lamination appears to have been preserved at the base of the underlying dark clay-rich diamicton where it has been deformed over a planar-bedded coarse sand raft, which in turn has been displaced over a stratified sand and mud melange (Figure 14.25c iii), the whole sequence demonstrating the highly contrasting styles of deformation intensity controlled by grain size and porosity.

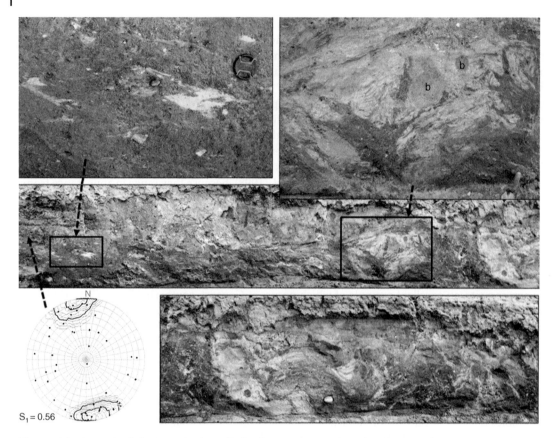

Figure 14.24 Example of glacitectonised stratified sediments cross-cut by soft-sediment deformation and burst-out structures, forming a melange that grades upwards into massive, matrix-supported diamicton with numerous sand boudins, attenuated lenses and stringers, Consort, southern Alberta, Canada. *Areas "b" are animal burrows.*

The glacitectonite case studies reviewed above provide clear illustration of the problems involved in genetic classification of subglacially deformed materials. Glacitectonites can inherit the sedimentary characteristics of parent materials and often modify them only slightly, so that glacitectonic disruption is visible only at microscale. As a glacitectonite can be manufactured out of any pre-existing material by glacier overriding, it probably qualifies as the most diverse sediment type on the planet! Hence, the potential for mis-identification and mis-classification is significant despite the availability of a number of diagnostic criteria that can be employed in determining glacitectonite genesis, as reviewed in the case studies above and in Chapters 4 and 6.

Most commonly, the pseudo-stratification/lamination or banding in glacitectonites, especially when associated with scattered clasts, can be mistaken for laminated glacilacustrine and glacimarine deposits with dropstones, but a deformation origin can be distinguished using the presence of characteristics such as lithological banding, attenuated rafts, asymmetric folds around pebbles and attenuated folds among the banding (Hart and Roberts, 1994; Ó Cofaigh and Dowdeswell, 2001; Roberts and Hart, 2005). Nevertheless, as we have seen above, the origin of banding or pseudo-stratification/lamination is more often than not inherited from pre-existing primary stratified deposits and this is often immediately apparent in obvious and laterally extensive compositional layering (Figure 10.13b). Interbedded sequences of stratified sediments and mass flow-derived

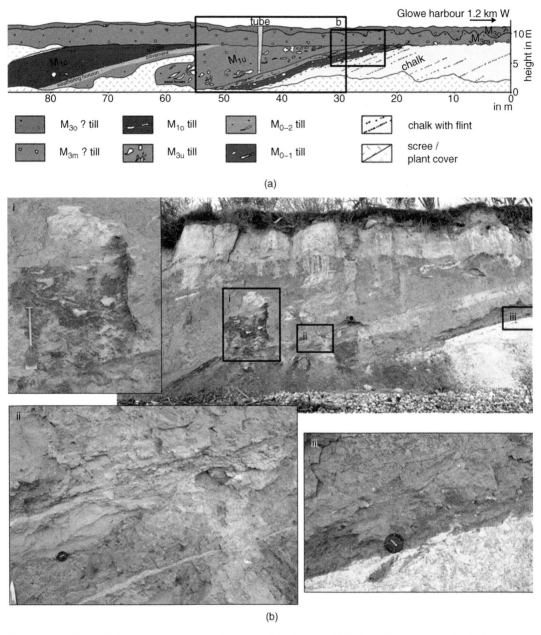

Figure 14.25 Stacked glacitectonites displaced over tilted chalk bedrock blocks at Glowe on the island of Rügen in northern Germany; (a) section sketch, with locations of photo-montages in (b) and (c) marked by black boxes. Stratigraphy compiled by Michael Kenzler, University of Greifswald.

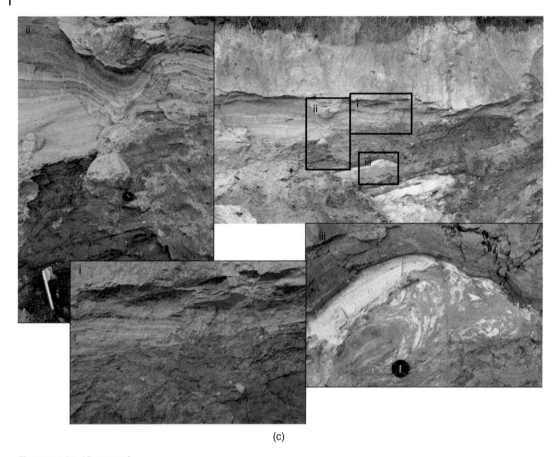

(c)

Figure 14.25 *(Continued)*

diamictons laid down in ice-proximal depo-centres (Figure 14.22) are prime materials for the development of glacitectonites that appear as pseudo-stratified diamictons (Figure 14.26). Often interpreted as melt-out tills because of the complex interbedding of diamicton and sand and gravel current bedforms (e.g. Shaw, 1982, 1987), these deposits are not always located in an overall geomorphological, stratigraphic and architectural context that is conducive to melt-out preservation, and in detail the high degree of sorting and bedform development are not typical of modern melt-out tills *per se* (cf. Lawson, 1979a, b; Larsen *et al.*, 2016); instead a glacitectonite interpretation invariably involves less assumptions.

For example, the case study depicted in Figure 14.26 from Drayton Valley, Alberta, Canada, lies at the westernmost extent of bedrock thrust block displacements by the Laurentide Ice Sheet and in the suture zone of the Cordilleran and Laurentide ice sheets, where changing ice flow dominance resulted in overprinted signatures (Klassen, 1989; Rains *et al.*, 1999). Thrust blocks of local Paskapoo Formation sandstone with younger, overlying preglacial Saskatchewan Sands and Gravels have been pushed over and into underlying Bearpaw Formation shale and dark grey massive, matrix-supported diamicton, the latter forming diapirs between bedrock blocks. This glacitectonic mélange is draped by a coarsening-upward, stratified diamicton (Dms) comprising interbedded sand/silt/clay laminae, diamictons and minor matrix-supported gravelly lenses, and displaying only localised soft-sediment

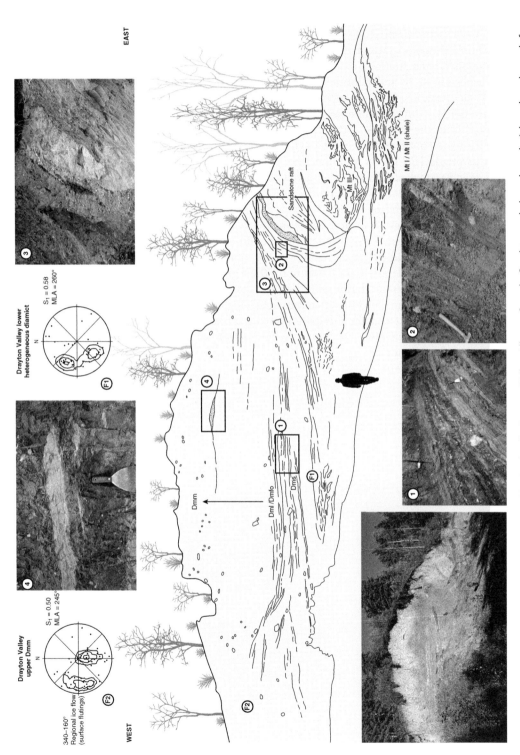

Figure 14.26 Stratified diamictons (glacitectonites) and bedrock rafts, Drayton Valley, Alberta, Canada. Sandstone, shale and preglacial sands and gravels form a melange with matrix-supported and stratified diamictons, capped by laminated to fissile diamicton (Dml/Dmf) and a massive, matrix-supported diamicton (Dmm). Clast macrofabrics are located in the upper diamictons.

deformation, thrust faulting and dislocated recumbent folds. It is truncated by a thin zone of laminated to fissile diamicton (Dml/Dmf) which lies at the base of a brown, massive, matrix-supported diamicton (Dmm) containing attenuated lenses of bedded sands, silts and clays that resemble the laminae of the underlying stratified diamicton and hence are likely to be rafts. A clast macrofabric in the Dms displays westerly and south-westerly dips, conforming to the enclosing bedding dip, and contains 35% shield erratics. The Dms appears to have originated as a subaqueous infill of ponds lying between thrust blocks, with laminated sands, silts and clays and latterly gravity mass flow diamictons being reworked from the margins of the Paskapoo and Bearpaw bedrock and flowing downslope towards the west and southwest, likely fed by drainage from the Laurentide Ice Sheet margin over and through its proglacially thrust moraine. The top of the Dms was then sheared to produce the Dml/Dmf zone and a subglacial till with attenuated sand rafts was emplaced over the whole sequence, as evidenced by the vertical increase in homogenisation towards the Dmm. A macrofabric from this Dmm is weak ($S_1 = 0.498$) and retains the westerly and south-westerly signature of the underlying Dms, but importantly contains no shield erratics and hence must have been emplaced by Cordilleran ice flowing from the northwest, as recorded by the drumlinised land surface in this area.

In a number of case studies reviewed above, the non-subglacial genetic interpretations of diamictons and associated deposits are based upon their gravity mass flow and subaqueous characteristics. Hence, it is inherent that a wider knowledge of the sedimentology of poorly sorted sediments (diamictons and mélange) is crucial to achieving a systematic and objective analysis of glacigenic deposits. The next chapter therefore provides a reasonably comprehensive overview of the process–form regimes involved in diamicton production in settings other than those of the subglacial traction/shear zone.

15

Glacial Diamictons Unrelated to Subglacial Processes

Glaciolacustrine … diamict lithofacies are widespread but are currently disguised in the literature as 'tills'.

Eyles and Eyles (1984b, p. 188)

The appropriate and less appropriate situations in which the term 'till' should be used were briefly discussed in Chapter 3, stemming from Lawson's (1979a, 1981a, b) proposition that a till is a sediment that has been deposited directly from or by glacier ice and has not been subject to subsequent disaggregation and resedimentation. What Lawson (1989) regarded as 'secondary deposits' or those which have undergone reworking by non-glacial processes are straightforward to recognise in their modern settings, for example, where gravity mass flows are being generated on downwasting, debris-covered glacier surfaces or from the collapsing fronts of active push moraines (Figure 3.2), but significantly more problematic and often controversial to classify genetically in the geological record. Hence, this chapter deals specifically with the sedimentology of diamictons created in glacial depositional settings but not related to subglacial processes, previously widely classified as 'flow tills', after Hartshorn's (1958) seminal definition, not only in early glacial research but also perpetuated in the literature of the modern era, despite the fact that a close association with a pre-existing till is often impossible to demonstrate in ancient stratigraphies.

A tour by a geomorphologist around a modern, receding glacier margin will rapidly formulate in their mind the impression that gravity mass flows are the most ubiquitous diamicton-forming process in glaciated basins (Figure 3.2). This is apparent not only in the more dynamic supraglacial settings (e.g. Boulton, 1972a; Eyles, 1979; Lawson, 1979a, b, 1989; Kjær and Krüger, 2001), but also on the freshly deposited morainic materials undergoing extensive paraglacial reworking (e.g. Eyles and Kocsis, 1988; Ballantyne and Benn, 1994, 1996; Harrison and Winchester, 1997; Curry and Ballantyne, 1999; Menzies and Zaniewski, 2003). It is surprising then that greater confidence appears to be demonstrated in identifying tills in ancient stratigraphic exposures than mass flow deposits. Some attempts to redress that imbalance have been delivered but more often than not result in controversial debates on the wider depositional setting, a good example being that of the relatively high deglacial sea levels required to generate subaqueous deposits rather than tills around the southern margins of the former British–Irish Ice Sheet (cf. McCabe, 1986; McCabe *et al.*, 1987; Eyles and McCabe, 1989a, b; Eyles *et al.*, 1989; McCarroll, 2001; Scourse and Furze, 2001). Importantly, some fundamental flaws in till identification and classification are exposed by such debates and hence the initiators (i.e. Eyles and McCabe, 1989b; Eyles *et al.*, 1989) should be credited for erecting outrageous geological hypotheses

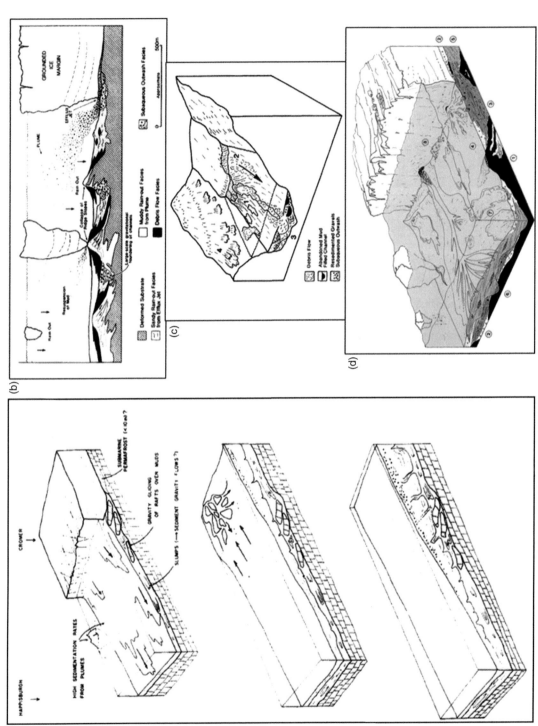

that have proven to be extremely constructive in glacial sedimentology (Figure 15.1). For example, in the case of the southern British–Irish Ice Sheet, the role of subaqueous sedimentation is now more widely acknowledged in glacial land system compilation and interpretation (cf. Thomas and Summers, 1982, 1983; Thomas and Kerr, 1987; Lunkka, 1994; Ó Cofaigh and Evans, 2001a, b; Evans and Ó Cofaigh, 2003; Thomas and Chiverrell, 2006, 2007; Phillips *et al.*, 2008; Ó Cofaigh *et al.*, 2011; Evans *et al.*, 2012), with the attendant recognition that sediment gravity flows, both subaqueous and subaerial, and ice-rafted (dropstone) diamictons are ubiquitous in glaciated basins even though precise origins are debated (cf. Boulton, 1977; McCarroll and Harris, 1992; Thomas *et al.*, 1998; Rijsdijk *et al.*, 1999, 2010; Rijsdijk, 2001; Hiemstra *et al.*, 2005, 2009; Knight, 2014; Evans *et al.*, 2015b; Figures 14.22 and 15.2). Hence, our approach to diamicton description and classification now focuses on those characteristics that have the potential to differentiate the parent process–form regimes, a task that will always be challenging because: (a) mass flow diamictons can be misinterpreted as tills and vice versa and (b) diamictic glacitectonites, and potentially fully homogenised tills, can be derived from mass flow deposits.

As was discussed in Chapter 3, 'till' has long been recognised as an inappropriate term for subaqueous sediments (cf. Evenson *et al.*, 1977; Dreimanis, 1979; Gravenor *et al.*, 1984; Powell, 1984), because it fails to acknowledge the disaggregation and/or remobilisation of glacigenic material once released from the glacier–bed interface into deep water. Instead, sedimentology recognises iceberg rafting and suspension settling (ice rafted debris; IRD) with terms such as 'dropstone diamicton', 'undermelt diamicton', 'iceberg contact deposits' and 'ice-keel turbate', and gravity mass flows with the terms 'subaqueous fall deposits', 'grain flows', 'olistostromes', 'subaqueous slumps' and 'slides', 'subaqueous debris flows' and 'turbidites'. The same argument has been made with respect to the applicability of the term 'flow till' in subaerial settings (e.g. Lawson, 1982, 1989). In order to compile a protocol for identifying diamictic IRD and mass flow deposits, both subaqueous and subaerial, we must first turn our attention to modern process–form regimes and deposits of known origin.

Sediment gravity flow behaviour and its implications for the nature of resultant deposits has been intensively studied (e.g. Hein, 1982; Lowe, 1982; Postma, 1986; Nemec, 1990; Major, 1997, 2000; Mulder and Alexander, 2001), and hence we can draw upon a firmly established knowledge base with respect to diamictons with mass flow origins. Diamicton forming flows are likely to be generated by the mobilisation of high-concentration, predominantly laminar flows, which contrast with low-concentration and turbulent or fluid flows. Whereas the latter generate fluid bedforms and are typically characterised by turbidites in subaqueous environments, high-concentration flows can deposit thick and massive beds, subaerially and subaqueously, and can also display faint, sub-horizontal shear laminae and water escape structures (Hein, 1982; Lowe, 1982; Postma,

---

Figure 15.1 Examples of schematic reconstructions of glacigenic depositional environments where a subaqueous (glacimarine) setting has been proposed for stratigraphic sequences around the southern margins of the former British–Irish Ice Sheet: (a) a glacimarine depositional model for the complexly glacitectonised Quaternary deposits of East Anglia, based upon the cliffs between Happisburgh and Cromer (from Eyles *et al.*, 1989); (b) schematic reconstruction of the style of deposition associated with a retreating tidewater ice margin based on observations around the Irish Sea Basin (from Eyles and McCabe, 1989a); (c) proposed model for glacimarine sedimentation along high-relief coasts during rapid deglaciation, based upon observations from valleys surrounding the northern Irish Sea Basin (from Eyles and McCabe, 1989a); (d) schematic diagram to show the processes and sediments that accumulate at the grounding line of a glacier terminating in an ice cliff (from Eyles and McCabe, 1991). (1) glacitectonised marine sediments, (2) subglacial till, (3) coarse-grained dropstone diamicton with evidence of current activity, (4) fine-grained dropstone diamicton, (5) proximal bouldery outwash, (6) subaqueous fan sands and gravels, (7) sediment gravity flow deposits, and (8) dumped supraglacial debris.

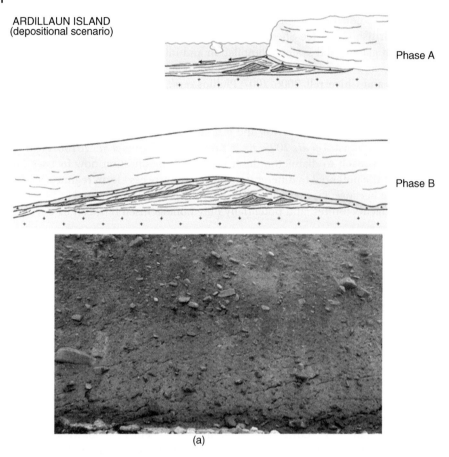

(a)

Figure 15.2 Case studies from the southern British–Irish Ice Sheet in which the role of subaqueous sedimentation, specifically in glacilacustrine environments, has been acknowledged in the production of thick diamicton sequences: (a) Reconstruction of the depositional sequence at Ardillaun Island, Connemara, western Ireland (from Evans *et al.*, 2015b). Phase A involved intermittent sub-marginal till wedge development (green) and subaqueous failure of the resulting push moraine to create sediment gravity flows (yellow), punctuated by periods of subaqueous fan sedimentation (orange). Photograph shows crudely stratified diamictons with shallow dipping bedding indicative of sediment gravity flows. (b) Reconstruction of the sequence of depositional events recorded at Waterville, southwest Ireland, where diamictons and associated sediments (LF1 and LF2) were laid down in grounding line fans dominated by sediment gravity flow, soft-sediment deformation and suspension settling during advance (green) and recession (blue) (from Evans *et al.*, 2012b). Supraglacial debris is red. A minor ice marginal oscillation (black) created glacitectonic structures and clastic dykes due to hydrofracture fill. Panel A represents the glacier maximum limit and panel D represents a subsequent readvance during overall recession. Panel E represents infilling of the freshly drained proglacial lake margins by glacifluvial outwash (LF 3). Upper photograph shows crudely to well-stratified mass flow diamictons. Lower photograph shows the 20-m-high section containing LF 1 subaqueous fan gravels and sands overlain by LF 2 gravity mass flow diamictons. (c) Stratigraphic architecture of the glacial sediments from Screen Hills area of the southeast Irish coast (from Thomas and Summers, 1983), showing the juxtaposition of subglacial tills, ice-contact subaerial and subaqueous fan sediments and mass flow deposits emplaced both subaerially and subaqueously, all associated with glacier marginal oscillations. Photographs show the sediments that typify this setting including: (left) the stratified diamicton of the Ballinclash Member, interpreted as subaqueous mass flow deposits; (right upper) broader-scale architecture of the Ballinclash Member, showing locally cross-bedded, sub-horizontal stacks of stratified to laminated diamictons; (right lower) basal Ballinclash Member, showing pseudo-laminated structure and sand boudins and discontinuous stringers, created by glacitectonisation of underlying Screen Member outwash deposits.

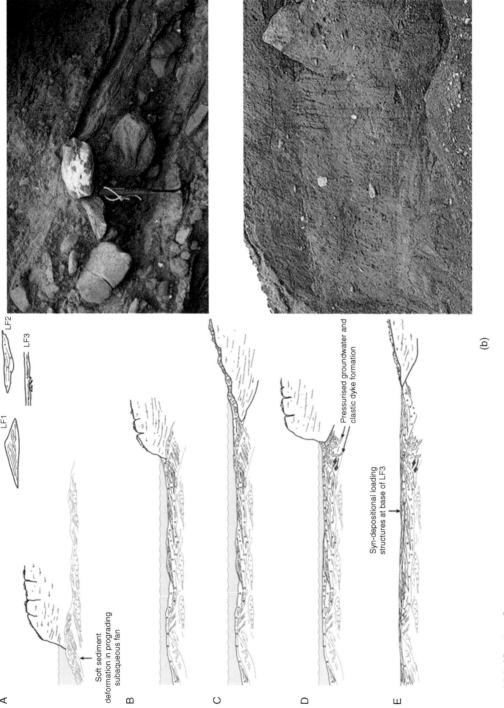

Figure 15.2 (*Continued*)

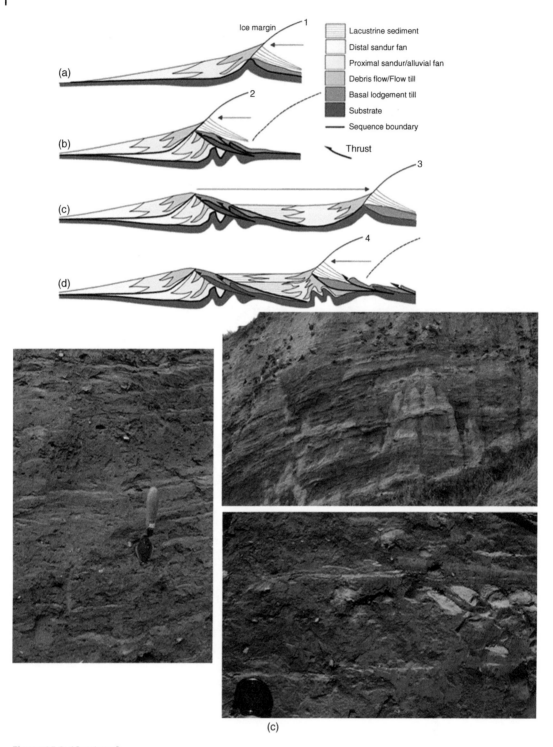

Figure 15.2 (*Continued*)

1986; Figure 15.3). Pre-existing materials that are subject to mass failure may remain cohesive or be arrested before they disaggregate and hence are classified as creeps, slides and slumps, resulting in predominantly low-strain and soft-sediment deformation structures. Increasing levels of disaggregation produce olistostromes or melange-type deposits and hyperconcentrated density flows (paraconglomerates), and then debris flows (diamictons), before the sediment–water mixture generates fluidal flows or grain flows. The importance of such mass flow processes in the generation of diamictons in glaciated basins has been likened to a cement mixer by Eyles and Eyles (2000; Figure 15.4a). The various sediment sources depicted in this analogy combine in slumps (cf. Boulton, 1968; Figure 15.4b) to produce increasingly more homogenised diamictons with greater mixing, creating a continuum from well-stratified or heterogeneous diamictons through weakly stratified or laminated diamictons to massive diamictons. The massive appearance of debris flows or debrites can be modified, as highlighted above, by shearing to produce centimetre to decimetre thick, bed-parallel laminae (Postma, 1986) and even striations have been reported from sediments underlying some landslide deposits (Gee *et al.*, 2005). The shearing that takes place at the base of mass flows (Figure 15.4c) can be directly contrasted with that taking place in subglacial deforming materials (cf. Figure 6.21), specifically in the reversal of the cumulative deformation profile, whereby the greatest displacement is observed towards the base of the deforming medium in debris flows (van der Wateren, 1995a; Hiemstra *et al.*, 2004).

The mass flow classification scheme of Lawson (1979a, b, 1982; Figure 15.5a), derived from supraglacial debris flows, is instructive in that it provides clear sedimentological signatures which relate to the varied subaerial flow processes, controlled by water content, in materials typical of glacial debris loads (Figure 15.5b). Type I flows create clast-rich, matrix-supported and massive, poorly sorted diamictons which may include blocks or intraclasts of pre-existing materials eroded and incorporated in the flow, and display concentrations of larger clasts at flow margins. Structures include zones of short, curving shear planes near flow bases and thin silt or sand horizons near the flow tops created by post-depositional water washing due to water expulsion from the flow; clast macrofabrics are very weak ($S_1 = 0.49-0.55$) with many vertically dipping clasts. Type II flows are subdivided into plug and shear zones, with the plug zone being massive, poorly sorted diamictons and the sheared zone containing clast concentrations in finer matrix due to sinking through the flow. The sheared zone also appears layered due to its textural differences with the overlying plug as well as including smeared silty clay inclusions. Clast macrofabrics are again predominantly weak ($S_1 = 0.50-0.65$) with numerous vertical clasts but relatively stronger than those of Type I flows. Type III flows are crudely stratified as a result of being matrix to clast supported, often lacking fine-grained matrix and displaying clear basal gravel zones. Hence, they appear as interbedded diamictons, clast concentrations and washed horizons as a result of pulsed flow, and individual clasts may be thicker than the whole deposit, projecting above the flow surface. Surges in flows will result in lateral variations in flow unit thickness. Clast macrofabrics are moderate to strong ($S_1 = 0.60-0.70$) with flow parallel and transverse orientations. Type IV flows are composed entirely of silt and sand matrix with the exception of their bases, which comprise granule gravels or coarse sand. This reflects the high water content and low matrix strength of the flow type, which also dictates the typically thin individual flow units (0.02–0.1 m). These units may build up thicker sequences of laminae over time. From this exhaustive assessment of debris flows, Lawson (1979a, b) was able to identify six distinct sedimentary zones, different combinations of which occur in each flow type (1–6 on sediment logs in Figure 15.5a): (1) texturally heterogeneous with increased tractional gravel content, massive to graded, with weak to absent macrofabric; (2) massive, texturally heterogeneous but with absence of large clasts due to settling, and weak macrofabric; (3) massive, texturally distinct and can contain structured sediments, with no obvious macrofabric but vertical clasts common; (4)

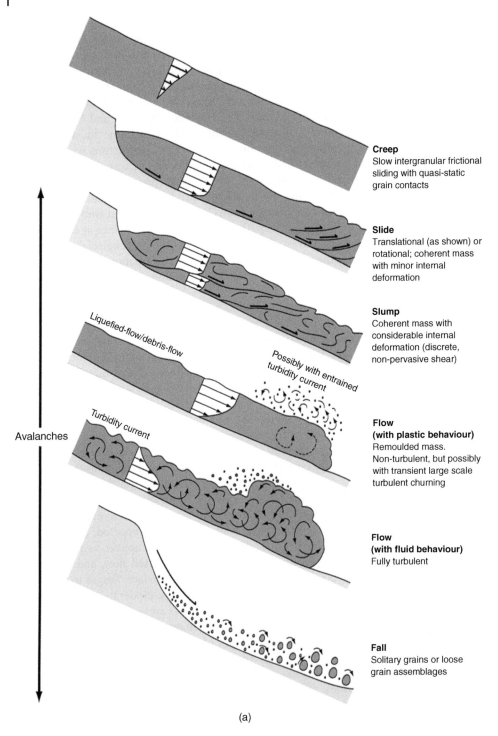

(a)

Figure 15.3 Classification schemes for mass flow types: (a) the range of gravity-driven sediment transport processes on subaqueous slopes (from Nemec, 1990); (b) the range of subaqueous sedimentary density flows, classified according to the criteria of dominant grain support mechanism, velocity profile, flow shape and schematic sedimentary logs (from Mulder and Alexander, 2001).

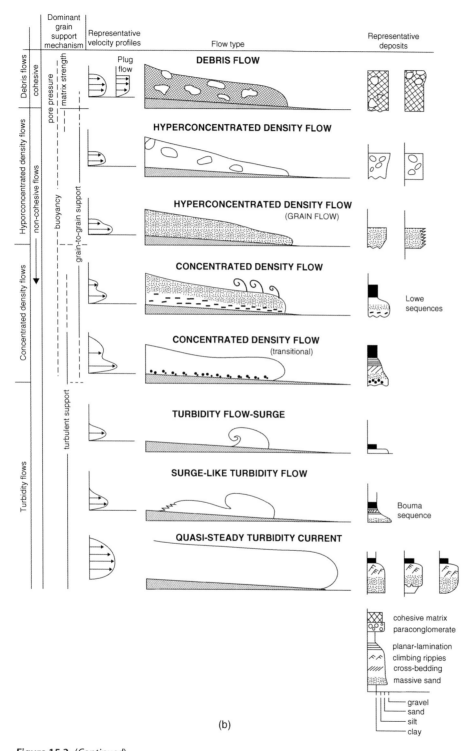

(b)

Figure 15.3 (*Continued*)

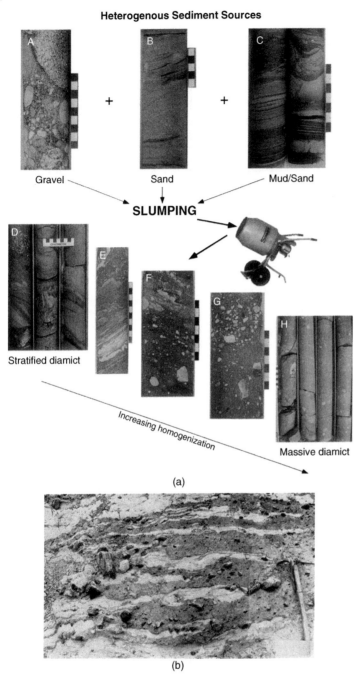

(a)

(b)

Figure 15.4 Mass flow processes and the generation of diamictons: (a) the 'cement-mixer' analogy for diamicton generation by downslope debris flow (from Eyles and Eyles, 2000); (A–C) various sediment sources feed material to slumps; (D) well-stratified diamicton of crudely intermixed and folded gravel, sand and mud; (E) well-stratified diamicton with folding of sand- and mud-rich units; (F) stratified diamicton with greater degree of homogenisation of sand- and mud-rich sediment; (G) weakly stratified diamicton with clast-rich horizon; (H) massive diamictons in which complete homogenisation of source materials has been achieved; (b) photograph of supraglacially derived, stratified mass flow diamictons observed on a Svalbard glacier snout by Boulton (1968); (c) summary sketch of the structures typically observed in the basal shear zones of debris flows (from van der Wateren, 1995a).

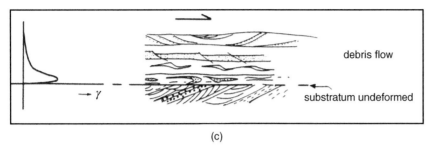

(c)

Figure 15.4 (*Continued*)

massive, fine-grained and lacks large clasts due to settling; (5) stratified to diffusely laminated silts and sands of meltwater flow origin; (6) massive to partially or fully graded silty sand with no obvious fabric. A range of basal contacts are evident between flows, including conformable, unconformable, sharp, gradational and deformed.

Unlike subaerial flows, subaqueous debris flows, when they are cohesionless, can display considerably more primary stratification due to water sorting and fluidal behaviour (e.g. Lowe, 1976a, b, 1982; Postma *et al.*, 1988; Sohn *et al.*, 1997; Talling *et al.*, 2004, 2012), facilitating genetic interpretation based upon diagnostic sedimentological signatures in the geological record (Figure 15.6). More problematic in terms of differentiating mass flow from till origins for diamictons are cohesive subaqueous debris flow deposits and potentially also hyperconcentrated debris flow deposits or paraconglomerates (Figure 15.3b). As these flows contain cohesive material or a clay matrix, they act as a fluid with cohesive strength and hence are characterised by a pseudo-plastic rheology (Johnson, 1965, 1970; Hampton, 1972, 1975, 1979; Rodine and Johnson, 1976; Lowe, 1979; Johnson and Rodine, 1984; Mulder and Alexander, 2001). The resulting deposits have been classified according to process–form regime by Lowe (1982) as either massive to inversely graded conglomerates, created by frictional 'freezing', or massive, matrix-supported pebbly mudstones, formed by the process of cohesive freezing. Also common are large clasts, boulders and blocks or intraclasts of pre-existing sediment which have been rafted along at high levels in the flow. Similar to Type I subaerial flows, the upper parts of cohesive subaqueous debris flows commonly consist of a semi-rigid plug riding passively over a thin basal shear zone, but their upper materials can become diluted due to turbulent mixing with the overlying water column, which can introduce an upwards increasing tendency to become stratified; such stratification may appear as normally graded and horizontally bedded gravels, sands and silts deposited from turbulent flows (turbidites). The basal shear zones can appear as attenuated clay or silt wisps, shear planes and normal or inverse grading, but can also develop into more substantial brecciated zones and melange diamictons (Visser *et al.*, 1984; Figure 15.7). In terms of architecture, cohesive debris flow deposits tend to form sheet-like vertically accreted beds, each bed typically decimetres to several metres thick and displaying planar or slightly scoured bases, and can extend for thousands of metres on, for example, glacially influenced submarine slopes (e.g. Laberg and Vorren, 2000; Dowdeswell *et al.*, 2008c; Ottesen *et al.*, 2008; Talling *et al.*, 2012; Talling, 2014). In more ice-proximal settings, 'mud aprons' (Kristensen *et al.*, 2009; Johnson *et al.*, 2013) can build up through repeat subaqueous mass flows generated on the tops of proglacially thrust glacimarine sediments, often appearing internally similar to glacitectonites and indeed being developed into glacitectonites where the ice-margin overrides apron tops. The presence of intraclasts derived from the substratum

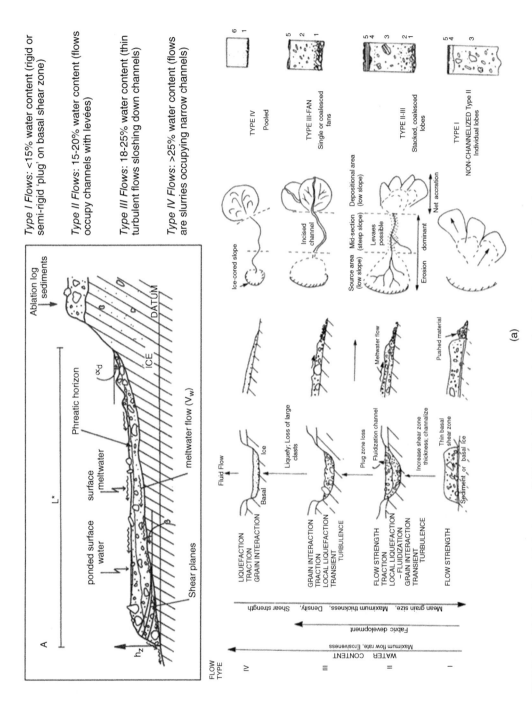

Type I Flows: <15% water content (rigid or semi-rigid 'plug' on basal shear zone)

Type II Flows: 15–20% water content (flows occupy channels with levées)

Type III Flows: 18–25% water content (thin turbulent flows sloshing down channels)

Type IV Flows: >25% water content (flows are slurries occupying narrow channels)

Figure 15.5 Subaerial debris flows: (a) the Lawson (1979a, b) classifcation scheme for debris flows observed on the Matanuska Glacier; (b) debris flow sediments recently deposited at the margins of Icelandic glaciers. Top panel shows fissile, compact subglacial till overlain by loose, friable and open framework, massive gravelly diamicton (large clast lies at junction of two units). The upper diamicton can be traced to a recently emplaced gravity mass flow derived from the melting cliff of the nearby Hrutarjökull snout. Middle panel shows compact and densely fissile subglacial till (below lens cap) overlain by crumbly, loose textured and massive gravelly diamicton created by a sediment gravity flow derived from the surface of the same till emerging from beneath the glacier snout visible in the background. Lower panel shows two separate exposures at the margin of Kvíárjökull in which crudely stratified gravelly diamictons document

(b)

Figure 15.5 (*Continued*)

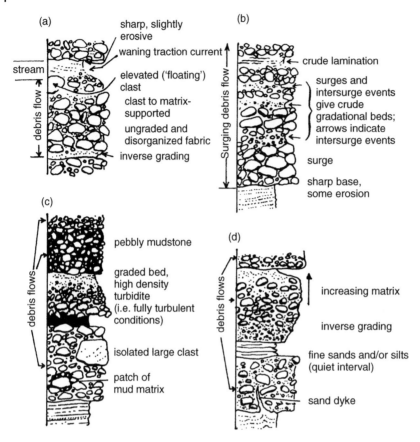

Figure 15.6 Examples of the different structures created by debris flows in subaerial (a, b) and subaqueous (c, d) settings (from Nemec, 1990).

are indicative of flow initiation by the downslope disaggregation of subaqueous slumps and slides. Incorporation of underlying materials during the passage of the flow may also modify a homogeneous, conglomoratic plug into a heterogeneous and chaotically bedded diamicton (Figure 15.8) or may lead to spatially variable matrix characteristics and/or the development of clast concentrations and discontinuous layers (Mulder and Alexander, 2001). The plug zones are otherwise typically composed of massive, matrix-supported diamicton or muddy sands and gravels (e.g. Carto and Eyles, 2012a).

Although the shear zones developed at the base of debris flows can be contrasted with those in subglacially deforming materials at macroscale (compare Figures 6.21 and 15.4c), the features developed at microscale are not as clearly diagnostic. Working on the Matanuska Glacier debris flows, Lachniet *et al.* (1999, 2001) identified a range of microscale features indicative of sediment deformation, fluid escape and incomplete mixing of source materials. They identified the fine lamination commonly observed at macroscale and related it to the development of a laminar flow fabric during ductile deformation, also evident in folds, pressure shadows, clast orientations around core stones, basal shear zones and imbrication and flow fabrics. Brittle styles of deformation were evident in shear faults and brecciation (Figure 15.9). A comparison between these microscale signatures and those of

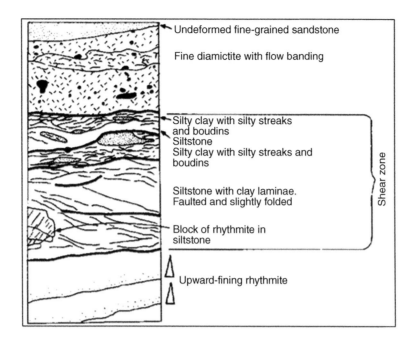

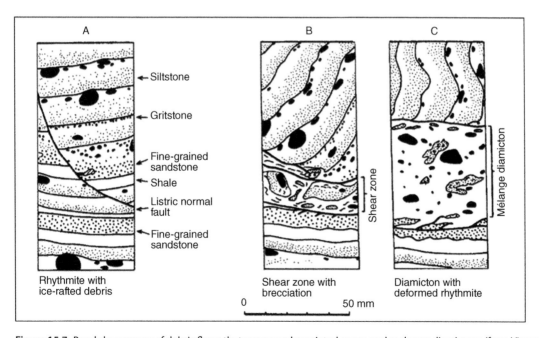

Figure 15.7 Basal shear zones of debris flows that appear as brecciated zones and melange diamictons (from Visser *et al.,* 1984).

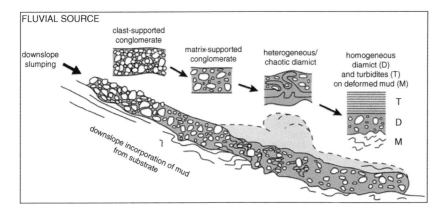

Figure 15.8 The origin of subaqueous debris flow diamictons and associated turbidite facies involving the episodic downslope slumping and mixing of underlying materials during the passage of the flow to produce a heterogeneous and chaotically bedded diamicton or homogenous diamictons (from Carto and Eyles, 2012a).

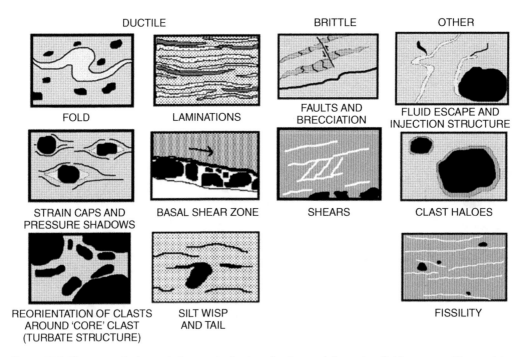

Figure 15.9 The range of microscale features indicative of sediment deformation, fluid escape and incomplete mixing of source materials in Matanuska Glacier debris flows (from Lachniet *et al.*, 2001).

subglacial materials reported by Menzies and Maltman (1992) and van der Meer (1993) allowed Lachniet *et al.* (2001) to isolate only basal shear zones and laminar flow fabrics as diagnostic of subaerial debris flows. Further diagnostic criteria were isolated by Menzies and Zaniewski (2003) based upon their comparisons between an ancient subglacial diamicton and a modern debris flow developed in it due to cliff erosion; these are 'tiled' and 'marbled' structures (Figures 4.9 and 10.17), thought to relate to flow deceleration and dewatering, although Kilfeather and van der Meer (2008) have related marble bed structures to pore space modification in shearing till (see Chapter 10). The interactions of soft-sediment plug deformation, basal shearing, sediment injection and water escape in debris flow emplacement have been investigated by Phillips (2006) in the micromorphological analysis of an ancient Lawson Type I debris flow (Figure 15.10). The range of features developed in the material underlying the flow deposit includes folds, thrusts and shears, as well as 'rotated' to slightly attenuated diamicton pebbles derived from the flow. The basal contact of the debris flow is not everywhere sharp but characterised also by elongate 'flames' of the substrate material separating lobate or pendant structures of the debris flow diamicton, all tilted downflow. Some detached 'flames' or ribbons of the substrate material occur in the flow basal material, as do indicators of rotational deformation, such as circular, arcuate and galaxy-like grain arrangements. Although none of these features is particularly diagnostic of a debris flow versus subglacial deformation origin, Phillips (2006) highlights that the macroscale context is clearly crucial to identifying the sediment origin. Additionally, the fluidisation of a thin basal layer of the substrate (Figure 15.10) is instructive in that it creates a zone of bedding-parallel hydrofracture fills.

The clast macrofabrics developed within debris flows clearly reflect localised clast alignment parallel or transverse to flow (e.g. Lawson, 1979a, b, 1982; Rappol, 1985; Eyles and Kocsis, 1988; Owen, 1991) to a strength compatible with subglacial tills but nonetheless reflecting in their high dip angles a significant degree of extreme particle rotation. This disturbance of particle alignment has been linked to the momentum exchange in debris flows by Iverson *et al.* (2008; cf. Iverson, 1997). In subaerial debris flows, macrofabrics range from isotropic in plug zones to moderately strong in basal shear zones. Clasts will also align themselves parallel with lateral and frontal margins of flows due to the compressive stresses experienced in those areas. Clast macrofabric shapes, when entered on ternary plots, reveal a similar range of signatures to those of subglacial tills with cohesionless subaqueous flows tending towards stronger, cluster-type fabrics (Figure 15.11).

In subaqueous settings, all of the above sediment gravity flows give rise to 'stratified diamictons' (Figure 15.12), a term that often appears somewhat contradictory because 'diamictons' are defined as poorly sorted materials. However, diamictons that appear crudely bedded at outcrop scale or locally contain laterally extensive beds of stratified sediment, accounting for more than 10% of the individual facies being classified, are often termed stratified diamictons, following the criteria set down by Eyles *et al.* (1983a). Using the same criteria, a massive diamicton is one that displays less than 10% stratification and is relatively rare outside subglacial settings. However, once stratified diamicton facies are sub-classified according to individual beds, massive diamicton units become more prevalent, and moreover some ice rafting and iceberg turbation processes can create significant thicknesses of apparently massive diamicton (Dowdeswell *et al.*, 1994). Such diamictons are problematic in glacial sedimentology because of their similarities to subglacial tills (e.g. Licht *et al.*, 1999), as demonstrated by the significant debate on the subglacial versus subaqueous origins of the extensive Great Lakes diamictons (cf. Evenson *et al.*, 1977; Gibbard, 1980; Eyles *et al.*, 1983b, 2005; C.H. Eyles and Eyles, 1983b; comments by Karrow, Dreimanis, Sharpe, Gravenor and reply by Eyles and Eyles, 1984b; Dreimanis *et al.*, 1987; Kelly and Martini, 1986; Hicock, 1992, 1993; Eden and Eyles, 2001; Dreimanis

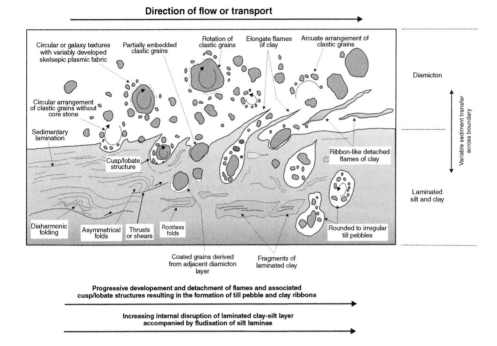

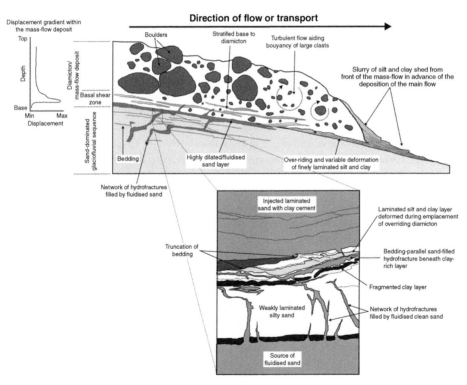

Figure 15.10 Details of sedimentary structures identified in a diamictic debris flow derived from glacigenic materials by Phillips (2006): Upper figure is a schematic cross-section sketch through the debris flow deposit showing the stratified basal shear zone and the network of sand and silt-filled hydrofractures that penetrate the underlying sand, the details of which are depicted in the inset box. Lower figure is a schematic summary sketch of the range of microstructures developed within the stratified base of the diamicton and the progressive development of cusp and flame structures and inclusion of till pebbles within the highly disrupted and homogenised clay–silt bands.

Figure 15.11 Clast macrofabric shape ternary diagram for cohesionless subaqueous flows and subaerial debris flows (from Benn, 2004a).

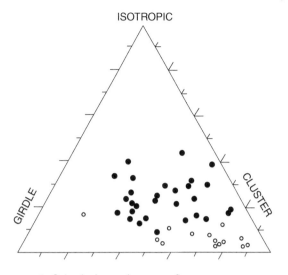

○ Cohesionless subaqueous flows
● Subaerial debris flows

Figure 15.12 Stratified diamictons: (top left) thin, clast-rich diamictons separated by clast lags and poorly sorted gravel beds, Coosderrig, western Ireland, deposited by repeated sediment gravity flows; (top right) crudely stratified boulder-rich diamictons overlying horizontally bedded matrix-supported gravels, laid down by ice-proximal debris flow-fed fans, Lake Pukaki, New Zealand; (bottom left) crudely stratified to pseudo-laminated diamicton, Filey Bay, Yorkshire, England, likely deposited as ice-proximal subaqueous mass flows; (bottom right) heterogeneous diamicton related to ice-proximal sediment gravity flow deposition, Southern Alberta, Canada.

and Gibbard, 2005). Ice rafted debris (IRD), for example, can create a range of deposits from laminated muds with dropstones to stratified diamictons and massive diamictons. Where a subaqueous origin can be proven, for example, in a stratigraphic context that displays conformable aggradation of deep-water sediments, such deposits are termed 'dropstone diamictons' (see Powell, 1984; Ó Cofaigh and Dowdeswell, 2001 and references therein). Dropstone diamictons are similar to, and indeed created in the same way as, suspension sediments or mud drapes with dropstones, hence Powell (1984) proposes that 10% clasts per unit area is the dividing line between the two classifications.

In addition to their association with stratified subaqueous beds, dropstone diamictons and subaqueous mass flow diamictons can be recognised by their internal sedimentological properties. For example, Domack and Lawson (1985), Smith (2000) and Evans *et al.* (2007; see Figure 4.14c) have presented clast macrofabrics for dropstones (Table 15.1) in an attempt to provide a diagnostic criterion for their recognition where other sedimentological characteristics are ambiguous. Such studies have confirmed previously identified tendencies for dropstones to display steep dip angles (e.g. Lavrushin, 1968; Griggs and Kulm, 1969; Spencer, 1971; Dalland, 1976; Gibbard, 1980), although this is not in every case the norm. This is because the fabric will be influenced by the character of the bottom sediment into which the clasts penetrate. For example, relatively stiff bottom sediments will allow the penetration of clasts that settle through the water column with their heavier ends pointing downwards. This will potentially preserve the high dip angles adopted by the clasts during their descent through the water column. In contrast, very soft bottom sediments, because they cannot hold clasts upright, will allow clasts to fall sideways after they impact the bottom. Alternatively, clasts may fall sideways because they cannot penetrate stiffer bottom sediments. Hence, dropstone diamictons may display high clast dip populations or girdle fabrics (Figure 15.13) and overall are difficult using macrofabric alone to distinguish from debris flows (cf. Figure 15.11).

Distinguishing subaqueous massive diamictons from subglacial matrix-supported tills is becoming increasingly undertaken using micromorphology. For example, Carr (2001, 2004) reports direct comparisons between glacimarine sediments and glacitectonite derived from those sediments as well as subglacial diamictons from the same tidewater glacier setting in Svalbard. The range of deformation structures regularly observed at microscale in subglacial materials (e.g. crushed quartz grains, grain lineations, pressure shadows, augen- or boudin-shaped instraclasts, unistrial and masepic fabric domains, rotational features and unidirectional fabrics; cf. Figures 4.9 and 10.15–10.17) could be contrasted with those in glacimarine sediments, such as winnowed plasma texture, dropstones,

Table 15.1 Clast macrofabric strengths (S_1 eigenvalues) for dropstones and dropstone diamictons from a range of field case studies (A-axis data unless listed otherwise).

Case study	Sediment type	Clast dip angles >45°	S_1 eigenvalue range	S_1 eigenvalue mean
Domack and Lawson (1985)	Fossiliferous glacimarine diamicton	4–32%	0.48–0.67	0.53
Smith (2000)	Glacilacustrine diamictons	15–61%	0.41–0.57	0.47
Evans *et al.* (2007)	Fossiliferous glacimarine diamictons and cyclopsams – A/B planes	0–26%	0.44–0.68	0.57
Evans *et al.* (2007)	Glacilacustrine diamictons and rhythmites – A/B planes	54%	0.53	N/A

Figure 15.13 Clast macrofabric shapes for glacilacustrine diamictons, using the data from Domack and Lawson (1985) for IRD and Lawson (1979a) for debris flows combined with diamictons from a high Arctic lake collected by Smith (2000).

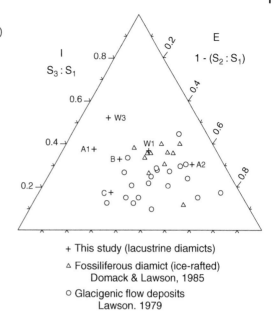

+ This study (lacustrine diamicts)

△ Fossiliferous diamict (ice-rafted)
  Domack & Lawson, 1985

○ Glacigenic flow deposits
  Lawson. 1979

in situ microfossils, a lack of shear deformation structures, graded bedding, a lack of plasmic fabric and high microfabric dip angles similar to those recognised in clast macrofabrics (Figure 15.14). Using macroscopically massive diamictons from a range of known sedimentary origins, Kilfeather *et al.* (2010) further identified specific micromorphological criteria for subaqueous sedimentation (Figure 15.15), including some apparently diagnostic features such as: (a) vertical lineations created by the passage of dropstones through soft bottom sediments; (b) the development of bimodal fabrics of near-horizontal and near-vertical alignments in debris flow deposits, relating to gravity-driven shearing and grain sinking through matrixes, respectively; and (c) laminated clay and silt coatings on clasts in debris flow deposits, relating to continuous rotation of clasts in buoyant, non-plug conditions. Kilfeather *et al.* (2010) also stress, however, that the deformation structures observed in iceberg turbated deposits (e.g. Woodworth-Lynas and Guigne, 1990; Dowdeswell *et al.*, 1994; Eden and Eyles, 2001; Eyles *et al.*, 2005) are predominantly impossible to differentiate from those of subglacial deformation based upon micromorphology alone. A good example of combining macro- and microscale analyses to circumvent this potential problem is that of Busfield and Le Heron (2013), who have provided a schematic model of subaqueous sediment gravity flow features in relation to a typical vertical deformation profile and contrasted this with what they observe in subglacially deformed diamictons in Neoproterozoic outcrops in Namibia (Figure 15.16).

Soft-sediment deformation at all scales can prove to be problematic in interpreting diamicton sequences. For example, soft-sediment deformation structures indicative of low strains were cited by Eyles and McCabe (1989a, b) and Eyles *et al.* (1989) in their models of subaqueous mass flow origins for stratified and pseudo-stratified diamictons (see also Thomas and Summers, 1982, 1983; Thomas and Kerr, 1987; Lunkka, 1994; Ó Cofaigh and Evans, 2001a, b; Evans and Ó Cofaigh, 2003; Thomas and Chiverrell, 2006, 2007; Phillips *et al.*, 2008; Ó Cofaigh *et al.*, 2011; Evans *et al.*, 2012a, b; Figures 15.1 and 15.2), but the rapid sedimentation rates that characterise ice-proximal glacigenic depo-centres can initiate significant large-scale, gravitationally induced loading and down-slope mobilisation of sediments in both subaerial and subaqueous environments. An excellent example is

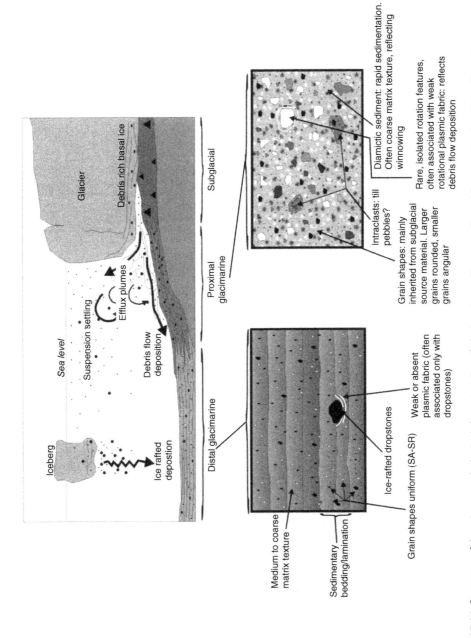

Figure 15.14 Summary of the micromorphological characteristics of glacimarine sediments collected from proximal and distal locations near a Svalbard tidewater glacier margin by Carr (2004).

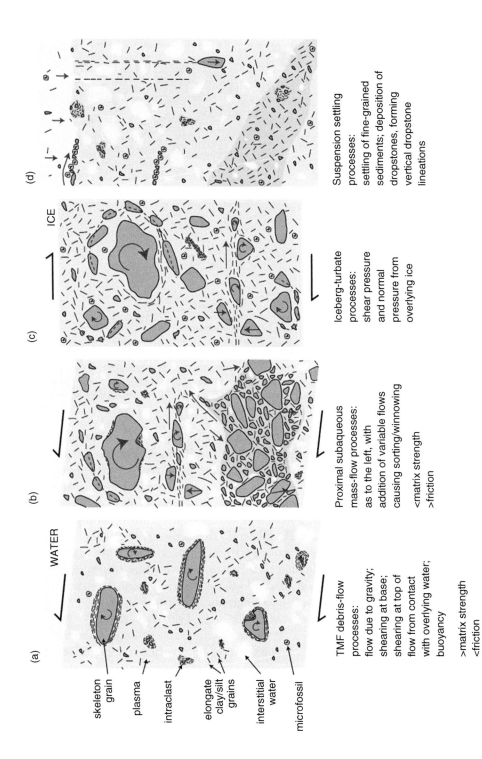

Figure 15.15 Micromorphological criteria that can be employed in differentiating subaqueous and other origins for massive diamictons (from Kilfeather *et al.*, 2010).

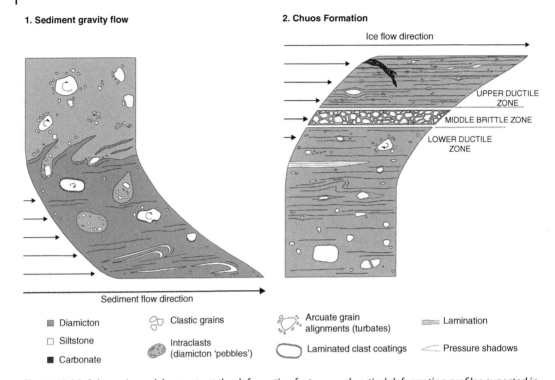

1. Sediment gravity flow

2. Chuos Formation

Ice flow direction

UPPER DUCTILE ZONE

MIDDLE BRITTLE ZONE

LOWER DUCTILE ZONE

Sediment flow direction

■ Diamicton   Clastic grains   Arcuate grain alignments (turbates)   Lamination

□ Siltstone   Intraclasts (diamicton 'pebbles')   Laminated clast coatings   Pressure shadows

■ Carbonate

Figure 15.16 Schematic model to contrast the deformation features and vertical deformation profiles expected in subaqueous sediment gravity flows and subglacial shearing zones based on the Neoproterozoic Chuos Formation, Namibia (from Busfield and Le Heron, 2013).

the extensive outcrop of distorted diamictons and gravelly stratified sediment at Traeth-y-Mwnt, southwest Wales (Figure 15.17), interpreted by Rijsdijk (2001) and Hiemstra *et al.* (2005) as spectacular syn-sedimentary gravity-induced loading, whereby aggrading glacifluvial outwash progressively loaded underlying low-viscosity, subaqueous deposited diamictons to produce synformal folding, teardrop- and block-shaped gravel rafts and clast stringers extending downwards into the diamicton. Similar pendant- and teardrop-shaped gravel bodies were described from the top of subglacial tills by Evans *et al.* (1995; Figure 15.18), who traced them laterally to less disturbed glacifluvial outwash deposits and hence interpreted them as gravity loading features produced by the increasing weight of the outwash over unconsolidated and saturated tills. This took place where tills were emplaced in a poorly drained sub-marginal environment similar to those observed around Icelandic glacier margins today where tills are squeezed upwards into longitudinal crevasses (cf. Price, 1970; Evans and Rea, 2003; Christoffersen *et al.*, 2005; Evans *et al.*, 2010, 2015a; Eyles *et al.*, 2015).

In the soft-sediment deformation cases above, stratigraphic context is critical in determining the origins of diamictons and associated sediments. When observed in restricted stratigraphic contexts, determining glacial versus non-glacial origins for stratified and massive diamictons is more challenging. In some settings, it may be simply determined by the lack of striated or subglacially shaped clasts (e.g. Carto and Eyles, 2012a). The nature of associated stratified deposits may also be diagnostic, even in restricted outcrops where, for example, turbidite facies lie between diamictons and hence clearly indicate a subaqueous mass flow or dropstone origin (e.g. Carto and Eyles, 2012b). Over restricted or larger outcrops, the stratigraphic signatures of subaqueous and subaerial debris

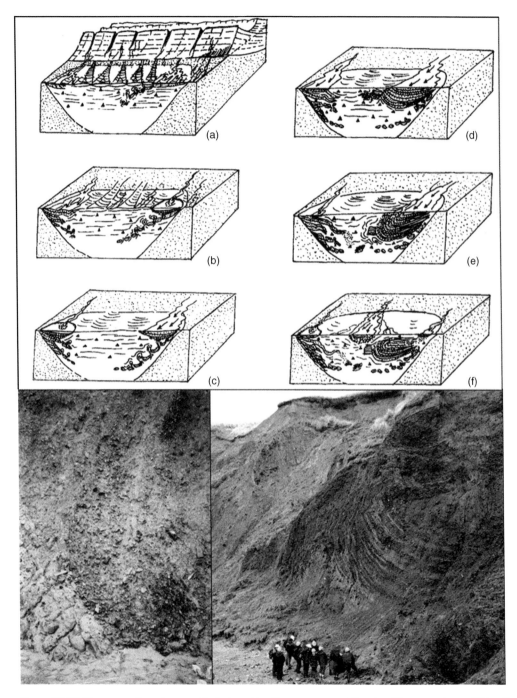

Figure 15.17 The syn-sedimentary gravity-induced loading structures at Traeth-y-Mwnt, southwest Wales (from Rijsdijk, 2001). Upper sequence of diagrams provides a reconstruction of the evolution of the structures: (a) filling of rock basin with glacigenic debris flows derived from stagnant melting ice and onset of density-driven deformation in gravels that cap the sequence; (b) density-driven faulting of gravel sheets and sinking of detached gravel blocks to form rafts; (c) lacustrine sedimentation in depression formed by sinking gravels; (d) syndeformational sedimentation in synclinal fold; (e) downward flexuring of gravels forming a fold; (f) final distal paraglacial redistribution of sediments. Lower left photograph shows a gravel tear-drop shaped structure that has sunk into underlying diamicton. Lower right photograph shows the large synclinal loading structure developed in stratified gravel outwash.

Figure 15.18 Pendant- and teardrop-shaped gravel loading features, overlain by rhythmites, at the top of the Skipsea Till, Yorkshire, England.

flows can be identified in vertical and lateral changes in facies characteristics that record flow dilution and/or transformation. This has been summarised by Sohn *et al.* (2002) based upon Cretaceous submarine channel deposits (Figure 15.19), where transformations between debris flow diamictons, turbidites, hyperconcentrated flow and stream flow deposits can be identified by facies changes. Difficulties may still arise where inter-diamict stratified layers resemble both subglacial canal fills (Figures 11.2–11.6) and debris flow surface stratification (Figure 15.5), although thick and laterally extensive sequences of multiple tills and canal fills are not likely to survive without undergoing at least some glacitectonic deformation (e.g. Figures 11.3 and 11.4).

At large or even basin-wide scales, ice rafted and subaqueous and/or subaerial mass flow origins for diamictons are more clearly demonstrated, as we shall now review before investigating the regional architecture of true tills and glacitectonites in Chapter 16. Extensive exposures containing diamictons of demonstrable subaqueous origin are documented from a range of large glaciated valley settings in which subglacial tills are rare. For example, in northwesternmost North America, Ferrians (1963), Eyles (1987) and Bennett *et al.* (2002) report thick and extensive stratified sediment infills that record glacigenic sedimentation in intermontane trenches, which include significant diamicton facies related to subaqueous sediment gravity flows. In the Upper Fraser River Valley of British Columbia, Eyles (1987) reports multiple units of thick diamictons and intervening but less substantial stratified sediments, all indicative of subaqueous failure of glacilacustrine sediment piles and resultant slump- and slide-generated debris flows (Figure 15.20). The diamictons are up to 10 m thick, unusual for subglacial tills, but locally display characteristics that are diagnostic of tills, including fluted basal scours, large rafts or intraclasts of pre-existing Tertiary materials and substrate deformation. A subaqueous origin for the diamictons is clear, however, in the occurrence of diamicton intercalations with stratified sediments characterised by laminations and dropstones, as well as typical debris flow features such as crude stratification at diamicton tops and attenuated intraclasts at their bases indicative of flow base laminar shear.

(a) Subaerial debris flow with flow dilution at the leading edge

(b) Subaerial debris flow with progressively dilute rear part

(c) Subaqueous nonhydroplaning debris flow

(d) Subaqueous hydroplaning debris flow

Debris-fall blocks

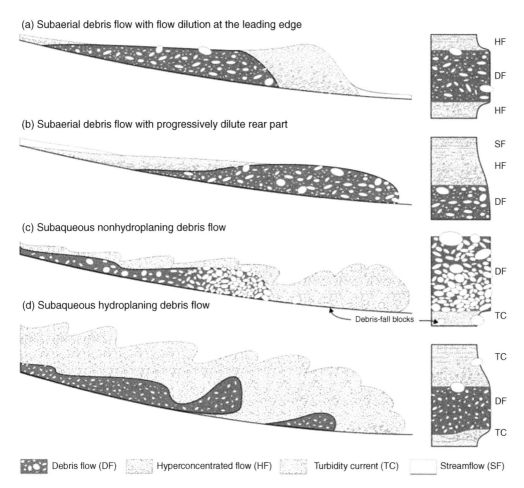

HF
DF
HF

SF
HF
DF

DF
TC

TC
DF
TC

Debris flow (DF)   Hyperconcentrated flow (HF)   Turbidity current (TC)   Streamflow (SF)

Figure 15.19 Schematic diagrams to show the four possible types of multiphase flows generated by the dilution of debris flows in subaerial and subaqueous settings; (a) subaerial debris flow, diluted at the leading edge by a streamflow, resulting in hyperconcentrated flow deposits overlain by debris-flow deposits; (b) incremental aggradation from subaerial debris flows, coarsest in the flow head and progressively more dilute and finer-grained toward the tail, resulting in a debris-flow deposit overlain by hyperconcentrated flow and streamflow deposits; (c) subaqueous debris flows, non-hydroplaning because of extremely permeable fronts, subject to mainly surface transformation. The surface-transformed suspended-sediment flows and debris-fall blocks can outpace the parental debris flows, resulting in outsized clast-bearing turbidites beneath debris-flow deposits; (d) subaqueous debris flows with impermeable fronts hydroplane and their flow fronts are repetitively detached and diluted to form voluminous turbidity currents. The turbidity currents outpace or are outrun by the debris flows, resulting in extensive turbidites beneath and above the parental debris-flow deposits.

The extensive and thick stratigraphic sequences in the Copper River Basin of Alaska have been ascribed a similar subaqueous origin also by Eyles (1987), based upon an architecture of multiple stacked diamictons and intervening lenses of turbidites and massive gravels. This location has more recently been described in detail by Bennett *et al.* (2002), who confirm a subaqueous origin for the basin infill and highlight various facies associations (FA), two of which in particular are critical to

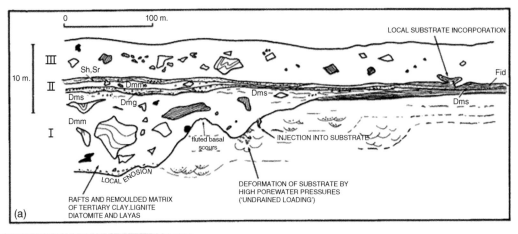

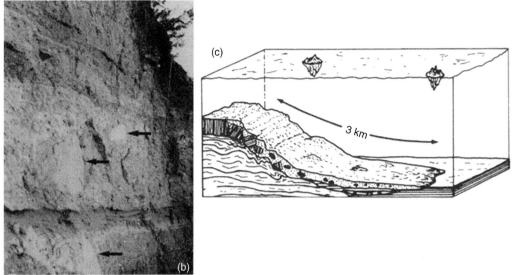

Figure 15.20 The products of slump- and slide-generated subaqueous debris flows in the Upper Fraser River Valley, British Columbia (from Eyles, 1987): (a) schematic section sketch of the main lithofacies I-III; (b) multiple units of thick diamictons separated by thin stratified sediment bodies; (c) reconstruction of the style of mass failure and re-sedimentation of the subaqueous sediment pile.

the interpretation of the diamicton-forming depositional environment (Figure 15.21). FA 1 includes massive and lenticular bedded diamictons with interbeds of stratified diamicton and laminated sand/silt and deformation structures. These components are interpreted as dropstone and subaqueous debris flow diamictons interrupted by bottom current activity and operating in an environments of suspension sedimentation and slumping. FA 2 comprises massive and graded diamictons, with matrix-supported gravels and rhythmites, together interpreted as the products of subaqueous debris flows that grade downslope into turbidites. In a basin-wide context, these facies associations are the more ice-proximal deposits within a stratigraphic sequence that documents subaqueous fan sedimentation, the diamictons being fed directly from subglacial tills emerging from beneath the

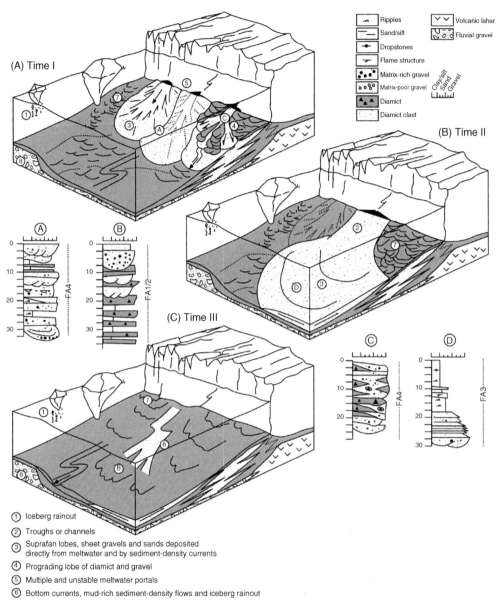

Ripples
Sand/silt
Dropstones
Flame structure
Matrix-rich gravel
Matrix-poor gravel
Diamict
Diamict clast
Volcanic lahar
Fluvial gravel
Clay/silt Sand Gravel

(A) Time I

(B) Time II

(C) Time III

① Iceberg rainout
② Troughs or channels
③ Suprafan lobes, sheet gravels and sands deposited
   directly from meltwater and by sediment-density currents
④ Prograding lobe of diamict and gravel
⑤ Multiple and unstable meltwater portals
⑥ Bottom currents, mud-rich sediment-density flows and iceberg rainout
⑦ Diamict sheet of tesselated debris flow lobes, derived
   from the release of sediment transported to the ice margin
   subglacially
⑧ Pre-glacial fluvial terrace

Figure 15.21 The subaqueous deposits of the Copper River Basin, Alaska (from Bennett *et al.*, 2002): (a) schematic reconstructions of the subaqueous depositional environment, identifying main facies associations; (b) photographs of the main diamictic deposits including: (i) two units of massive diamicton separated by a laminated fine sand interbed with channel form (Facies Association-1); (ii) lenticular and trough-shaped diamict units (Facies Association-1); (iii) soft-sediment clast within trough-shaped diamicton units (Facies Association-1); (iv) units of massive diamicton with variable clast concentrations (Facies Association-1); (v) graded and massive diamictons in stratigraphic succession (Facies Association-2); (vi) trough-shaped diamicton units (Facies Association-1).

(b)

Figure 15.21 *(Continued)*

glacier grounding lines. Similar basin-wide subaqueous diamicton architectures have been reported in intermediate upland relief (e.g. Benn, 1996; Figure 15.22a), ice sheet marginal glacimarine settings (e.g. McCabe, 1986; Ashley *et al.*, 1991; Figure 15.22b) and in active alpine settings (e.g. Mager and Fitzsimons, 2007; Evans *et al.*, 2013). In the latter case, the proximity of glacier ice, very high debris turnover and development of lakes that are at least partially supraglacial, combine to give rise to huge vertical and horizontal variability in diamicton characteristics, structures and architecture (Figure 15.23) due to accumulation in rapidly aggrading subaqueous debris flow-fed depo-centres with a background of continuous suspension settling and iceberg rafting and syn-sedimentary soft-sediment deformation (Evans *et al.*, 2013).

The large volumes of debris that are turned over at tectonically active alpine glacier margins such as those in New Zealand are important in the development of thick sequences of mass flow diamictons in large latero-frontal moraines and debris flow-fed ice-contact fans (Owen and Derbyshire, 1989; Owen, 1991; Benn *et al.*, 2003), which often constitute the only glacigenic diamictons (i.e. an almost total absence of till) in some glaciated basins. Examples of the mass flow diamictons and their stratigraphic architectural contexts created in such settings are well exposed in the latero-frontal moraine loops of the west coast of New Zealand's South Island, where thick sequences of clino-forms containing clast- and matrix-supported diamictons interbedded with poorly sorted gravels and boulder- to cobble-sized rubble beds and occasional discontinuous sandy gravel stratified lenses record aggradation of debris flows from debris-charged glacier snouts (Evans *et al.*, 2010b;

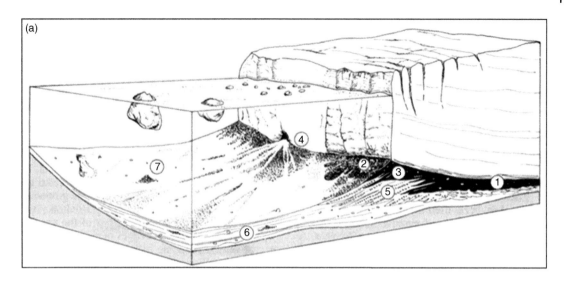

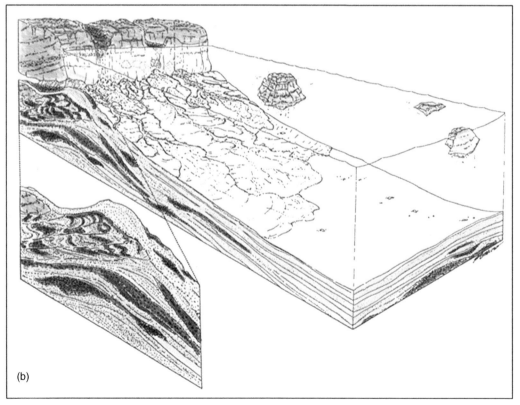

Figure 15.22 Reconstructions of subaqueous depo-centres in which diamictons are ubiquitous: (a) intermediate upland relief in the Scottish Highlands (from Benn, 1996); (b) ice sheet marginal glacimarine environment on the coast of Maine (from Ashley *et al.*, 1991).

Figure 15.23 Subaqueously deposited diamictons in the former proglacial/supraglacial lake of the Tasman Glacier, Lake Pukaki, New Zealand: (a) crudely stratified Dmm/Dms; (b) lenticular, cross-cutting units of Dms and gravel/sand interbeds, showing local soft-sediment deformation; (c) weakly stratified diamicton with thin silty, sand bed highlighted; (d) interbedded rhythmites and gravelly to matrix-supported mass flow diamictons with numerous dropstones. Camera lens cap for scales above vertically aligned clast over which a mass flow diamicton has been draped; (e) rhythmically bedded lake sediments with dropstones interbedded with stratified gravelly mass flow deposits and arranged in crude clinoforms and overlying massive to crudely bedded, clast-rich diamictons; (f) fine gravel lens in weakly stratified to massive diamicton.

Figure 15.24 Subaerially deposited mass flow diamictons in the latero-frontal moraines of the west coast of South Island, New Zealand: (a) crudely stratified, bouldery diamictons separated by discontinuous beds of laminated fines; (b) detail of sandy, silt lens between bouldery diamictons; (c) stacked sequence of crudely bedded, boulder to cobble gravel diamictons separated by stratified gravelly sand units; (d) clast-supported bouldery diamicton; (e) contorted silt laminae lens within a bouldery, clast-supported diamicton, indicative of surface ponding of expelled water in gravity mass flow that was later overrun by further mass flow material.

Figure 15.24). Such deposits commonly create a stratified appearance to lateral moraine outcrops (Boulton and Eyles, 1979; Small, 1983; Owen and Derbyshire, 1993; Lukas, 2012; Lukas *et al.*, 2013; Figure 15.25), where the term 'till' has traditionally, but inappropriately, been applied as a genetic classification. Even in lower-relief terrains, the sedimentological characteristics and stratigraphic architecture of moraines often indicate diamicton deposition by gravity mass flow (Lukas, 2005; Benn and Lukas, 2006; Reinardy and Lukas, 2009; Lukas and Sass, 2011; Figure 15.26), a process that

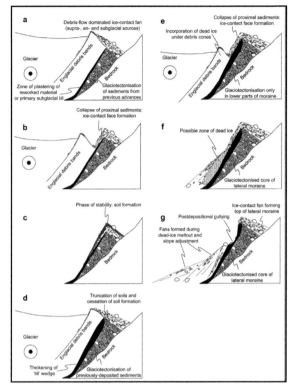

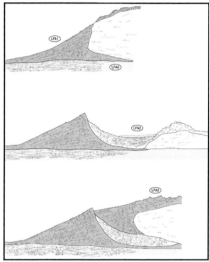

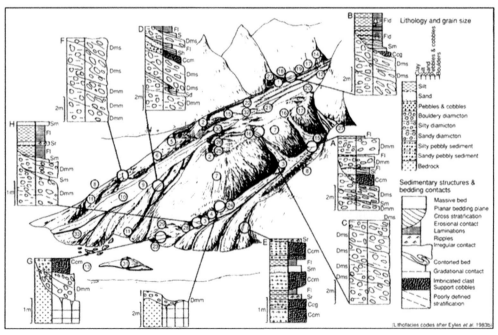

Figure 15.25 Models of latero-frontal moraine construction: (top left) conceptual model of lateral moraine formation based on observations at Findelengletscher by Lukas *et al*. (2012); (top right) reconstruction of the general depositional sequence involved in the production of the latero-frontal moraine loops of the west coast of New Zealand's South Island (from Evans *et al*., 2010). Upper panel shows initial glacier advance over proglacial outwash deposits (LFA 0) and deposition of debris flow-fed ice-contact fans (LFA 1). Middle panel shows ice recession and deposition of glacilacustrine sediments interdigitated with subaerial to subaqueous sediment gravity flow deposits (LFA 2) on the proximal faces of latero-frontal moraines. Lower panel shows glacier readvance, resulting in glacitectonic disturbance and hydrofracture filling of LFAs 1 and 2 and the deposition of debris flow-fed fans (LFA 3) sourced from monolithological supraglacial debris that originated as rock slope failure; lower figure) the landforms and sediments of the high-relief glaciated valley land system (from Owen and Derbyshire, 1993), in which diamictons produced by mass movements dominate the depositional signature: (1) truncated scree; (2) and (5) latero-frontal dump moraine; (3) laterally drained outwash channel; (4) glacifluvial outwash channel; (6) debris flow cones; (7) slide-modified lateral moraine; (8) abandoned lateral outwash fan; (9) meltwater channel; (10) meltwater fan; (11) abandoned meltwater fan; (12) bare ice; (13) trunk valley river; (14) debris flow; (15) flow slide; (16) gullied lateral moraine; (17) lateral moraine; (18) 'ablation valley' lake; (19) 'ablation valley'; (20) supraglacial lake; (21) supraglacial stream; (22) ice-contact terrace; (23) subglacial till exposure; (24) roche moutonnée; (25) flutings; (26) diffluence col; (27) high-level till remnant; (28) diffluence col lake; (29) fines washed out from supraglacial debris; (30) ice-cored moraines; (31) alluvium; (32) supraglacial debris; (33) dead ice.

Figure 15.26 Sedimentological characteristics and stratigraphic architecture of moraines dominated by gravity mass flow diamictons: (top) gravelly diamicton and poorly sorted gravel clinoforms at the core of a Younger Dryas moraine hummock, Seathwaite, English Lake District; (lower left) massive, clast-rich diamicton overlain by crudely stratified boulder gravel and clast-supported diamictons and stratified diamicton, Lake Coleridge, New Zealand; (lower right) crudely stratified, clast-rich diamictons overlain by horizontally bedded outwash fan gravels, Lake Coleridge, New Zealand. The Lake Coleridge diamictons were deposited in ice-contact debris flow-fed fans at the end of the last glaciation, and have been traditionally regarded as the local 'till'.

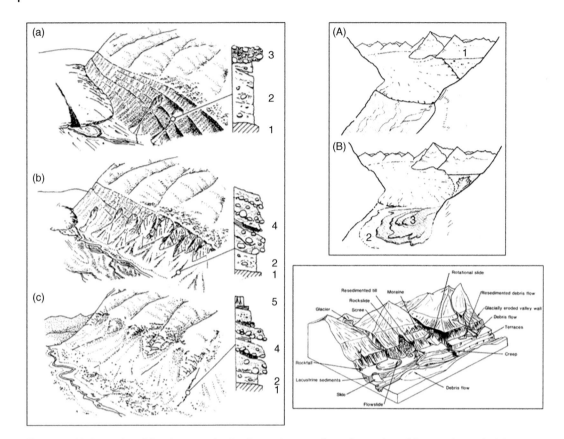

Figure 15.27 Examples of diamicton production by gravity mass flows due to slope failure and paraglacial denudation processes: (left panel) schematic temporal reconstruction of paraglacial landform modification and sediment reworking of drift-mantled slopes in freshly deglaciated terrain (from Ballantyne and Benn, 1996) – (a) initial slopes exposed by glacier recession, showing lateral moraines and the onset of gully incision; (b) advanced gully development and deposition of coalescing debris fans downslope; (c) exposed bedrock and stabilised, vegetated gullies and largely relict debris fans due to restricted debris supply. Facies are: (1) bedrock; (2) subaerial sediments relating to an earlier episode of paraglacial sedimentation; (3) ice-marginal deposits; (4) paraglacially reworked sediment (debris flows and intercalated slopewash deposits); (5) soil horizons; (upper right panel) reconstruction proposed for the thick sequences of diamictons in the Bow Valley, near Banff, Alberta, Canada, by Eyles *et al*. (1988b). (A) retreating valley glacier ponds water in tributary valley at location 1 and deposits proglacial outwash in the main valley; (B) glacier thinning releases floodwater from the tributary valley (2) followed by debris flows (3) generated by failures in the freshly exposed lake sediments; (lower right panel) schematic reconstruction of paraglacial adjustment processes and forms in the Karakoram Mountains by Owen (1991), showing the dominance of mass movements generated by the failure of valley-side drift accumulations, especially after they are incised by downcutting meltwater streams.

is immediately evident in the construction of modern push moraines and minor ice-contact fans (Figures 3.2 and 15.25).

Beyond these ice-contact depositional processes, another reason why the most ubiquitous diamicton-forming agencies in glaciated basins might be those associated with gravity mass flowage is that freshly deglaciated surfaces are subject to intensive paraglacial reworking (e.g. Eyles and Kocsis, 1988; Ballantyne and Benn, 1994, 1996; Harrison and Winchester, 1997; Curry and Ballantyne, 1999; Ballantyne, 2002a, b, 2003; Menzies and Zaniewski, 2003; Figure 3.2). Inherent instability in a range of glacigenic and non-glacigenic deposits can trigger gravity mass flows at local slope (Ballantyne and Benn, 1994, 1996; Curry and Ballantyne, 1999) to basin-wide (Eyles *et al.*, 1988b; Owen, 1989, 1991; Derbyshire and Owen, 1990; Owen and Sharma, 1998) scales, giving rise to potentially laterally extensive and thick, stacked sequences of massive to weakly stratified diamictons, separated by thin and often discontinuous stratified sediments created by intervening periods of debris flow surface fluvial modification (Figure 15.27). Such deposits have traditionally proven controversial, as both subglacial and mass flow origins have been argued for the same outcrops, especially where diamictons are predominantly massive and contain clasts with subglacial wear signatures (cf. Eyles *et al.*, 1988b, 1990; Mandryk and Rutter, 1990). Where attempts have been made to differentiate subglacial and paraglacial debris flow diamictons some diagnostic criteria have been isolated. At macroscale, simple sedimentary architectural features, such as significant unit thickness and slope parallel bedding, as well as smaller-scale details, such as downslope orientated macrofabrics, slope-parallel fluvial interbeds between diamictons and bimodal diamicton textures indicate a mass flow origin (cf. Eyles *et al.*, 1988a, b; Ballantyne and Benn, 1994, 1996; Owen, 1994; Harrison, 1996; Curry, 1999, 2000a, b; Curry and Ballantyne, 1999; Figures 3.2c and 15.28a). At microscale, Owen (1994) provides a range of criteria with which to differentiate tills and paraglacially modified tills (debris flow diamictons) in a Himalayan setting (Figure 15.28b).

Although most active in higher-relief terrains, as illustrated by the case studies above, paraglacial reworking is not an insignificant process in areas of lower relief, especially is permafrost environments where large volumes of buried glacier ice melt out and release debris very slowly (e.g. Mackay, 1956, 1959; Mackay *et al.*, 1972; Kaplyanskaya and Tarnogradskiy, 1986; Astakhov and Isayeva, 1988; Astakhov *et al.*, 1996; Waller, 2001; Dyke and Evans, 2003; Murton *et al.*, 2005; Lacelle *et al.*, 2007; Murton *et al.*, 2004, 2005; Waller *et al.*, 2009; Figure 6.23). Here, the activity of retrogressive flow slides on very-low-angle slopes ensures that the debris released is predominantly subject to reworking in sediment gravity flows, often developing slicken-sided boundaries between diamicton layers similar to those produced by subglacial processes (Figure 15.29).

(a)

(b)

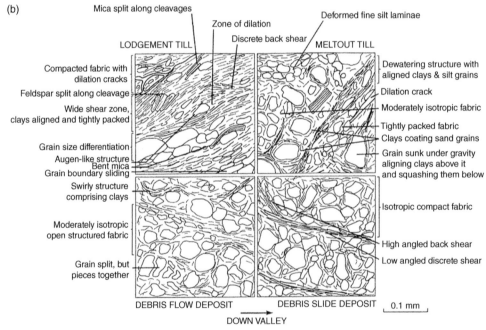

Figure 15.28 Examples of sediments and facies architecture in paraglacial deposits. (a) macroscale features, including: (i) crudely-to-well-bedded, largely clast-supported diamictons and poorly sorted gravel beds deposited by paraglacial slope reworking of local tills, Morfa Bychan, West Wales; (ii) gullied exposures through valley-side stacks of crudely stratified and clast-supported to massive matrix-supported diamictons, near Jasper, Alberta, Canada; (iii) crudely to well-bedded sequence of predominantly clast-supported diamictons, poorly sorted gravels and intervening pockets of sandy gravels, Lahul Himalaya; (iv) slope-parallel bedding in paraglacially reworked local till, Hayeswater, English Lake District; (b) microscale criteria proposed by Owen (1994) as critical to differentiating tills and paraglacially modified tills or debris flow diamictons.

Figure 15.29 Examples of paraglacial reworking of glacial deposits on Banks Island, in areas of lower relief characterised by permafrost and buried glacier ice where retrogressive flow slides operate: (a) flow slide in morainic topography; (b) reworking of debris-charged buried glacier ice; (c) retrogressive flow on <5° slope in glacilacustrine deposits; (d) slickensides developed on flowslide base in glacilacustrine deposits.

16

Till Spatial Mosaics, Temporal Variability and Architecture

The glaciotectonic end moraines squeezed from beneath frontal parts of the Pleistocene ice sheets are common in the European Lowlands. They are usually several tens of metres high. … They simply represent (glacio)tectonic structures composed of sediments of various, most often non-glacial origin. Why should they be called tills?

Ruszczynska-Szenajch (2001. p. 580)

This chapter summarises the nature of the geological record of glacigenic deposits and their appearance in glacial stratigraphies by highlighting specifically the vertical and lateral continua of tills and associated deposits. Beyond the valuable, but often ambiguous and sometimes contentious, genetic interpretations of diamictons based upon outcrop and microscale investigations, it is their wider stratigraphic context, architecture, geomorphic setting and geography that are all crucial in identifying the boundaries and assessing probabilities of the possible genetic outcomes. Hence, a holistic approach to glacial sedimentology is advocated here, with recommendations that we focus wherever possible on all scales, from the thin section to the basin.

At the outset it is important to emphasise our knowledge base with respect to the process–form relationships controlling till emplacement in modern glacial systems. In previous chapters the subglacial experiments at Breiðamerkurjökull (Boulton and Hindmarsh, 1987; Boulton *et al.*, 2001) have been highlighted as the base line for the identification and genetic interpretations of ancient tills, simply because they constitute a rare but unequivocal demonstration of till process sedimentology. The Breiðamerkurjökull case study is not without its problems however, mostly because it represents a sub-marginal rather than a fully subglacial scenario. This has been highlighted by van der Veen (1999), who raises the possibility that the sub-marginal environment in which the till is evolving is most likely being squeezed out from underneath the glacier rather than being subglacially deformed under the applied shear stress. This is certainly illustrated by other studies around the margins of Icelandic outlet glaciers where sub-marginal to marginal squeezing and till flowage, likely due to weak ice–bed coupling, is demonstrated by a range of phenomena (discussed in Chapters 8 and 10; Price, 1970; Evans *et al.*, 2010a, 2016; cf. Eyles *et al.*, 2015) including: (1) crevasse squeeze ridges; (2) sawtooth moraines; (3) till eskers; and possibly (4) weak clast macrofabrics. Numerical modelling of glacier advance over deformable till by Leysinger-Vieli and Gudmundsson (2010; Figure 16.1) also demonstrates this process, but in terms of till thickness patterns it importantly verifies that the till will thicken towards the margin to form 'till bulges' or 'propagating till waves'. Hence, investigations into till sedimentology in former ice–marginal tills will not be representative of subglacial process–form regimes. This is a very important message for till sedimentologists to digest, because throughout more than a century

Till: A Glacial Process Sedimentology, First Edition. David J A Evans.
© 2018 John Wiley & Sons Ltd. Published 2018 by John Wiley & Sons Ltd.

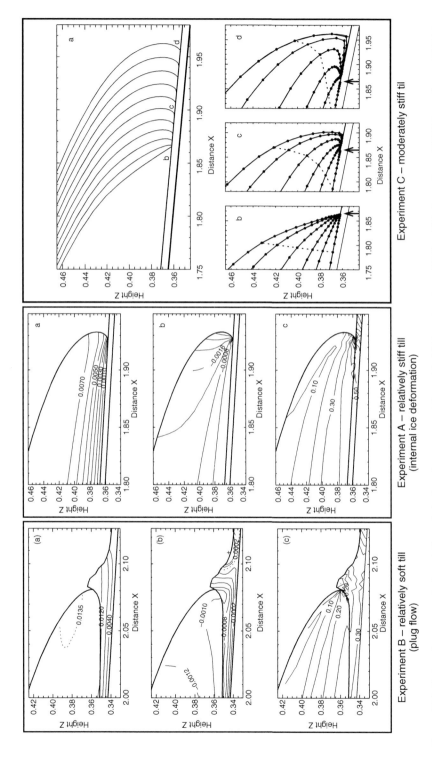

Figure 16.1 Output from numerical modelling of glacier advance over deformable till (from Leysinger-Vieli and Gudmundsson, 2010). In each case till thickens towards the glacier margin but it is a process that is most effective in situations where the till is soft. Experiments A and B show snapshots at dimensionless time $T = 15$ and represent (a) horizontal velocity, (b) vertical velocity and (c) effective stress. Experiment C shows (a) glacier surfaces at intervals of two dimensionless time units with selected front positions marked $c - e$ and replicated in the lower plots. Graph (b) shows the maximum height (H_b) of the sediment bulge.

Experiment A – relatively stiff till
(internal ice deformation)

Experiment B – relatively soft till
(plug flow)

Experiment C – moderately stiff til

of till investigations, in both ancient and modern settings, the focus has been predominantly on locations where tills are at their thickest and most complex. If we reflect on where such studies have been concentrated it is clear that they have clustered on former ice sheet marginal locations. The relative paucity of focus on ice sheet interiors is revealing in itself, as it demonstrates that tills are presumably not thick enough to attract our attentions!

The above reflections notwithstanding, it is important to review models of till emplacement, firstly in modern settings and then using theoretical constructs based upon regional till patterns. The construction of till-cored moraines as modelled by Leysinger-Vieli and Gudmundsson (2010; their 'till bulges') was initially demonstrated based upon field criteria by Price (1970) and then by Sharp (1984), but more recently conceptual models have increasingly recognised that there is a seasonal signature in the till accretion that takes place at soft-bedded glacier margins, at least in temperate settings (Krüger, 1993, 1995, 1996; Matthews *et al.*, 1995; Evans and Hiemstra, 2005; Chandler *et al.*, 2016). This complexity is claimed by Leysinger-Vieli and Gudmundsson (2010) to be reflected in their numerical model (Figure 16.1) by changing till hardness, thereby replicating Truffer *et al.*'s (2009) proposal that glacier flow in the Alaskan Taku Glacier changes from internal deformation in the winter to plug flow (experiment B in Figure 6.1) in the summer. As it applies to the same south Iceland glacier setting as that of the Breiðamerkurjökull experiments, the model of Evans and Hiemstra (2005; Figure 16.2) is now reviewed as a modern exemplar of sub-marginal till emplacement controlled by seasonal conditions. This model attempts to integrate the observed processes of ice–marginal till extrusion (Price, 1970; Sharp, 1984) and sub-marginal till freeze-on (Krüger, 1993, 1994, 1996; Matthews *et al.*, 1995) observed at temperate glacier margins by invoking till freeze-on and forward transport by glacier flow during the period of low ablation (winter), followed by summer thaw and squeezing/pushing of the till to form moraines that comprise a distal-thickening wedge. As moraine construction has been demonstrated to be predominantly annual in this area (e.g. Boulton, 1986; Evans and Twigg, 2002; Chandler *et al.*, 2016), the thickness of advected till can be assessed as in the order of 0.2–1.5 m per year. The construction of multiple till wedges/moraines was observed in real time in the early- to mid-1990s when the Icelandic south coast glacier snouts became stationary and stacked a succession of push moraines (Figure 8.12a). As each till layer was subject to subglacial shear stress throughout the period of forward glacier flow, it likely behaved in accordance with experiment A in Figure 16.1 during the winter and then with either experiments B or C in the summer depending upon local drainage conditions (Evans *et al.*, 2015a; Chandler *et al.*, 2016). The tills only occasionally display A and B horizon characteristics, and clast macrofabrics were only moderately strongly aligned with ice flow direction; at microscales, they contain ubiquitous water escape and sediment flowage features. These characteristics were interpreted by Evans and Hiemstra (2005) as the combined products of sub-marginal deformation, melt-out and flowage that emplaced till slabs over several seasonal cycles with each cycle (Figure 16.2) involving: (1) subglacial lodgement, bedrock and sediment plucking, subglacial deformation and ice keel ploughing in late summer; (2) freeze-on of subglacial sediment to the thin outer snout in early winter; (3) readvance and failure along a till decollement plane, resulting in the carriage of till onto the proximal side of the previous year's push moraine in late winter; and (4) melt-out of the till slab, initiating porewater migration, water escape and sediment flow and extrusion in early summer. The arrival of stratified debris-rich (supercooled) basal ice on the proximal slopes of push moraines could potentially lead to the local preservation of melt-out till (Figure 13.4).

Although multiple till sequences were demonstrably accreted in this way, likely by Leysinger-Vieli and Gudmundsson's (2010) 'propagating till waves' and 'till bulges', the repeated reworking of the resultant sub-marginal till wedges inevitably results in overprinted strain signatures and weakly developed clast pavements or clast lines. Additionally, any A and B horizon characteristics that may develop as a response to sub-marginal deformation will likely get truncated and superimposed,

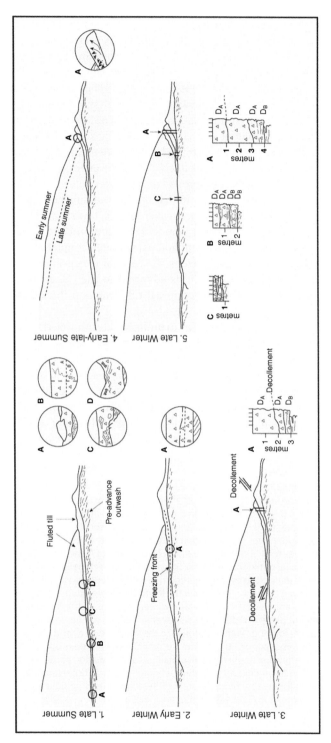

Figure 16.2 Schematic model of till slab emplacement over several seasonal cycles (from Evans and Hiemstra, 2005): (1) Situation in late summer at typical Icelandic glacier snout where subglacial processes include lodgement and sliding (A), bedrock and sediment plucking (B), subglacial deformation (C) and ice keel ploughing (D) in a temporally and spatially evolving process mosaic. (2) During early winter, the thin part of the glacier snout freezes onto part of the subglacial till. The till slab that freezes onto the ice sole is likely to be from the more porous A horizon (A). (3) The later winter readvance initiates failure along a decollement plane within the A horizon or at the junction with the more compact B horizon, resulting in the carriage of A horizon till onto the proximal side of the previous year's push moraine. (4) In the early summer, the melt-out of the till slab (A) initiates porewater migration, water escape and sediment flow (small arrows) and sediment extrusion due to glaciostatic and glaciodynamic stresses. (5) The late summer situation is again followed by winter freeze-on and marginal stacking of subglacial till produced by the reworking of existing subglacial sediments and fresh materials advected to sub-marginal locations from up-ice. Repeated reworking of the thin end of sub-marginal till wedges produces overprinted strain signatures and clast pavements.

as identified by Evans and Twigg (2002) and Evans and Hiemstra (2005) in their observations of overprinted strain signatures, where later deformation events (e.g. A and B horizons) are overprinted on earlier till horizons, giving rise to partially or totally eroded horizons and second-generation deformation structures (Figures 8.12 and 8.13); hence the diagnostic porous and bubbly texture of A horizons becomes compacted and sheared by later till overriding. Hence, incomplete stacks (e.g. Evans and Twigg, 2002; Evans and Hiemstra, 2005) should not be unexpected but modified stacks are possible, such as that described by Evans *et al.* (2016) from a former glacier sub-marginal setting at þorisjökull, central Iceland (Figures 8.6 and 16.3). At this site a vertical succession of alternating beds of massive and fissile diamictons display sub-boulder size clast macrofabric strengths indicative of shear strains too low for a steady state strain signature but strong boulder size macrofabrics reflecting high cumulative shear strains; these trends are explained as likely due to the effect of clast collisions in clast-rich till and the perturbations set up by the numerous large boulders (see Chapter 8). The alternating massive and fissile units are interpreted as A and B horizons of subglacial deforming layer couplets although the massive units do not possess the weaker fabrics, low shear strength and bubbly texture of traditional A horizons. Hence, applying the model of Evans and Hiemstra (2005), each couplet is hypothesised to record seasonal emplacement and partial inter-couplet modification in the form of B horizon superimposition on older A horizons. A seasonal interpretation of the A and B horizon couplets implies that <1 m of subglacial till is advected to the glacier margin per deformation event, a figure that is compatible with proven annual till advection thicknesses from Iceland's south coast.

A similar multiple till stack has been recently exposed in drumlinised terrain at the margin of the Icelandic surging glacier Múlajökull. An in-depth study of the till stratigraphy by Johnson *et al.* (2010) indicates that there are up to five tills, of around 40–120 cm thickness. They are interpreted as the products of combined deformation and lodgement at the ice–bed interface, with each till unit being emplaced by a separate surge. Hence, like the þorisjökull case study above, the till sequence at Múlajökull could be employed in an event stratigraphy linked to known glacier dynamics.

Potential overprinting of strain signatures in such multiple till sequences is just one reason why A and B horizons are not well represented in ancient till sequences even though multiple tills relating to

Figure 16.3 Case study of former glacier sub-marginal multiple till stack at þorisjökull, central Iceland (from Evans *et al.* 2016); (a) vertical profile log representing the vertical succession of alternating beds of massive and fissile diamictons (LF2–9), their interpretation as A and B horizon couplets, and their clast macrofabrics; (b) conceptual model to explain the development of the multiple subglacial tills, which assumes that seasonal conditions impact upon glacier sub-marginal processes and hence identifies the separation of spring-summer deformation events by a phase of winter freeze-on. During 'deformation event 1' a subglacial traction till comprising A and B horizons develops over a glacitectonite of former glacifluvial outwash, within which hydrofracture fills are commonly produced by elevated groundwater pressures. The first till developed over a glacitectonite will be characterised by a basal zone of sheared inclusions. Plucked blocks derived from bedrock steps below the icefall are delivered to the deforming layer by meltout of debris-rich basal ice. 'Deformation event 2' begins after winter freeze-on of the thin snout ice to the top of the A horizon, initiating a decollement plane and down-ice displacement of the top of the A horizon. This is followed by the advection of a new subglacial deforming layer in response to thawed conditions and elevated porewater pressures in the following spring-summer period. At this time the new B horizon is developed in the top of the old A horizon and deeper shear planes may develop in the older till units due to deformation partitioning. Specific processes identified widely in subglacial traction tills, including ploughing, clast lee-side matrix perturbations, lodgement and abrasion of large clasts, clast collisions and micro-shears (fissility) are also incorporated into the model. Note that the clast macrofabrics are examples from this study that are indicative of the various levels in the A and B horizons. The cumulative relative displacement curves are representative of the individual displacement events and therefore must be combined when assessing the total strain signature for a multiple till sequence. The impact of potential shearing at depth within a subglacial till is reflected in the alternative curves for deformation event 1.

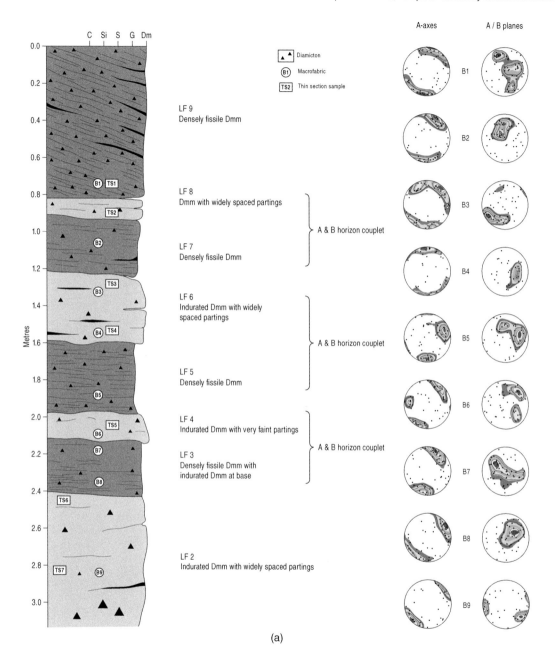

(a)

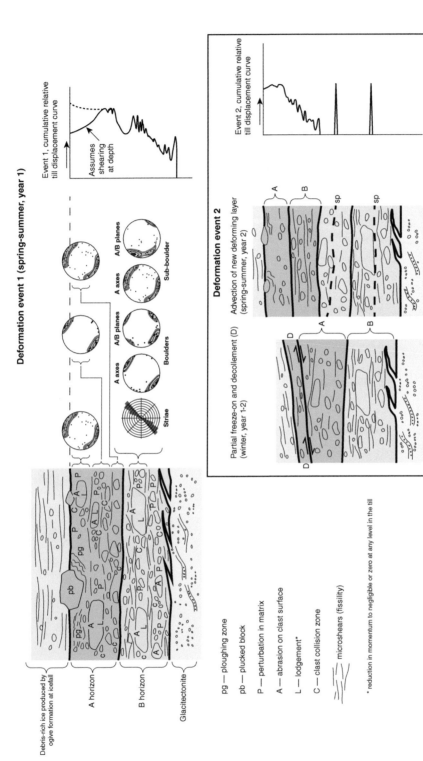

Deformation event 1 (spring-summer, year 1)

Event 1, cumulative relative till displacement curve

Assumes shearing at depth

A/B planes — Sub-boulder
A axes
A/B planes — Boulders
A axes
Striae

Debris-rich ice produced by ogive formation at icefall

A horizon

B horizon

Glacitectonite

pg — ploughing zone
pb — plucked block
P — perturbation in matrix
A — abrasion on clast surface
L — lodgement*
C — clast collision zone

///| microshears (fissility)

* reduction in momentum to negligible or zero at any level in the till

Deformation event 2

Event 2, cumulative relative till displacement curve

Advection of new deforming layer (spring-summer, year 2)

A

B

sp

sp

Partial freeze-on and decollement (D) (winter, year 1–2)

D

A

B

D

Top of A horizon displaced down ice

New B horizon partially developed in old A horizon + potential of deeper shear planes (sp)

(b)

Figure 16.3 *(Continued)*

changing processes have been proposed (e.g. Piotrowski *et al.*, 2006) and predominantly two-tiered subglacial till stratigraphies, comprising high-shear-strength diamictons overlain by thick (<5 m) low-shear-strength diamictons, have been related to ice stream subglacial deformation on the Antarctic continental shelf (e.g. Dowdeswell *et al.*, 2004; Ó Cofaigh *et al.*, 2005; see below). Additionally, the A horizons in particular are not very robust when it comes to resisting post-depositional denudation, as initially evaluated at the time of the Breiðamerkurjökull experiments by Boulton and Dent (1974). We also have to reflect on the observations made in Chapter 15 on paraglacial reworking and entertain the notion that A horizon classifications are being applied to diamictons that appear similar to subglacially deformed A horizons but which are instead localised mass flow diamictons; this might be manifest in some of the typically weak clast macrofabrics found in field settings (Table 8.1). Hence, there are four outcomes in terms of interpreting the weaker fabrics, low-shear-strength and bubbly texture of some sub-marginally emplaced diamictons: (1) they are A horizons as defined by Boulton (1979), Boulton and Jones (1979), Boulton and Hindmarsh (1987) and Benn (1995); (2) they are crevasse squeeze ridge deposits (e.g. Price, 1970); (3) they are mass flow diamictons produced by till flowage during deformation in a shear zone that becomes increasingly saturated as the melt season advances, as observed, for example, in the slurrying of a strongly aligned till in the fluted till of west Fláajökull by Evans and Hiemstra (2005); (4) they are mass flow diamictons created by post-depositional flowage of till on the surfaces of freshly constructed push moraines (Sharp, 1984). Clearly, the geomorphic setting, stratigraphic context and architecture of such diamictons is crucial to arriving at realistic interpretations of their genesis.

The modern glacier snout observations and numerical modelling reviewed above identify a marginal thickening of subglacially deforming till, a concept developed independently at a regional scale by Boulton's (1996a, b) numerical model of regional erosional and depositional zones and till architecture beneath ice sheets (Figures 9.1), supported notionally by Alley *et al.*'s (1997) assessment of subglacial sediment advection by all potential processes, not just deformation (Figure 1.4). Boulton's (1996a) theory of regional erosional and depositional zones on a deformable bed stems from the larger-scale operation of A and B horizons as identified in the south Iceland case studies (Figure 16.4a). Increasing ice flux leads to the mobilisation of the B horizon due to lowering of the A/B horizon interface, thereby increasing A horizon net flux and eroding the substrate. Conversely, a reduction in ice flux near the ice sheet margin leads to a drop in A horizon net flux. For a steady state ice sheet, this predicts a thickening of, and an increase discharge in, the deforming horizon towards a sub-marginal zone and hence the development of erosional and depositional (more correctly accretionary) zones (Figure 16.4b), providing also a solution to clast pavement development (Figure 9.4). Over the growth and decay stages of an ice sheet, sub-marginal thickening wedges of till are thus predicted to develop during advance and retreat (Figure 16.4c) and thereby develop advance and retreat phase till wedges whose architecture will vary according to ice margin temporal and spatial oscillation patterns (Figure 16.4d). Although this model over-emphasises the role of A and B horizon subglacial deformation, the regional architecture predicted for till thickness in particular appears to be born out in ancient glacigenic stratigraphies where tills and glacitectonites thicken towards former ice margins (see below). Removal of soft substrates to create erosional zones was explained in Chapter 10 in terms of the 'excavational' deformation of Hart *et al.* (1990), Hart and Boulton (1991) and Hart (1995). This has been related by Boyce and Eyles (1991) to the operation of an erosional shear zone cutting downwards into substrates or an 'erodent layer' by Eyles *et al.* (2016; Figure 10.6). The 'erodent layer hypothesis' (ELH) invokes wearing surfaces (i.e. ploughing clasts and/or frozen rafts that protrude from the deforming layer base) to explain subglacial surfaces characterised by lodgement, deformation and ploughing (Eyles and Boyce, 1998; Figure 6.6), processes that repeatedly

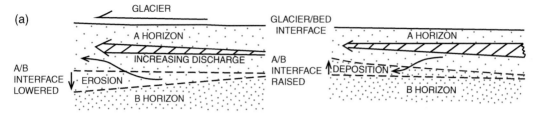

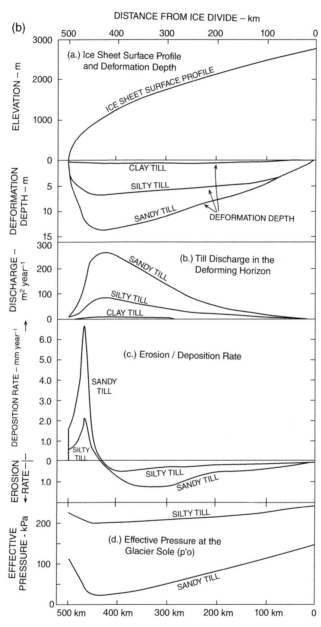

Figure 16.4 Selected diagrams from Boulton's (1996a) theory of regional erosional and depositional zones on a deformable bed: (a) relationship between ice flux and A and B horizon thicknesses. Left panel shows increasing ice flux associated with higher shear stress and increasing sediment discharge, hence a lowering A/B interface. Right panel shows decreasing ice flux and longitudinal compression and rising A/B interface.

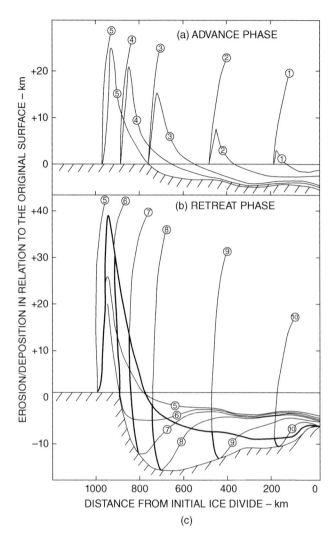

Figure 16.4 (b) the pattern of till discharge and hence deposition in relation to grain size and associated porewater pressure and deformation depth beneath a steady state ice sheet; (c) the changing patterns of till wedge production over the growth and decay stages of an ice sheet; (d) the spatial and temporal pattern of till deposition in relation to ice sheet oscillations. Left side shows an ice sheet with a prolonged period of standstill during advance and hence a substantial thickness of till is deposited during the standstill which is not all eroded during subsequent overriding. Right side shows ice sheet that readvances to its maximum, thereby producing stacked tills with intervening erosion surfaces (prime locations for clast pavements). Each ice margin contains a record of advance and retreat tills.

rework and advect the subglacial till to produce overprinted strain signatures and clast pavements at the thin end of sub-marginal till wedges.

The ELH has been invoked by Eyles *et al.* (2016) as an alternative to the 'instability mechanism' of subglacial bedform (particularly drumlin) construction advocated by Fowler (2000, 2009, 2010), Dunlop *et al.* (2008), Clark (2010) and Stokes *et al.* (2011, 2013), which hypothesises that subglacial bedforms arise from the deformation and local thickening of subglacial till (Figure 16.5). The instability

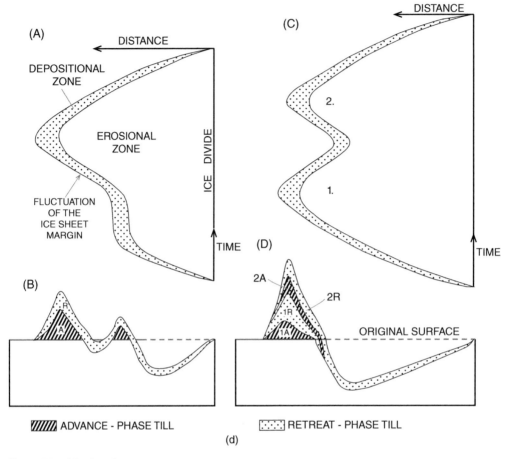

(d)

Figure 16.4 (*Continued*)

Figure 16.5 The instability mechanism for drumlin formation and subglacial deforming till layer evolution: Left panel shows a schematic diagram of parameters and underlying principles of the instability theory. Ice and sediment deformation and sliding at the ice–till interface (A) creates a system prone to the development of an along-flow instability which in turn creates waveforms (bedforms) at the ice–till surface (B). These then emerge as drumlins of dominant wavelength (C). The parameters identified in A relate to the flow of ice as a Newtonian viscous material, a sliding law relating basal shear stress (τ) to basal velocity (u), basal effective pressure (N), and sediment flux (q). The assumption is that granular till will only deform if $\tau > \mu N$, where μ is a coefficient of friction. Also assumed is that τ increases with u and N, while q increases with τ but decreases as N increases. Importantly, because N in the till increases with depth below the ice–till interface, till deformation is limited to a thin mobile layer of tens of centimetres to metres, typical of those observed in modern ice sheets. Right panel shows how drumlins emerge and till thins through time due to inhibited till continuity (i.e. till transport > till supply). Instability at the ice–till interface is unstable and becomes wavy but the wavy interface erodes downwards due to inhibited till supply but drumlin emergence requires a pre-existing, metres thick layer of till (A). As till is removed from the system and thins, some drumlins are anchored by bedrock perturbations that might act as cores (B). Further till exhaustion leads to the creation of more obvious drumlin cores (C), crag-and-tail features (D), and eventually a till free bedrock surface (E).

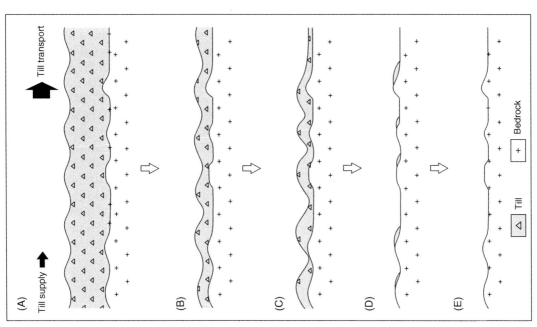

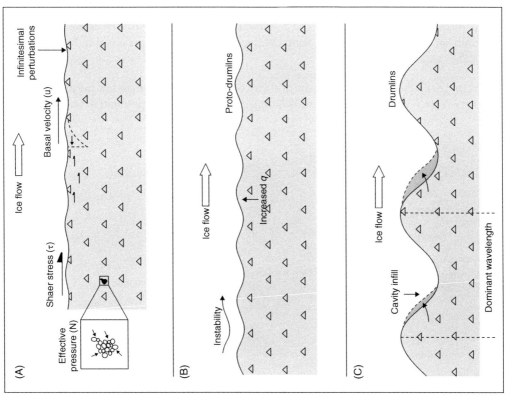

mechanism provides a potential explanation for thick tills on former ice sheet beds, but Eyles *et al.* (2016) highlight a problematic aspect of the theory in that the stratigraphy of drumlin cores in particular are often not entirely composed of till and, furthermore, appear to be streamlined residuals of pre-existing sediment over which a thin till cap has been smeared (e.g. Shaler, 1889; Alden, 1905, 1918; Armstrong, 1949; Dean, 1953; Gravenor, 1953, 1957; Jewtuchowicz, 1956; Flint, 1957; Clayton and Moran, 1974; Krüger and Thomsen, 1984; Whittecar and Mickelson, 1977, 1979; Stephan, 1987; Newman and Mickelson, 1994; Colgan and Mickelson, 1997; Möller, 2006; Schomacker *et al.*, 2006; Kerr and Eyles, 2007). Additionally, Eyles and Doughty (2016) use the often-reported juxta-position of rock-cored (Figure 16.6) and sediment-cored drumlins to propose a common erosional origin, rather than regard them, as well as non-till-cored drumlins, as anomalies or 'clones' of the 'emergent' instability-formed drumlins (cf. Clark, 2010). It is important to emphasise, however, that the instability theory encompasses not just the building of till-cored drumlins but also the downward excavation of the streamlined interface right down to the bedrock substrate in situations where till supply is restricted. This 'till continuity control' (defined as the balance between till transport in the deforming layer and till supply from substrate erosion and advection from up ice) over deforming bed thickness is essentially the same process as that envisaged by the ELH. Important research questions arise from these discussions, including:

(1) How common are till-cored drumlins?
(2) Do they contain tills indicative of the perceived instability process, and indeed what is sedimentologically diagnostic of the instability process?
(3) Where do till-cored drumlins occur, and does their locational pattern fit alternative theories of till accretion, such as sub-marginal incremental thickening, followed, presumably, by glacier over-riding and streamlining?

Stokes *et al.* (2011) attempt to address the problem of drumlin composition but the glacial research community will likely always be significantly hampered by a lack of representative samples in terms

Figure 16.6 Rock-cored drumlin in the Tweed Valley palaeo-ice stream, exposed by bedrock quarrying near Coldstream, Scotland. Note the different-coloured multiple till units of <1.5 m thickness.

of drumlin exposures. We will revisit these research questions below, specifically with a review of the small number of till-cored drumlin case studies.

The advection of subglacial material by 'excavational deformation' or erodent layer development can be compared with the 'propagating till waves' modelled by Leysinger-Vieli and Gudmundsson (2010), a process that has the potential to build up multiple till units in sub-marginal stacks, as observed in modern settings (e.g. Evans and Hiemstra, 2005) and inherent within numerical and theoretical models (e.g. Leysinger-Vieli and Gudmundsson, 2010; Boulton, 1996a, b). In stratigraphic terms, this 'sub-marginal incremental thickening' (Evans and Hiemstra, 2005) has been likened to the process–form regime of 'punctuated aggradation' or repeated episodic deposition of one facies on another by Eyles *et al.* (2011; cf. Brett and Baird, 1986). This has been demonstrated in an ancient setting by Eyles *et al.* (2011) using the 16.5-m-thick sequence of Wildfield Till in the Trafalgar Moraine on the shores of Lake Ontario, Canada. The till thins gradually on its proximal side to 2-m-high flutings overlying shale bedrock and hence forms a ramp or wedge shape deposit comprising multiple till units separated by deformed silty-clay interbeds or sharp unconformities (Figure 16.7). This model, and indeed the outcomes of the instability theory illustrated in Figure 16.5, are compatible with common reports of drumlin fields being best developed immediately up ice of the moraine constructed at the time as their formation (e.g. Mooers, 1989; Boyce and Eyles, 1991; Patterson and Hooke, 1996; Colgan *et al.*, 2003; Kerr and Eyles, 2007) and are compatible also with genetically linked fluting and push moraine associations at modern glacier snouts (Price, 1970; Sharp, 1984; Evans and Twigg, 2002; Evans and Hiemstra, 2005; Evans *et al.*, 2015a, b).

Punctuated aggradation/sub-marginal incremental thickening has been proposed as an explanation for the occurrence of thick multiple till sequences ('till moraines' of Chapman and Putnam, 1951; cf. Hansel and Johnson, 1987; Colgan *et al.*, 2003; Patterson *et al.*, 2003) along the former margins of the southern Laurentide Ice Sheet by Boulton (1996a, b), Jensen *et al.* (1996), Eyles *et al.* (2011) and Evans *et al.* (2012a, b, 2014). Although these tills are locally characterised by a macroscopically massive appearance and have been selected for laboratory shear tests, because they are considered to be subglacial tills (e.g. Iverson *et al.*, 1996, 1998; Moore and Iverson, 2002; Thomason and Iverson, 2006; see Chapter 7), marginal thickening appears to have been especially effective where ice margins have advanced into or oscillated in proglacial lake sediments, the implications of which will be reviewed below. Some marginal till stacks appear to display stratigraphies similar to those of modern glacier margins, in that massive diamictons are separated merely by partings between till units (Figures 8.12a and 16.7; e.g. Eyles *et al.*, 2011). At increasingly larger scales, multiple tills are separated by stratified interbeds (Figures 8.12b and 11.2–11.6) in laterally extensive and vertical sequences, indicative of changing subglacial conditions during till accretion. The stratified interbeds record periods of subglacial meltwater sheet flow or canal infilling and soft-bed sliding (Chapter 11) between phases of ice–bed coupling and till accretion, during which the stratified sediments are at least partially, and often intensively, glacitectonised. This is illustrated by the multiple till sequence reported from central Poland by Piotrowski *et al.* (2006; Figure 8.3), in which three till units display phases of alternating deformation and hydraulic decoupling (Figure 16.8). Further examples of such changing subglacial sedimentation regimes are presented by Boyce and Eyles (2000) and Meriano and Eyles (2009) for the Northern Till in Ontario, Canada (Figure 11.5) and the tills of the Northumberland coast, England by Eyles *et al.* (1982; Figure 11.2). More substantial stratified intra-till beds can represent the widening of canal fills into subaqueous fans in marginal cavities near glacier grounding lines, as has been described on the eastern English coast by Davies *et al.* (2009) and Evans and Thomson (2010), where they form wedge-shaped

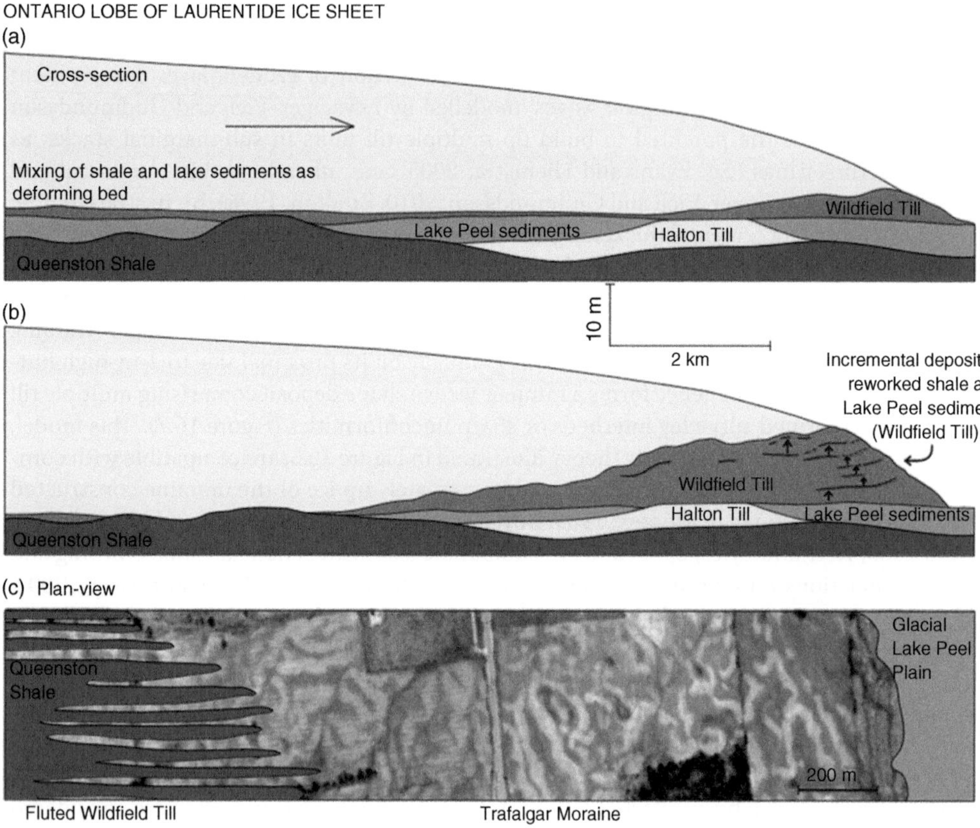

ONTARIO LOBE OF LAURENTIDE ICE SHEET

(a)

Cross-section

Mixing of shale and lake sediments as
deforming bed

Lake Peel sediments   Halton Till

Wildfield Till

Queenston Shale

(b)

10 m

2 km

Incremental deposition of
reworked shale and
Lake Peel sediments
(Wildfield Till)

Wildfield Till

Halton Till   Lake Peel sediments

Queenston Shale

(c) Plan-view

Glacial
Lake Peel
Plain

Queenston
Shale

200 m

Fluted Wildfield Till   Trafalgar Moraine

Figure 16.7 A generalised depositional reconstruction (vertical scale exaggerated) for the Trafalgar Moraine and the Wildfield Till invoking sub-marginal incremental thickening by the deformation mixing and advection of lake sediments and shale to the ice front (from Eyles *et al.*, 2011): (a) initial overriding and mobilisation of sediment as a subglacial deforming bed; (b) sub-marginal incremental thickening of Wildfield Till; (c) aerial image of the moraine/till surface, showing extent of flutings in Wildfield Till and their passage into shale bedrock. Note also the 'brain-like' ridge patterns on the till surface indicative of late-stage settling/pressing of stagnant ice.

components at the outer edges of retreat-phase tills and thin up flow to the canal fills (Figures 11.3, 11.4 and 16.9).

It is becoming increasingly clear from both modern processes observations and sedimentologically based theory, as reviewed in Chapter 6, that these complex till sequences are the product of the operation of subglacial mosaics of deforming and sliding bed conditions that vary in space and time, especially near the termini of glaciers and ice sheets. Additionally, it has been proposed that the locus of deformation may change spatially and temporally so that different deformation events are partially superimposed, as illustrated by the concept of deforming spots (Figure 16.10a, b) proposed by Piotrowski and Kraus (1997) and Piotrowski *et al.* (2004). A similar spatially and temporally changing, anastomosing rather than patchy, pattern of till deformation has been proposed by Shumway and Iverson (2009; Figure 16.10c) to explain spatially variable fabric alignments that do not record a bed-parallel simple shear regime. More recently, Phillips *et al.* (2013a, in press) have employed localised microscale structures within tills, such as alternating layers of diamicton with laminated silt and clay and 'vinaigrette' patterns to propose that localised liquefaction events (e.g. Figure 6.12b) had

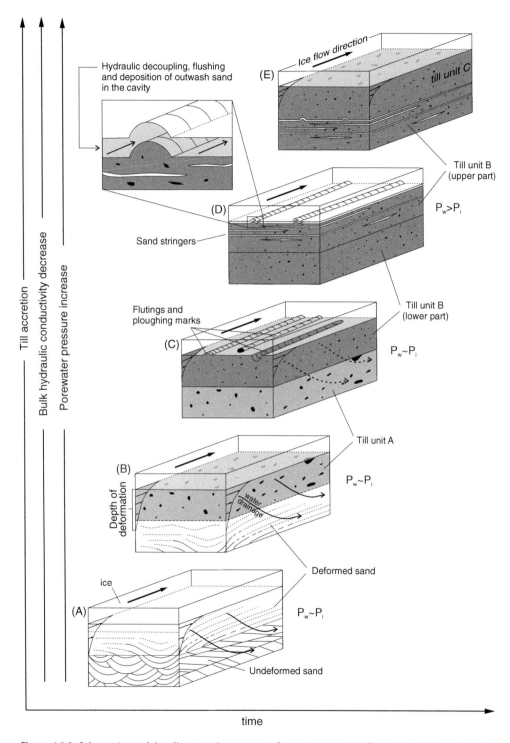

Figure 16.8 Schematic model to illustrate the concept of time-transgressive formation of till and associated meltwater deposits based upon the Kurzetnik section in Poland (from Piotrowski *et al.*, 2006). The processes identified are bed deformation (A, B), lodgement with ploughing (C), and basal decoupling and water film operation and winnowing (D).

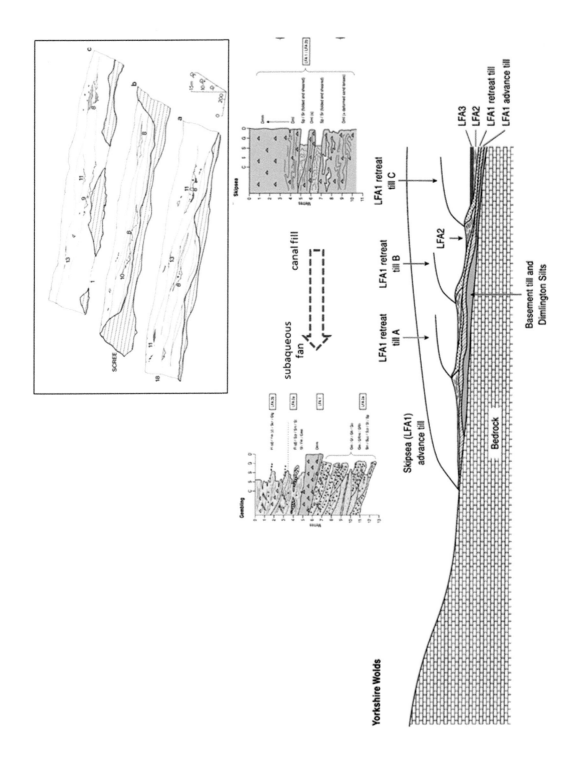

injected the till with pressurised water, disrupting the matrix (vinaigrette) and momentarily inducing basal sliding (Figure 16.11); laminated sediments were deposited by water trapped along the ice–bed interface after the ice had recoupled with its bed and hence the interbedded diamict and stratified layers record intermittent ice–bed separation and recoupling. Attenuation of such deposits after their emplacement could be envisaged as being an effective way of creating glacitectonic lamination and pseudo-stratified diamictons at the ice–bed interface, particularly as the rafting and attenuation model depicted in Figure 6.22a does not satisfactorily explain the 'stratification' in such materials. The possibility of localised liquefaction and flowage or at least dilation of till at the ice–bed interface in this way was employed by Evans *et al.* (2006b), to develop a model of spatially and temporally variable deformation (Figure 16.12) in which patches of mobile A horizon moved through the bed whenever the till-matrix framework broke down (Chapter 10; Figure 10.1). Such events, if they took place immediately prior to final till emplacement, would be effective in weakening clast macrofabrics as well as introducing the water escape features now widely observed, especially at microscale, in tills.

These notions of changing till properties and behaviour, including the potential viscous slurry-like behaviour of tills, were elucidated by Menzies (1987, 1989) in his classification of hard, mobile and quasi-mobile zones (H, M and Q beds), the spatial and temporal migration of which were developed by van der Meer *et al.* (2003; Figure 16.13). The hard or 'H beds' are characterised by high meltwater activity at their surface (dendritic channels, linked cavities, water films; 4, 5 and 7 in Figure 6.13) and represented by either rock or sediment that is frozen or of low conductivity. This could include till that has effectively stabilised due to processes such as strain hardening or constructional deformation. Mobile or 'M beds' are composed of saturated till wherein free meltwater flow is limited and hence is transferred only by bulk movement in a deforming layer (1 in Figure 6.13), giving rise to advective pervasive flow of soft saturated debris. Quasi-mobile or 'Q beds' represent a combination of the H and M bed types but both are spatially and temporally transitory at a localised scale. The Q bed type is regarded as the most common beneath glaciers and ice sheets by van der Meer *et al.* (2003), who consider the deforming bed to be a matrix of variations in composition, water content, shear strength and applied shear stress levels which interact with variations in thickness and velocity. Such a scenario has been proposed to explain vertical changes in till properties by Lian and Hicock (2000), in a till evolutionary model that involves lodgement followed by till thickening and elevated porewater pressures, in turn leading to deformation and even flowage as a viscous slurry; dewatering and stiffening then lead to brittle deformation being superimposed on early ductile structures.

As a consequence of these perceived spatial and temporal changes, inter- and intra-till properties will inevitably reflect complex formational histories or constitute genetic hybrids (e.g. Dreimanis *et al.*, 1987; Hicock, 1990; Hicock and Dreimanis, 1992a, b). Additionally, multiple tills, often but

---

Figure 16.9 Examples of the changing architecture of stratified intra-till beds where former canal fills moved across the ice–till interface and also widened into subaqueous fans near glacier grounding lines. Upper panel shows the three-dimensional architecture of canal fills in the Northumberland tills of northern England as reported by Eyles *et al.* (1982) from the repeatedly quarried face of an open cast coal mine. The till units are bounded by erosional surfaces and the line of section lies transverse to former ice flow direction. Main features include: 1 – striated rockhead; 8 – channel fill; 9 – diapiric till intrusion up into the base of a channel; 10 – channel sidewall failure; 11 – upper surfaces of cut and fill channels partially eroded by ice flow, thereby resulting in deformed inclusions in overlying till; 13 – slickensided bedding plane; 18 – base of postglacial weathering profile. Note that the sections a–c represent sequential faces exposed by quarrying and that this allows the visualisation of the ribbon-like cut and fill channels orientated sub-parallel to ice flow and wedging-out of certain till units. Lower panel shows the regional stratigraphy of subaqueous fans (LFA 2) deposited in association with the Skipsea Till (LFA 1) at the margin of the North Sea Lobe on the eastern English coast (after Evans and Thomson, 2010). The wedge-shaped fans grade up-ice into canal fills in the Skipsea Till.

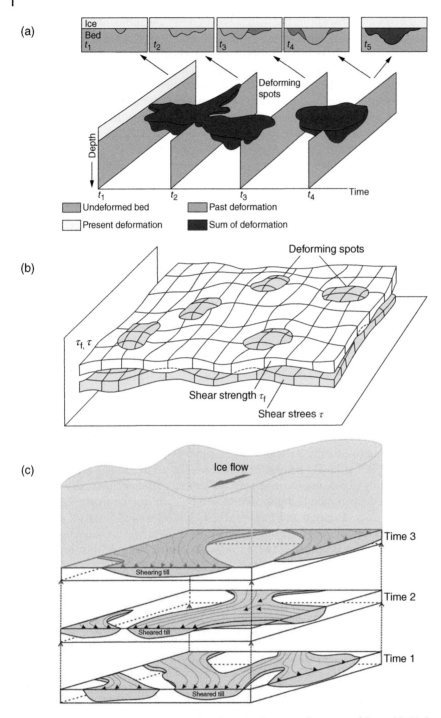

Figure 16.10 Schematic reconstructions to illustrate the general concept of the stable/deforming subglacial bed mosaic: (a) Diagram to show how the position of deforming spots on the bed changes spatially and temporally. This produces overprinted deformation signatures in the subglacial material (from Piotrowski *et al.*, 2004). (b) Subglacial deforming/sliding bed mosaic proposed by Piotrowski and Kraus (1997), which compares subglacial shear stress (τ) with sediment shear strength (τ_f). The shear stress is lower than the sediment shear strength wherever high basal water pressures cause ice–bed decoupling. Elsewhere, the bed develops deforming spots. (c) Model of anastomosing rather than patchy pattern of till deformation proposed by Shumway and Iverson (2009).

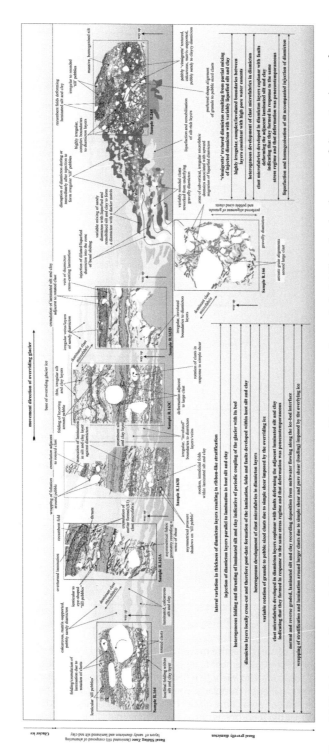

Figure 16.11 Schematic diagram linking microscale structures in specific thin sections from subglacial tills to the development of alternating layers of diamicton with laminated silt and clay and 'vinaigrette' patterns (pseudo-stratified diamictons), as a product of localised liquefaction events at the ice–bed interface (from Phillips et al. in press).

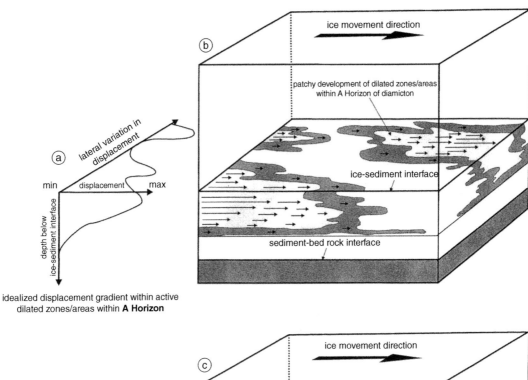

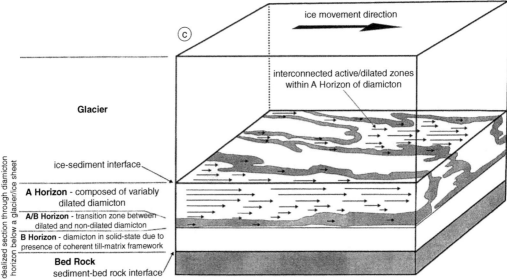

Figure 16.12 A model for the operation of areas of active movement at the ice–bed interface due to dilating subglacial sediment (from Evans *et al.*, 2006b). Movement takes place either in isolated patches (a and b) or as a more coherent network of interconnected zones (c). Arrow length reflects the relative displacement magnitude throughout the deforming sediment due to lateral variations in subglacial conditions (e.g. water content, porewater pressure, drainage efficiency). The net effect is that one area of the glacier bed will be moving but an adjacent area may be stationary. This model also proposes that the A horizon is not a tabular zone that stretches across the whole glacier bed.

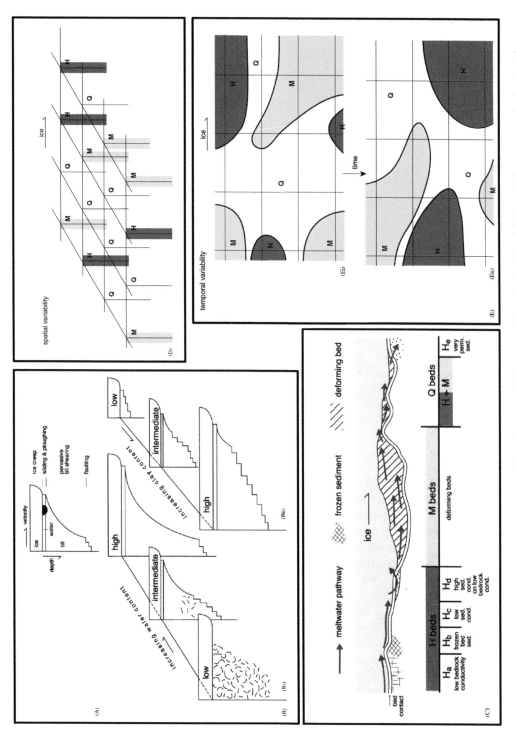

Figure 16.13 Diagrams to convey the concept of hard, mobile and quasi-mobile zones (H, M and Q beds) and their spatial and temporal migration (from van der Meer *et al.*, 2003): (A) and (B) various styles of deformation applied to Alley's (1989a) subglacial profile diagram, including changes in style and intensity of deformation with increasing water content (Bi) and clay content (Bii); (C) the variable styles and intensity in deformation in 'H', 'Q' and 'M' beds; (D) theoretical spatial distribution of 'H', 'Q' and 'M' beds under a glacier; (E) theoretical maps at two time slices (Ei and Eii) of the possible distribution of 'H', 'Q' and 'M' beds, showing the potential temporal variability of the deforming bed.

not exclusively with distinctly different provenances, have also been explained as the superimposed subglacial imprints of competing ice lobes within ice sheets (e.g. Broster and Dreimanis, 1981; Rappol and Stoltenberg, 1985; Stea and Brown, 1989; Hicock and Fuller, 1995; Hicock and Lian, 1999), which based upon the marginal thickening concepts developed above it is possible to envisage as a vertical accretion of lithologically and/or sedimentologically and structurally distinct till units. Boulton (1996a) has provided an explanation of vertically stacked tills of distinct lithological compositions based upon an understanding of the ice travel distance over specific bedrock outcrops (Figure 16.14), emphasising the importance of locally derived materials. This illustration that a till is a blend of all the bedrock lithologies that it travels over must be modified, however, in situations where lowland glacier lobes have moved over and reworked substantial depo-centres that already contain blends of lithologies derived from regions previously traversed by ice sheets, good examples being marine and large lake basins. An example of where this has likely taken place is presented by Boston *et al.* (2010), based upon the East Yorkshire tills and associated deposits of the former North Sea Lobe of the British–Irish Ice Sheet. Here, the geochemical signatures of folded and stacked till units do not everywhere reflect a gradual transition in lithological composition, as predicted by Boulton's (1996a) model, but are instead repeated vertically throughout exposures through morainic ridges. This vertical repetition is thought to reflect the independent structural evidence that the till units have been deposited through the folding, attenuation and stacking of offshore marine deposits and proglacial lake sediments as a result of competing ice flow units within the North Sea Lobe moving onshore and initiating glacitectonic transposition (Figure 16.15).

The example of the East Yorkshire depositional setting briefly described above demonstrates that marginal thickening of deposits traditionally classified as 'tills' is not always achieved just by till wave propagation/advection but also by predominantly ductile styles of glacitectonic construction such as in folds and fold nappes. The strain markers that appear ubiquitous in such marginal sediment wedges indicate not only their style of construction but also their primary origin, often not as true tills but as stratified deposits modified into glacitectonites. The complex array of depositional processes operating at glacier margins, as reviewed in Chapter 15, is clearly significant in the generation of source materials for the development of such sub-marginal glacitectonite and till assemblages, especially at oscillating ice margins (e.g. Figure 15.2c). Indeed, low–moderate strain deformation is recorded in many classic 'multiple till' stratigraphies, some that are clearly sub-marginal (e.g. the 'North Sea Drifts' of East Anglia in England and the 'Irish Sea Till' of the southern Irish Sea Basin; Eyles *et al.*, 1989; Ó Cofaigh and Evans, 2001a, b; Evans and Ó Cofaigh, 2003; Phillips *et al.*, 2008; Figure 16.16) and others that are likely to have been (e.g. the Southern Alberta Ice Stream deposits of the Canadian prairies; Evans *et al.*, 2008, 2012a, 2014; Figure 16.17).

In the Irish Sea Basin and Canadian prairies examples, ice sheet advance resulted in the damming of regional drainage networks, thereby creating vast glacilacustrine sediment sinks into which a wide array of subaqueous deposits, including mass flow and dropstone diamictons, accumulated in front of advancing and oscillating ice margins (Figures 15.20–15.23). On the southern Irish coast, the 'Irish Sea Till' comprises glacitectonite derived from pre-existing periglacial, offshore marine and glacilacustrine deposits (Figure 16.16). The incorporation of periglacial deposits in the 'till' is evidenced by the presence of folded, thrust, attenuated, and boudinage type inclusions or rafts of the local 'head'. Additionally, prominent transition or amalgamation zones occur between the head and overlying shelly diamicton (classic 'Irish Sea Till'), as defined by pseudo-laminated or stratified diamicton composed of attenuated inclusions of head and shelly diamicton. A vertical succession of deformation intensity is also apparent in the sequences, comprising undisturbed or lightly deformed

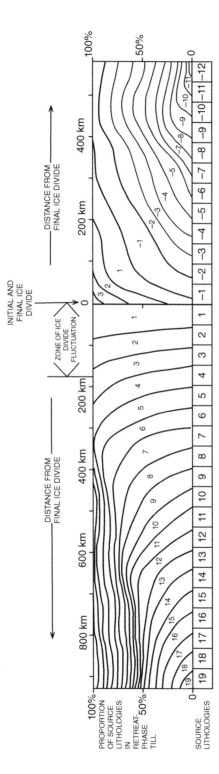

Figure 16.14 A theoretical diagrammatic explanation of vertically stacked tills of distinct lithological compositions (from Boulton, 1996a). This shows bulk lithological composition of retreat-phase tills in relation to their source lithologies, which are shown along the bottom, where 1–19 are located south of the initial and final ice divide and −1 to −12 are situated to the north. Proportions of individual source lithologies are shown in relation to their distance from the ice divide. Material is transported across the ice divide to the north due to the southwards migration of the divide during maximum glaciation. Retreat-phase till deposited within 300 km south of the ice divide is predicted to be dominated by local lithologies, whilst beyond this up to 50% is predicted to be made up of far-travelled material. This is due to high ice velocities and erosion rates in the outer zone during ice expansion.

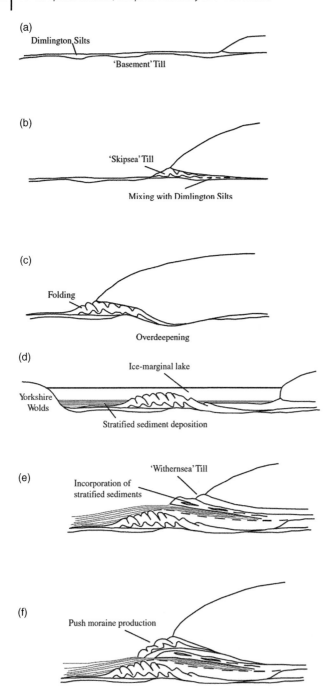

(a)
Dimlington Silts
'Basement' Till

(b)
'Skipsea' Till
Mixing with Dimlington Silts

(c)
Folding
Overdeepening

(d)
Ice-marginal lake
Yorkshire Wolds
Stratified sediment deposition

(e)
'Withernsea' Till
Incorporation of stratified sediments

(f)
Push moraine production

Figure 16.15 Reconstruction of North Sea ice lobe dynamics and deposition of the traditional till units of eastern England as an explanation of till lithological signatures (from Boston *et al.*, 2010): (a) initial ice advance over the 'Basement' Till, created ponding and the deposition of the Dimlington Silts; (b) incorporation of the Dimlington Silts into the subglacial traction layer and mixing with more distal sediments; (c) compression and folding of the 'Skipsea' Till at the ice margin and over-deepening in the sub-marginal zone; (d) ice margin retreat across Holderness and sedimentation in Glacial Lake Holderness; (e) overriding of the stratified sediments by readvancing ice margin and the cannibalisation of some stratified sediments into the 'Withernsea' Till; (f) production of a push moraine at Dimlington High Land during a major ice advance.

head, overlain by more heavily deformed head, overlain by a transitional zone of deformed and attenuated head and diamicton, overlain in turn by massive diamicton with deformed sediment inclusions.

On the Canadian prairies, the lobate southern margins of terrestrially terminating palaeo-ice streams are demarcated by large arcuate assemblages of complex glacigenic sediment sequences, containing diamictons up to 20 m thick. As we have seen in previous chapters, such thicknesses are not typical of subglacial tills but can be reconciled with models proposing the sub-marginal accretion of deforming layers or stacking of glacitectonites, especially as the sediment sequences display clear evidence of glacitectonic deformation and stacking, such as large-scale fold and shallow reverse fault structures associated with deformed bedrock rafts and intra-clasts. The individually thick units or sequences of diamicton are explained by Evans *et al.* (2012a) using a combination of two sets of processes (Figure 16.17). First, glacier-marginal till thickening, which produces stacked till sequences or wedges in areas where the ice stream margin was stationary for short periods of time, a scenario documented also for the southern Laurentide Ice Sheet lobes by Johnson and Hansel (1999; Figure 16.18) and Eyles *et al.* (2011; Figure 16.7). The occurrence of multiple subglacial tills, because their accretion is known to be annual in some modern settings, potentially represents a record of annual sub-marginal incremental thickening; this scenario is entertained by Evans *et al.* (2012a), specifically because densely spaced recessional push moraines also document the early retreat history of the Canadian prairie ice streams, landforms clearly associated with annual till propagation in many modern temperate glacier snouts (Evans *et al.*, 2008). Second, accumulation of mass flow diamictons, interdigitated with glacilacustrine deposits and created by gravity-induced slumping of ice–marginal till aprons directly into lake waters occupying ice-dammed preglacial valleys. These till aprons and diamictons were fed directly by the advection of sub-marginal till over valley thalwegs. In addition to these two processes, glacitectonites are ubiquitous due to the extensive overriding and deformation/cannibalisation of the substantial thalweg infills. The widespread occurrence of thrust bedrock masses in the tills and glacitectonites of the region have long been explained as the products of elevated porewater pressures created by ice flow against preglacial valley-side scarps orientated transverse to ice flow (Tsui *et al.*, 1989); the liberation of bedrock blocks from pre-existing valley floor badlands is also thought likely by Evans *et al.* (2012a) to be a potential source of rafts. Finally, the overprinting and stacking of regional till sheets has been a characteristic of competing palaeo-ice stream lobes whose dominance has switched during advance and recessional oscillations and has given rise to changing sediment provenance through time, as reflected in the traditional nomenclature of the regional tills (i.e. Stalker, 1963, 1969, 1983; Stalker and Wyder, 1983).

The spatial architecture of tills as outlined in the case studies reviewed above has been explained largely as a function of sediment advection from erosional to depositional zones in sub-marginal settings, augmented at some ice margins with glacitectonic folding and thrust stacking. Critical to the functioning of the process of incremental thickening, punctuated aggradation or propagating waves is 'till continuity' (Figure 16.5). In addition to subglacial deformation (Boulton, 1996a; Evans and Hiemstra, 2005; Eyles *et al.*, 2011), the feeding of sub-marginal till stacks is achieved at the same time by Alley *et al.*'s (1997) integrated system of sediment transfer (Figure 1.4), illustrated above by examples of meltwater networks and glacitectonic folding and thrusting but also involving ice-marginal processes such as supercooling and melt-out till formation. If a till is to be maintained elsewhere on an ice sheet or glacier bed, it must be replenished by processes operating up flow, hence the concept

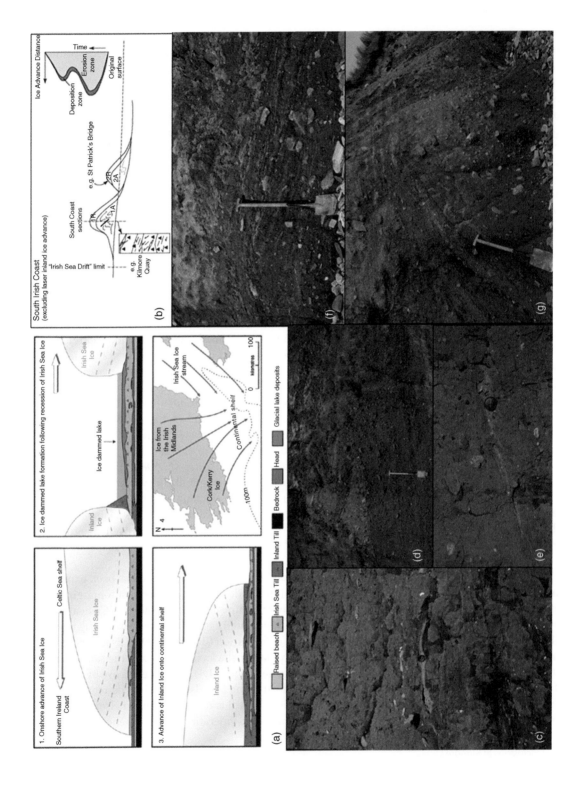

(a)

1. Onshore advance of Irish Sea ice

Southern Ireland Coast

Celtic Sea shelf

Irish Sea Ice

Inland Ice

2. Ice dammed lake formation following recession of Irish Sea ice

Ice dammed lake

Irish Sea Ice

Inland Ice

Irish Sea Ice stream

Ice from the Irish Midlands

Cork/Kerry Ice

Continental shelf

N

100m

0   100

kilometres

3. Advance of Inland Ice onto continental shelf

Inland Ice

Raised beach | Irish Sea Till | Inland Till | Bedrock | Head | Glacial lake deposits

(b)

South Irish Coast
(excluding laser inland ice advance)

Ice Advance Distance

Time

Deposition zone

Erosion zone

Original surface

e.g. St Patrick's Bridge

2A

South Coast sections

1A

"Irish Sea Drift" limit

e.g. Kilmore Quay

(c)

(d)

(e)

(f)

(g)

of till continuity (Alley, 2000; Iverson, 2010). In Chapter 6, it was demonstrated that material was added to the subglacial traction zone in a variety of ways, including regelation and melting, ice keel ploughing and shear zone development, making up for till loss by down-ice advection. However, once a till reaches a critical threshold thickness, it will seal off the source of further till-forming ingredients. Cuffey and Paterson (2010) provide an assessment of how deforming layer tills are maintained or sustained over time at any one location, specifically identifying the time variation in till thickness (h) as dependent on debris supply and outflow:

$$\frac{\partial h}{\partial t} = \dot{s}_i + \dot{s}_b - \nabla \cdot q_b - \nabla \cdot q_w,$$

where $\dot{s}_i$ is till deposition rate from the ice, $\dot{s}_b$ is the rate of erosion from the underlying substrate, q_b is the till flux conveyed by shear, and q_w is the sediment flux transported by water.

Hence, Iverson (2010) highlights that wherever a till layer of more than a few decimetres thickness occurs at the ice–bed interface, $\dot{s}_b$ will tend to be small, thereby initiating a depletion in the till layer unless areas of the substrate are exposed up flow and can be eroded to provide material to till further down flow (Cuffey and Alley, 1996; Alley, 2000). Till continuity can be compounded in hard substrate settings because meltwater drainage systems can flush out subglacial sediments, although, as we have seen above, deforming tills and subglacial meltwater channels appear to be able to co-exist and indeed operate as a coupled system of sediment advection (Figure 6.13) to produce sliding bed deposits (Chapter 11).

Although the thickening and stacking/overprinting of multiple tills has been related above to former sub-marginal environments, and such environments have been acknowledged as likely being over-represented in till stratigraphic studies, there is a growing knowledge base on the nature and likely behaviour of tills in more subglacial locations. The inaccessibility of modern ice sheet beds was highlighted in earlier chapters as representing a severe challenge to the establishment of a more confident set of definitive predictions on subglacial process–form regimes, but recent successes in remotely sensing large areas of ice sheet beds have given us a tantalising glimpse of the spatial and temporal operation of deforming beds. Repeat observations have enabled Smith *et al.* (2007) and Smith and Murray (2009) to monitor the changing mosaic of sliding/erosion and subglacial bedform construction beneath the Rutford Ice Stream, Antarctica. Bed elevation changes between 1991 and 2004 indicated a broadly small thickening or thinning of a deforming layer overlying a

Figure 16.16 The sub-marginal stratigraphy of the 'Irish Sea Till' and associated deposits of the southern Irish Sea Basin (after Ó Cofaigh and Evans, 2001a, b; Evans and Ó Cofaigh, 2003): (a) reconstructions of the ice dynamics and associated deposition of glacigenic sediments along the Irish south coast (from Ó Cofaigh *et al.*, 2012) – (1) onshore advance of the Irish Sea Ice Stream onto the coastline, depositing the Irish Sea Till as a glacitectonite derived from offshore stratified deposits, (2) withdrawal of Irish Sea ice from the coastline and formation of ice dammed lakes and sedimentation in coastal embayments, (3) offshore flow of inland ice from southwest and central Ireland onto the continental shelf forming a series of inland tills which cap the stratigraphic sequence, (4) ice advance onto the continental shelf; (b) idealised cross-section through the sediment–landform assemblages of the southern Irish coast in the context of the time–distance diagram of erosional and depositional zones and advance and retreat tills proposed by Boulton (1996a). The 'advance tills' and 'retreat tills' in this location are glacitectonically disturbed proglacial sediments and glacitectonites (Irish Sea Till), as typified by Kilmore Quay; (c)–(g) examples of the glacitectonites that comprise the Irish Sea Till, showing pseudo-stratified or pseudo-laminated diamictons, with attenuated and folded inclusions of pre-existing stratified sediments cannibalised from the Irish Sea Basin, including: (c) glacitectonic lamination at Kilmore Quay; (d) periglacial 'head' drawn upwards into Irish Sea Till; (e) attenuated gravel inclusions at Kilmore Quay; (f) laminated diamicton overlain by homogenised massive diamicton at Whiting Bay; (g) glacitectonic lamination at Whiting Bay.

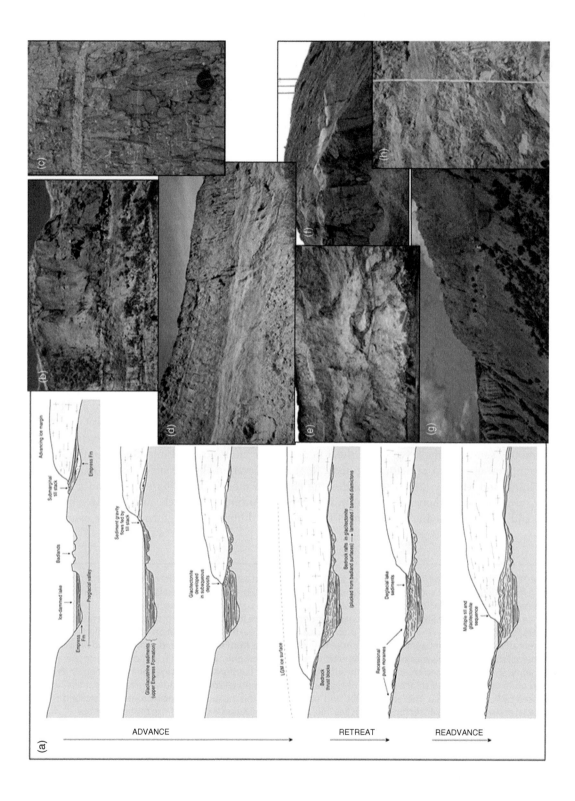

(a)

ADVANCE

RETREAT

READVANCE

Advancing ice margin

Empress Fm

Subaqueous till stack

Ice-dammed lake

Bedrock

Preglacial valley

Empress Fm

Glaciolacustrine sediments (upper Empress Formation)

Sediment gravity flows fed by till stack

Glaciotectonite developed in subaqueous deposits

LGM ice surface

Bedrock thrust blocks

Bedrock rafts in glaciotectonite (plucked from bedded surfaces) → laminated / banded diamictons

Recessional push moraines

Deglacial lake sediments

Multiple till and glaciotectonite sequence

non-deforming substrate, thickening being greatest at the location of emerging drumlins with mean heights of 22 m and many extending for up to 2 km along flow (Figure 16.19). This verifies not only the models of till mosaics described above but also that pervasive deformation operates up to at least a few metres into the bed and can persist over an area of many square kilometres (cf. King *et al.*, 2007, 2009). Interestingly, evidence of canals operating at the ice–till interface was also identified by King *et al.* (2004) and Murray *et al.* (2008), verifying theoretical models of alternating deforming and sliding subglacial regimes.

The existence of a deforming and streamlined, presumed dilatant, till layer up to a few metres thick overlying a thicker substrate of presumed stiff till, is typical of extensive areas of the deglaciated continental shelf around Antarctica. In those ancient tills, basal diamictons with high shear strengths are capped by low-strength diamictons up to 5 m thick with surfaces displaying well-developed MSGL. The latter are thought to be a record of coupling of ice streams and their deforming till layer, with all meltwater being evacuated through the till matrix (e.g. Dowdeswell *et al.*, 2004; Ó Cofaigh *et al.*, 2005). As was discussed in Chapter 8, the origins of such diamictons and their associated bedforms are problematic in terms of modes of till emplacement; for example, clast macrofabrics indicated that there is a contrast between Spagnolo *et al.*'s (2016) apparently strongly sheared MSGL tills and the apparently non-pervasively deformed MSGL tills of Ó Cofaigh *et al.* (2013), a contrast that may be related to grain size variability but potentially also the strength of coupling between the shearing medium and the overriding ice. The occurrence of relatively thick (<5 m) low-shear-strength diamictons with strongly fluted surfaces are more difficult to reconcile with the four outcomes used at the beginning of this chapter to explain similar sub-marginally emplaced diamictons, with the exception of outcome (1) that they are A horizons as defined by Boulton (1979), Boulton and Jones (1979), Boulton and Hindmarsh (1987) and Benn (1995) and potentially outcome (3) that they are mass flow diamictons produced by till flowage during deformation in a shear zone that becomes increasingly saturated as the melt season advances. Both scenarios are potentially enhanced by the fast flow of marine-based ice streams as well as their operation close to buoyancy. The impact of ice–bed decoupling and sediment moulding during the later stages of till emplacement have been invoked by Eyles *et al.* (1999, 2011) and Boone and Eyles (2001), applying earlier concepts of Stalker (1960), to explain till surface features such as 'brain-like patterns', 'humdrums', 'rim-ridges' and various hummocky landforms that grade up-flow into drumlins and flutings (Figures 16.7 and 16.20). This model of relatively low-strain ice pressing may not be always manifest in such well-developed

Figure 16.17 The sub-marginal 'multiple till' stratigraphies of the Southern Alberta Ice Stream on the Canadian prairies (after Evans, 1994; Evans *et al.*, 2008, 2012a, 2014): (a) Idealised reconstructions of the ice dynamics and sedimentary processes hypothesised as the origins of the thick diamicton sequences of southern Alberta. This communicates two, often co-existing, depositional scenarios, including: (1) glacier-marginal till thickening and stacking of till wedges at temporarily stationary ice stream margins and (2) proglacial lake and valley infilling with glacilacustrine rhythmites and mass flow diamictons associated with the advection of sub-marginal till into preglacial bedrock depressions. Added to this are the processes of glacial overriding/deformation of valley infills and the dislocation and entrainment of bedrock rafts, leading to the widespread development of glacitectonites and megablocks. (b) Multiple massive to laminated and deformed diamictons overlying preglacial gravels and including rafts of Cretaceous bedrock, Wolf Island. (c) Laminated diamicton and sandy gravel lens, Bain Bluff, Medicine Hat. (d) Massive diamicton overlying heterogeneous and stratified diamictons, One Tree Creek, Dinosaur Provincial Park. (e) Heavily contorted massive diamicton in the middle of heterogeneous/stratified diamicton, One Tree Creek, Dinosaur Provincial Park. (f) Cretaceous bedrock rafts in diamictons at Fort Whoop Up, Lethbridge. (g) Multiple diamictons, including stratified, laminated and massive varieties, Bain Bluff, Medicine Hat. (h) Heterogeneous diamicton or melange, Little Sandhill Creek, Dinosaur Provincial Park.

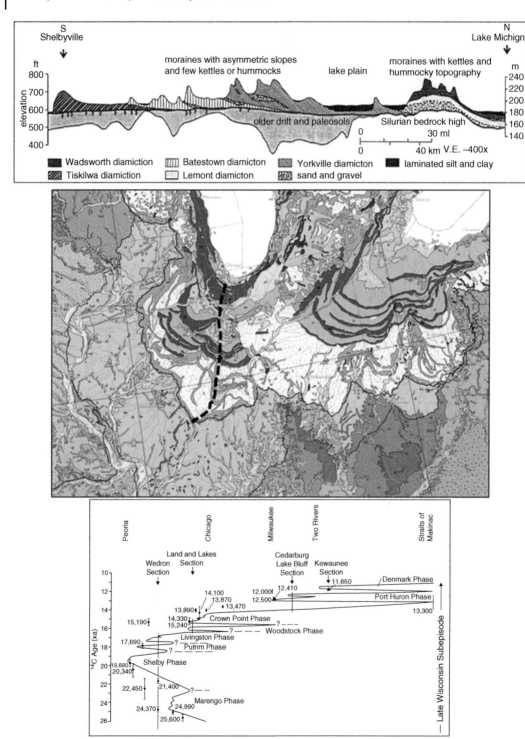

Figure 16.18 The products of ice sheet marginal till thickening, in the form of stacked till sequences or wedges that mark the oscillating lobes of the southern Laurentide Ice Sheet in Illinois (after Johnson and Hansel (1999). Upper panel shows the architecture of the superimposed till wedges along the transect marked by the dashed line in the middle panel/map of the moraines in the region (from Fullerton *et al.*, 2003). Lower panel shows the chronology of ice marginal oscillations along the transect.

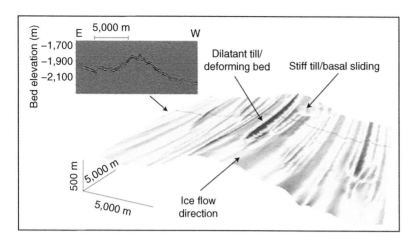

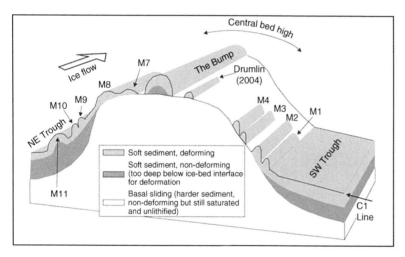

Figure 16.19 Selected results from the remote sensing of evolving subglacial bedforms beneath the Rutford Ice Stream, Antarctica: Upper panel shows a three-dimensional image of the bed looking in the downstream direction and colour shaded based on the difference between the short-wavelength topography and a long-wavelength trend surface. The topography is dominated by MSGLs. Inset box shows an example radar profile (from King *et al.*, 2009). Lower panel shows a similar view with specific bedforms highlighted (from Smith and Murray, 2009).

landform generation, as depicted in Figure 16.20, but instead in weak fabrics, localised till flowage and gravity loading structures associated with subsequent outwash sedimentation (e.g. Figures 15.17 and 15.18).

Where till continuity is restricted on glacier beds, most usually in hard-bedded terrains, tills will thin out, especially over bed protuberances. This can take place even in lowland settings where bedrock protrudes above the feather ends of sub-marginally thickening wedges. Consequently, relatively intensive bedrock quarrying can proceed even late on in a glacial cycle due to the gradual thinning or even complete removal of a protective deforming layer. In Chapter 14, this was proposed as a mechanism for introducing freshly plucked bedrock blocks to subglacial till layers (Figure 14.5), especially where well-jointed bedrock is prone to till/glacitectonite being squeezed into its crevices

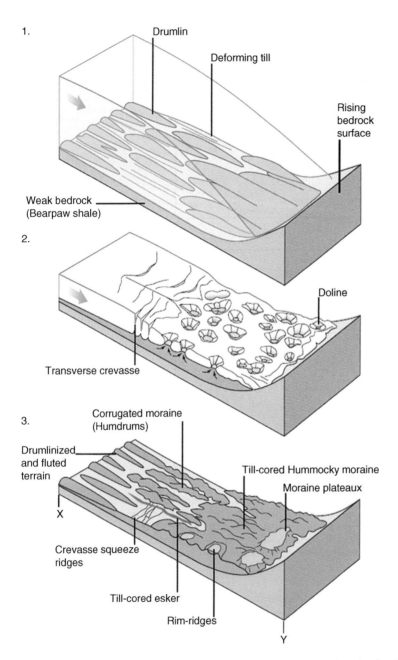

Figure 16.20 Conceptual model proposed by Eyles *et al.* (1999) to explain the development of subglacially moulded (squeezed) terrain comprising till-cored hummocks, plateaux, rim ridges and corrugated ridges or 'humdrums'. Note also the association with till eskers and crevasse-squeeze ridges.

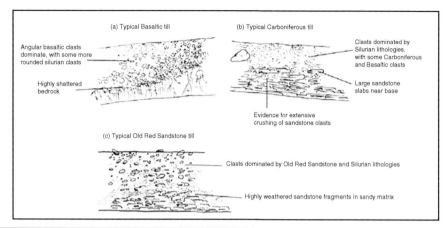

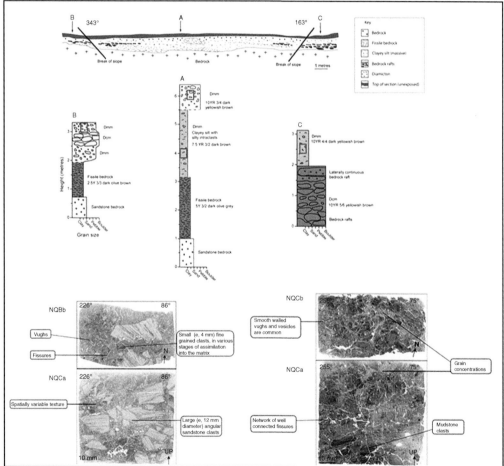

Figure 16.21 Details of the subglacial materials in a relatively sediment-depleted zone of the Tweed palaeo-ice stream bed. Upper panel shows some typical thin till profiles overlying various bedrock types on the ice stream bed (from Kerr, 1978). Lower panel shows section sketch and vertical profile logs of the outcrop depicted in photograph in Figure 16.6 and displays the nature of the glacitectonised sandstone bedrock and overlying amalgamation zone of bedrock rafts and till. Thin section images show the details of the inheritance of bedrock fragments in the till at this site rendering it similar to pulverised bedrock/melange (from Channon, 2012).

(Broster *et al.*, 1979; Harris, 1991; Evans *et al.*, 1998; Figure 14.5). Alternatively, in the absence of a deforming layer till, shear zones may migrate downward by exploiting sub-horizontal jointing or bedding (Knill, 1968; Money, 1983; Harris, 1991; Hiemstra *et al.*, 2007; Evans and Ó Cofaigh, 2008). In smaller mountain glacier systems (e.g. Evans *et al.* 2016), an abnormally wide range of clast angularity values in subglacial tills may reflect the localised input of freshly plucked and hence relatively highly angular blocks to the deforming layer, a characteristic of stepped bedrock profiles. On lowland ice stream beds, the juxtaposition of rock-cored and sediment-cored drumlins has been explained by the instability theory and the ELH (Figures 10.6 and 16.5), wherein the downward excavation of the streamlined interface can reach the bedrock substrate if till supply is restricted. Hence, till continuity controls both deforming bed thickness and the localised cannibalisation and or mobilisation of some bedrock lithologies. This is well illustrated on the bed of the Tweed palaeo-ice stream in Scotland where rock-cored drumlins are overlain by patchy or thin tills with locally sourced bedrock rafts (Kerr, 1978; Channon, 2012); the upper zone of the bedrock is commonly glacitectonised, plucked of angular clasts and partially amalgamated into the till (Figures 16.6 and 16.21), thereby constituting a vertical glacitectonite – till continuum in a zone of an ice stream where till supply was close to exhaustion.

17

Concluding Remarks: The Case for a Simplified Nomenclature

Out of intense complexities, intense simplicities emerge.

Winston Churchill

Diamictons are one of the most ubiquitous deposits on the Earth's surface and in the bedrock record. Although they are mostly associated with glacial and/or paraglacial processes in glaciated basins, the role of gravity mass flows in both subaerial and subaqueous settings and rafting (sea ice and seaweed in addition to iceberg rafting; cf. Reimnitz and Kempema, 1988; Gilbert, 1990) are significant in diamicton production in environments beyond any direct glacial influence. Our expanding knowledge base on such non-glacial diamictons has fostered an appreciation that large volumes of such material will inevitably dominate over subglacial diamictons ('tills' as they are defined in this book) in glaciated basins. Moreover, the differentiation of tills and other glacigenic diamictons requires intensive sedimentological investigation at a range of scales from thin section right up to regional architecture. Such investigations allow us to differentiate between subglacial versus supraglacial and other ice-contact deposits and hence isolate diagnostic criteria of sediment production in the subglacial traction zone. Traction zones are critical to the genetic classification of many deposits, because they relate sedimentological process to form, whether it be on river beds or on aeolian dune surfaces. Hence, Evans *et al.* (2006b) proposed the term 'subglacial traction till' for diamictons created within the traction zone (i.e. the ice–bed interface) of glaciers and ice sheets.

Simplified nomenclatures have been proposed on a number of occasions throughout the history of till research. For example, it was explained in Chapter 3 that the umbrella term 'subglacial till' was entertained by Dreimanis (1989), based upon numerous acknowledgements (e.g. Anderson *et al.*, 1980, 1986; Kemmis, 1981; Bergersen and Garnes, 1983; Dreimanis, 1983; Lundqvist, 1983; Stephan and Ehlers, 1983; Ringberg *et al.*, 1984; van der Meer *et al.*, 1985; Rappol, 1985; Hansel and Johnson, 1987) that the sedimentological signatures of lodgement, melt-out, deformation and undermelt could not be unequivocally diagnosed in till deposits. Nevertheless, glacial sedimentologists have invested much effort in generating a genetic classification scheme for till, especially subglacial till, that purveys a sense of precision that we now view as unlikely to be realistic, based upon recent developments of our understanding of glacier beds to be spatial and temporal mosaics.

Such spatial and temporal variability was acknowledged by Hicock's (1990, 1992) concept of a genetic 'till prism' or 'spaghetti prism' (Figure 17.1), in which the broad range of tills and related diamictons could be viewed three-dimensionally as a continuum and even used to depict the deformation history/changing rheology of materials emplaced at one location by different ice lobes. The spaghetti prism in particular engaged with the changing ductile to brittle styles of deformation as

Till: A Glacial Process Sedimentology, First Edition. David J A Evans.
© 2018 John Wiley & Sons Ltd. Published 2018 by John Wiley & Sons Ltd.

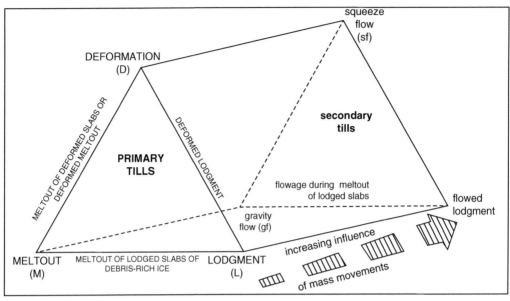

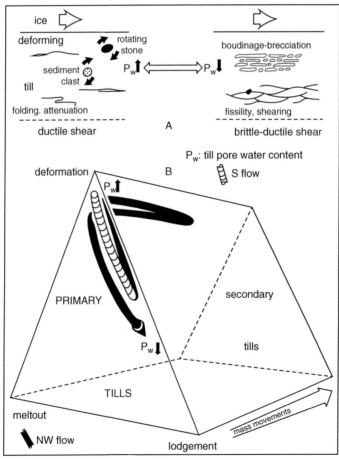

Figure 17.1 Hicock's (1990, 1992) concepts of the genetic 'till prism' and 'spaghetti prism'.

well as the tendency for a till to be subject to flowage, hence the divergence of the till behaviour arrow away from deformation and lodgement modes towards mass movements or 'secondary tills'. The wider spatial implications of the overprinting of tills, as identified by Hicock's (1992) model, are captured in Boulton's (1996a, b) migrating erosional and depositional zones over time and the concomitant development of partially superimposed advance and retreat tills. The continuous evolution of the subglacial meltwater system and its interaction with, and dynamic control of, the deforming till bed must also be incorporated into such spatio-temporal models of till genesis. Because the dynamics of the overriding ice must also be taken into account, for example, in situations where ice motion is driven more by mechanisms external to the deforming/sliding bed such as surging, calving and drawdown and ice quakes, Phillips *et al.* (in press) have compiled a summary model of ice–bed interface processes (Figure 17.2) that evaluates the combined roles of deformation, meltwater and ice fracture (quakes). This creates an environment that they term the 'transient mobile zone' (also 'active layer' of Evans *et al.*, 2006b) in which spatially and temporally restricted liquefaction events lower the cohesive strength of tills and thereby cause them to accommodate the shear imposed by the overriding ice by a process called 'soft-bed sliding'. Either side of this state, both temporally and spatially, raised or reduced water contents can bring about basal sliding or sticky spot states, respectively; the potential for overprinted sedimentological and structural signatures in till deposits is therefore infinite, rendering classifications specific to individual processes – such as 'lodgement', 'squeeze flow' and 'comminution', for example – at best premature but most likely untenable. All such deposits have certainly been deformed, but 'deformation till' is also an inappropriate term, largely because it is redundant, but more specifically because the alternative 'lodgement till' relates to only one process component involved in the emplacement of subglacially sheared materials.

A three-fold classification scheme for subglacial materials was proposed in Chapter 3, following recommendations by Evans *et al.* (2006b), and includes 'subglacial traction till', 'glacitectonite' and 'melt-out till'. The process–form regimes that are inherent within the largely overprinted depositional and structural signatures in subglacial traction tills include lodgement, cavity fill, deformation, soft-bed sliding/ploughing and sliding bed meltwater flow (see Chapters 6, 9–11). A depositional continuum exists between subglacial traction tills and glacitectonites where the latter are progressively homogenised after being liberated by deformation processes from pre-existing materials. Benn and Evans (2010) argued for the inclusion of melt-out till in the subglacial traction till classification scheme, largely to accommodate the important observations of Ham and Mickelson (1994) at Burroughs Glacier, Alaska, where melt-out till appears to underlie a subglacial till surface and hence was likely overrun before it had melted out. The implications of these observations are that the tills have undergone multiple cycles of emplacement and that the melt-out/debris-rich ice facies are part of an overridden apron and hence are inextricably linked with subglacial processes as a type of glacitectonite. Hence, it is important to maintain melt-out till as a separate end member, despite its poor preservation potential (Chapter 13).

The term 'subglacial traction till' may ultimately be regarded as somewhat cumbersome, in that the deliberations on genetic labelling as presented in the preceding chapters of this book have aligned the term 'till' specifically and exclusively with the subglacial traction zone and hence till cannot be produced in any other location. Hence, the words 'subglacial traction' are strictly redundant but retained at this juncture in order to emphasise the specificity of the process–form regime to which 'till' is allocated. 'Subglacial traction till' is defined as:

> Sediment deposited at a glacier sole, the sediment having been released directly from the ice and/or liberated from the substrate and then disaggregated and completely or largely homogenised during transport.

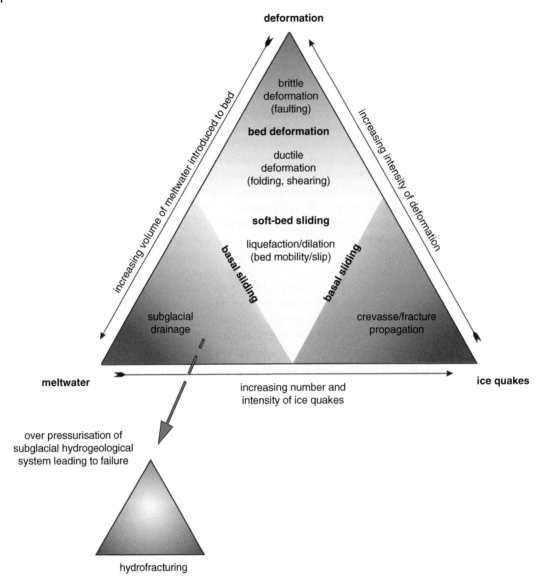

Figure 17.2 Schematic ternary diagram showing the relative effects of deformation, increased meltwater and ice quakes as potential triggers for soft-bed sliding versus bed deformation versus basal sliding as the main mechanism for glacier motion (from Phillips *et al.*, in press).

This definition encompasses products that may be at least partially recognisable in specific, but commonly overprinted, sedimentological or structural signatures, including: (i) ploughing and lodgement; (ii) localised lee-side cavity filling; (iii) stratified sediments deposited by water films and channels associated with sliding bed processes; (iv) localised pseudo-lamination or vinaigrette structures created by stick-slip/liquefaction or soft-bed sliding.

The raw materials for subglacial traction till are derived from melt-out from the ice base, quarrying and abrasion of hard rock surfaces where they protrude into the shear zone, and the excavation/

cannibalisation of particles or particle aggregates (e.g. tectonic slices) from weaker substrates. Within the traction zone, sediment can be continuously reworked in water films and canals at the ice–sediment interface or within the till. High till porewater contents can also result in mixing through liquefaction. Subglacial deformation can alternate with ice–bed decoupling and sliding, debris freeze-on and melt-out, especially in sub-marginal settings. Deformation takes place down to only a few metres at most, often at depths of less than 1 m, and in more heterogeneous materials will be partitioned according to sediment rheological properties. In homogenous tills, zones of maximum strain will migrate vertically in response to changing porewater pressures, even on a diurnal basis, to create complex cumulative strain patterns (Figure 17.3). In any location, the production of subglacial traction till relies upon continuity of debris supply and hard rock beds are less likely to maintain that supply as well as soft substrates and/or pre-existing sediment piles. Hence, lowland terrains are prime locations for the maintenance of subglacial deforming layers, comprising first glacitectonites and subsequently subglacial traction tills. As Iverson (2010) demonstrates, it is also meltwater that plays a significant role in modulating till continuity, and hence we should expect till stratigraphies to display significant stratified components. With these controls on subglacial traction till production in mind, Boyce and Eyles (2000) provided a simplified sedimentological and stratigraphic approach to describing and interpreting the spatial and temporal mosaics of typical subglacial sequences based on architectural element analysis (Figure 17.4). Superimposed subglacial deformation events, punctuated by phases of ice–bed separation and subglacial meltwater sheet flow typically produce three architectural elements with characteristic internal structures, separated by and including seven orders of bounding surface. The 'deformed zone' (DZ) comprises glacitectonised pre-existing sediment or glacitectonite. This is commonly overlain by multiple 'diamict elements' (DE), which can be separated by erosional surfaces, clast pavements and 'inter-bed elements' (I) of stratified sediment.

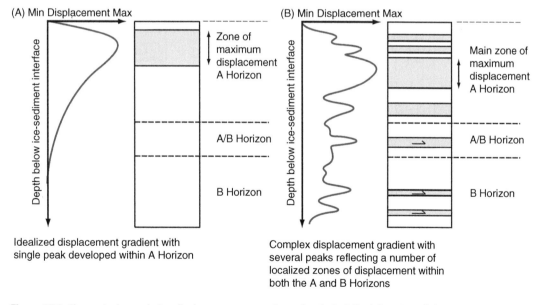

Figure 17.3 Theoretical cumulative displacement curves through subglacially deforming till showing (left) a simple vertical decrease in displacement, as portrayed in the Breiðamerkurjökull case study and (right) a complex curve depicting multiple failure loci due to deformation partitioning in which the displacement curve represents the differential horizontal displacement with depth. Areas shaded blue are those in which significantly greater numbers of failure events take place.

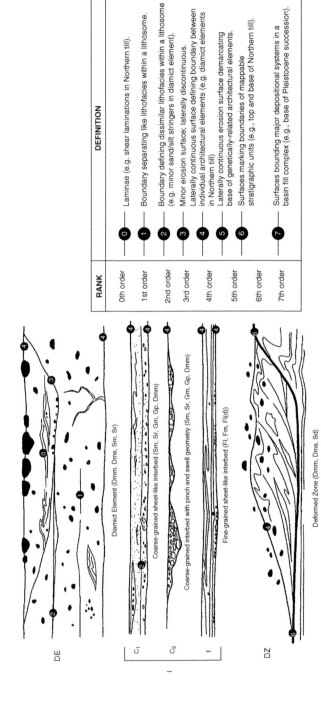

RANK	DEFINITION
0th order	Laminae (e.g. shear laminations in Northern till).
1st order	Boundary separating like lithofacies within a lithosome.
2nd order	Boundary defining dissimilar lithofacies within a lithosome (e.g. minor sand/silt stringers in diamict element).
3rd order	Minor erosion surface; laterally discontinuous.
4th order	Laterally continuous surface defining boundary between individual architectural elements (e.g. diamict elements in Northern till).
5th order	Laterally continuous erosion surface demarcating base of genetically-related architectural elements.
6th order	Surfaces marking boundaries of mappable stratigraphic units (e.g., top and base of Northern till).
7th order	Surfaces bounding major depositional systems in a basin fill complex (e.g., base of Pleistocene succession).

DE — Diamict Element (Dmm, Dms, Sm, Sr)

c_1 — Coarse-grained sheet-like interbed (Sm, Sr, Gm, Gp, Dmm)

c_2 — Coarse-grained interbed with pinch and swell geometry (Sm, Sr, Gm, Gp, Dmm)

f — Fine-grained sheet-like interbed (Fl, Fm, Fl(d))

DZ — Deformed Zone (Dmm, Dms, Sd)

ARCHITECTURAL ELEMENTS IN NORTHERN TILL		CODE	OUTCROP (2-D) GEOMETRY	APPROX. SCALE	L/T	LITHOFACIES ASSEMBLAGE	INFERRED PROCESSES	FIGS.
Diamict Element		DE	tabular diamict beds, planar to gently undulating bounding contacts; boulder pavement often marking upper surface	>100 (L); <10 m (T); $>10^3$ m² (A)	>10	Dmm,Dms, Dms, Sm, Sr	subglacial aggradation of deformation till units	4:6, 4:9A, 4:9B
Interbed	Coarse	$I\text{-}c_1$	laterally continuous, sheet-like sands and gravels separating diamict elements	>100 m (L); <5 m (T); 10^3 m² (A)	>25	Sm, Sr, Sp Gm, Gp, Dmm	ice-bed separation; erosion and deposition by subglaciofluvial meltwater sheet-flow	4:4, 4:6, 4:12A
		$I\text{-}c_2$	laterally discontinuous sand gravel body, With pinch and swell geometry	<10 m (L); <1 m (T); 10^2 m² (A)	>10	Sm, Sp, Gm, Gcm	ice-bed separation; localized inclusion by subglaciofluvial meltwater sheet-flow	4:3, 4:6, 4:10
	Fine	$I\text{-}f$	laterally continuous tabular silt and mud units separating diamict elements	>10 m (L); <1 m (T); 10^2 m² (A)	>10	Fl, Fm, Flr, Fl	ice-bed separation; low energy sedimentation in subglacial water body	4:6, 4:7B
Deformed Zone		DZ	undulatory zone of deformed till and thrusted sediments at base of till sheet; variable thickness and spatial extent	variable; <10 m (T); 10^2 m² (A)	?	Dmm, Dms + Included sub-till sediments	subglacial deformation of pre-existing strata	4:6, 4:912B

1 m

Figure 17.4 Schematic illustration and classification tables of architectural element types with typical bounding surfaces as represented by interbeds (from Boyce and Eyles, 2000).

The conceptual model of Evans *et al.* (2006b) is now used in a modified form (Figure 17.5) to summarise the sedimentology, stratigraphy and architecture of subglacial traction till, glacitectonite and associated ice–bed interface deposits. Vertical profile logs A-G show the typical sequences of subglacial deposits in relation to different types of dominant ice–bed interface processes, but not all are necessarily coeval. The graphs to the right of each vertical profile log indicate the likely relative differential pattern of cumulative horizontal displacement/strain created by deformation partitioning. The light blue shaded area of the glacier bed represents the subglacial traction till or till sequence.

Sequence A represents glacitectonite developed in pre-advance lake sediments (DZ) capped by a subglacial traction till acting as an erodent layer. Such localised pockets of soft substrate are critical to till continuity in upland landscapes and areas of thin tills over scoured bedrock (e.g. Figures 14.4b and 14.23). Sequence B represents lee-side cavity fill diamicts and crudely stratified sediments capped by subglacial traction till where the ice has re-coupled with the bed due to cavity closure and sheared the upper-cavity fill beds (e.g. Figure 9.5b). In both scenarios A and B, the tops of the sub-till sediments are classified as glacitectonite which is locally cannibalised and dragged into the base of the deforming till layer after disruption due to ice-keel ploughing, clast ploughing and/or migrations in the locus of failure planes in subglacial sediment, aided by cyclical changes in porewater pressure. This produces a pseudo-laminated diamicton in which banding is composed of discontinuous stringers or tectonic slices derived from the pre-existing sediment piles. Over distance, such lamination is gradually destroyed by various processes of homogenisation into A and B horizon tills.

The family of sequences in box C represents sheared tills with partially to intensively glacitectonised stratified intra- and inter-beds produced over hard beds: (i) multiple diamict elements (DE) separated by stratified inter-beds ($I - c_1$ and $I - c_2$) are continually aggraded due to continuous replenishment of deformable sediment from up-ice to create multiple subglacial traction tills and ice–bed separation deposits such as canal fills (e.g. Figure 11.4); (ii) progressive lodgement of large clasts to produce a clast pavement at the base of a thin subglacial traction till (erodent layer) due to finer-grained material being continuously advected down-ice (e.g. Figure 10.7); (iii) a single subglacial traction till (erodent layer) with associated glacitectonised canal fill at a site where aggradation is restricted by poor till continuity (e.g. Figure 11.7); (iv) regelation of subglacial till (light blue shade on log) leads to the stacking of tills which later melt out to form melt-out till or are released to the subglacial traction zone down flow. The example in C1 reflects the localised influence of liquefaction and flow processes due to stick-slip events and/or ice quakes.

Sequence D depicts pre-advance or remobilised melt-out till overlain by subglacial traction till. If the melt-out sequence is not fully de-iced, this might represent a scenario similar to that reported for the Burroughs Glacier by Ham and Mickelson (1994). The top of the melt-out till or melt-out sequence is classified as glacitectonite and is locally cannibalised and dragged into the base of the deforming till layer after folding/thrusting is induced in its upper layers. If not de-iced, it can also be effectively re-coupled with the deforming layer and glacier ice to produce an apron incorporation effect (*sensu* Shaw, 1977; Evans, 1989; Fitzsimons, 1990).

Sequence E represents glacitectonised soft bedrock grading upwards into subglacial traction till with bedrock rafts, created in areas where subglacial till thins over bedrock high points and/or is depleted by poor till continuity. Sequence E1 shows how this may also display a vertical continuum of dislocated bedrock overlain in turn by fractured and then pulverised bedrock capped by homogenised, largely mono-lithological diamicton (e.g. Figure 14.6–14.9). Short travel distances are often reflected in the freshly plucked angular bedrock blocks and the deforming layer till may become injected into rock crevices to aid in the plucking process (Figure 14.5).

The two sequences in box F represent glacitectonite, derived from glacially overridden lake sediments, overlain by subglacial traction till, which are stacked in a sub-marginal setting:

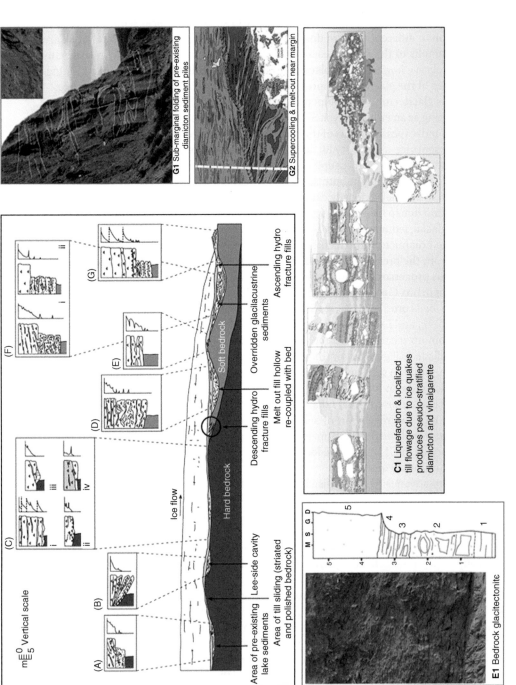

Figure 17.5 Conceptual model of the full spectrum of till production and depositional mechanisms resulting in a continuum of till formation (modified from Evans et al., 2006). The processes depicted are those that would be expected beneath a large, predominantly temperate outlet glacier with a seasonally frozen snout, although supercooling (G2) might be in operation in some settings. Larger, ice-sheet-scale processes of ice-marginal deformation and stacking of pre-existing deposits is represented by case G1. The shaded area of the glacier bed represents the subglacial traction till (STT). The final thickness of a till deposit does not reflect the thickness of the subglacial traction/deforming layer at any one point in time. Not all the process–form products would necessarily be coeval and signatures may be overprinted due to glacier advance and retreat cycles. The graphs to the right of each vertical profile log indicate the likely relative pattern of cumulative strain or more specifically differential horizontal displacement with depth.

(i) glacitectonite locally cannibalised and dragged into the base of the deforming layer after folding/ thrusting is induced at its upper interface with the till. Thick sequences of glacitectonite are common in such settings and can display strong inheritance of primary bedding; (ii) glacitectonite is eroded by the base of the till so that the contact between the two sediment bodies is essentially a fault gouge (see Figures 8.12b, 10.9, 10.13 and 13.9).

Sequence G represents a typical sub-marginal to marginal stack of glacifluvial outwash or lake sediment, glacitectonised in its upper layers and overlain by subglacial traction tills, often displaying A and B horizons which are commonly overprinted (Figures 6.9a, 8.6a, 8.12 and 8.13). Each till unit is emplaced by the sub-marginal incremental thickening process observed at modern temperate glacier snouts, where stacking can result from quasi-stationary glacier margins and slabs of till are delivered over a seasonal cycle of freeze-on, deformation and melt out. Note that some tills are thin and contain clast pavements due to their development at the up flow ends of till wedges. Sequence G1 represents a similar ice-marginal thickening of tills and glacitectonites but where substantial accumulations of mass flow diamictons, fed by gravity-induced slumping of ice-marginal till aprons into either glacilacustrine or subaerial fans, have been folded and stacked (e.g. Figures 16.16 and 16.17). Sequence G2 depicts the stratigraphy created in situations where bodies of debris-charged supercooled ice have melted out above subglacial traction tills (e.g. Figure 13.4). The long-term preservation potential of such loosely packed sediments is very low indeed, and hence interpretations of large exposures of pseudo-stratified diamictons in the geological record as melt-out till must be able to not only account for the unusual circumstances of their preservation but also confidently dismiss alternative, simpler explanations such as glacitectonite production.

References

Aario, R., 1977. Classification and terminology of moraine landforms in Finland. *Boreas* **6**, 87–100.

Aber, J.S., 1982. Model for glaciotectonism. *Bulletin of the Geological Society of Denmark* **30**, 79–90.

Aber, J.S., 1985. The character of glaciotectonism. *Geologie en Mijnbouw* **64**, 389–395.

Aber, J.S., Croot, D.G., Fenton, M.M., 1989. *Glaciotectonic landforms and structures*. Kluwer, Dordrecht.

Alden, W.C., 1905. The drumlins of Southeastern Wisconsin. *USGS Bulletin* **273**, 44 pp.

Alden, W.C., 1918. The Quaternary Geology of Southeastern Wisconsin: with a chapter on the Older Rock Formations. USGS Professional Paper **106**, 356 pp.

Alley, R.B., 1989a. Water pressure coupling of sliding and bed deformation: 1.Water system. *Journal of Glaciology* **35**, 108–118.

Alley, R.B., 1989b. Water pressure coupling of sliding and bed deformation: II. Velocity-depth profiles. *Journal of Glaciology* **35**, 119–129.

Alley, R.B., 1991. Deforming-bed origin for southern Laurentide till sheets? *Journal of Glaciology* **37**, 67–76.

Alley, R.B., 1992. How can low pressure channels and deforming tills coexist subglacially? *Journal of Glaciology* **38**, 200–207.

Alley, R.B., 1993. In search of ice stream sticky spots. *Journal of Glaciology* **39**, 447–454.

Alley, R.B., 2000. Continuity comes first: recent progress in understanding subglacial deformation. In: Maltman, A.J., Hubbard, B., Hambrey, M.J. (Eds.), *Deformation of Glacial Materials. Geological Society*, London, Special Publication, vol. 176, pp. 171–179.

Alley, R.B., Blankenship, D.D., Bentley, C.R., Rooney, S.T., 1986. Deformation of till beneath ice stream B, West Antarctica. *Nature* **322**, 57–59.

Alley, R.B., Blankenship, D.D., Bentley, C.R., Rooney, S.T., 1987a. Till beneath ice stream B: 3. Till deformation: evidence and implications. *Journal of Geophysical Research* **92**, 8921–8929.

Alley, R.B., Blankenship, D.D., Rooney, S.T., Bentley, C.R., 1987b. Till beneath ice stream B: 4. A coupled ice-till flow model. *Journal of Geophysical Research* **92**, 8931–8940.

Alley, R.B., Cuffey, K.M., Evenson, E.B., Strasser, J.C., Lawson, D.E., Larson, G.J., 1997. How glaciers entrain and transport basal sediment: physical constraints. *Quaternary Science Reviews* **16**, 1017–1038.

Alley, R.B., Lawson, D.E., Evenson, E.B., Strasser, J.C., Larson, G.J., 1998. Glaciohydraulic supercooling: a freeze-on mechanism to create stratified debris-rich basal ice: II. Theory. *Journal of Glaciology* **44**, 562–568.

Alley, R.B., Strasser, J.C., Lawson, D.E., Evenson, E.B., Larson, G.J., 1999. Glaciological and geological implications of basal-ice accretion in overdeepenings. In: Mickelson, D.M., Attig, J.W. (Eds.), *Glacial Processes Past and Present*. Geological Society of America Special Paper, vol. 337, pp. 1–9.

Till: A Glacial Process Sedimentology, First Edition. David J A Evans.
© 2018 John Wiley & Sons Ltd. Published 2018 by John Wiley & Sons Ltd.

Altuhafi, F.N., Baudet, B.A., Sammonds, P., 2009. On the time-dependent behaviour of glacial sediments: a geotechnical approach. *Quaternary Science Reviews* **28**, 693–707.

Åmark, M., 1986. Glacial tectonics and deposition of stratified drift during formation of tills beneath an active glacier – examples from Skåne, southern Sweden. *Boreas* **15**, 155–171.

Anandakrishnan, S., Alley, R.B., 1994. Ice Stream C, Antarctica, sticky spots detected by micro-earthquake monitoring. *Annals of Glaciology* **20**, 183–186.

Andersland, O.B., Alnouri, A., 1970. Time-dependent strength behaviour of frozen soils. *Journal of the Soil Mechanics and Foundations Division SM4*, 1249–1265.

Anderson, F.W., 1967. Robert George Carruthers. *Proceedings of the Geological Society, London* **1636**, 186–187.

Anderson, J.B., Kurtz, D.D., Domack, E.W., Balshaw, K.M., 1980. Glacial and glacial marine sediments of the Antarctic continental shelf. *Journal of Geology* **88**, 399–414.

Anderson, J.B., Goldthwait, R.P., McKenzie, G.D. (Eds.), 1986. *Observed processes of glacial deposition in Glacier Bay, Alaska.* Columbus Institute of Polar Studies, Ohio State University.

Andrews, J.T., 1971. *Techniques of Till Fabric Analysis.* British Geomorphological Research Group Technical Bulletin, Geo Abstracts.

Andrews, J.T., Shimizu, K., 1966. Three-dimensional vector technique for analysing till fabrics: discussion and Fortran program. *Geographical Bulletin* **8**, 151–165.

Andrews, J.T., Smithson, B.B., 1966. Till fabrics of the cross-valley moraines of north-central Baffin Island, NWT, Canada. *Bulletin of the Geological Society of America* **77**, 271–290.

Armstrong, J.E., 1949. Fort St. *James map area. Geological Survey of Canada Memoir* **252**.

Armstrong, J.E., Brown, W.L., 1954. Late Wisconsin marine drift and associated sediments of the Lower Fraser Valley, British Columbia, Canada. *Bulletin of the Geological Society of America* **65**, 349–364.

Ashley, G.M., Boothroyd, J.C., Borns, H.W., 1991. Sedimentology of late Pleistocene (Laurentide) deglacial-phase deposits, eastern Maine: an example of a temperate marine grounded ice-sheet margin. In: Anderson, J.B., Ashley, G.M. (Eds.), *Glacial marine sedimentation; paleoclimatic significance: Geological Society of America Special Paper*, **261**, pp. 107–125.

Astakhov, V.I., Isayeva, L.L., 1988. The 'ice-hill': an example of 'retarded deglaciation' in Siberia. *Quaternary Science Reviews* **7**, 29–40.

Astakhov, V.I., Kaplyanskaya, F.A., Tarnogradskiy, V.D., 1996. Pleistocene permafrost of West Siberia as a deformable glacier bed. *Permafrost and Periglacial Processes* **7**, 165–191.

Atre, S.R., Bentley, C.R., 1993. Laterally varying basal conditions beneath Ice Stream B and C, West Antarctica. *Journal of Glaciology* **39**, 507–514.

Atre, S.R., Bentley, C.R., 1994. Indication of a dilatant bed near downstream B camp, Ice Stream B, Antarctica. *Annals of Glaciology* **20**, 177–182.

Baligh, M.M., 1972. Applications of plasticity theory to selected problems in soil mechanics. Unpublished PhD thesis, California Institute of Technology, Pasadena.

Ballantyne, C.K., 2002a. Paraglacial geomorphology. *Quaternary Science Reviews* **21**, 1935–2017.

Ballantyne, C.K., 2002b. A general model of paraglacial landscape response. *The Holocene* **12**, 371–376.

Ballantyne, C.K., 2003. Paraglacial landsystems. In: Evans, D.J.A. (Ed.), *Glacial Landsystems.* Arnold, London, pp. 432–461.

Ballantyne, C.K., Benn, D.I., 1994. Paraglacial slopes adjustment and resedimentation following recent glacier retreat, Fåbergstølsdalen, Norway. *Arctic and Alpine Research* **26**, 255–269.

Ballantyne, C.K., Benn, D.I., 1996. Paraglacial slope adjustment during recent deglaciation: implications for slope evolution in formerly glaciated terrain. In: Brooks, S., Anderson, M.G. (Eds.), *Advances in Hillslope Processes.* Wiley, Chichester, pp. 1173–1195.

Banham, P.H., 1975. Glaciotectonic structures: a general discussion with particular reference to the contorted drift of Norfolk, In: Wright, A.E., Moseley, F. (Eds.), *Ice Ages: Ancient and Modern*. Seel House Press, Liverpool, pp. 69–84.

Banham, P.H., 1977. Glaciotectonites in till stratigraphy. *Boreas* **6**, 101–105.

Banham, P.H., 1988. Polyphase glaciotectonic deformation in the Contorted Drift of Norfolk, In: Croot, D. (Ed.), *Glaciotectonics: Forms and Processes*. Balkema, Rotterdam.

Beaumont, P., 1971. Break of slope in particle size curves of glacial tills. *Sedimentology* **16**, 125–128.

Beeler, N.M., Tullis, T.E., Blanpied, M.L., Weeks, J.D., 1996. Frictional behaviour of large displacement experimental faults. *Journal of Geophysical Research* **101**, 8697–8715.

Benediktsson, Í.Ö., Ingolfsson, Ö., Schomacker, A., Kjær, K.H., 2009. Formation of sub-marginal and proglacial end moraines: implications of ice flow mechanism during the 1963–64 surge of Bruarjökull, Iceland. *Boreas* **38**, 440–457.

Benn, D.I., 1994a. Fluted moraine formation and till genesis below a temperate glacier: Slettmarkbreen, Jotunheimen, Norway. *Sedimentology* **41**, 279–292.

Benn, D.I., 1994b. Fabric shape and the interpretation of sedimentary fabric data. *Journal of Sedimentary Research A* **64**, 910–915.

Benn, D.I., 1995. Fabric signature of till deformation, Breiðamerkurjökull, Iceland. *Sedimentology* **42**, 735–747.

Benn, D.I., 1996. Subglacial and subaqueous processes near a glacier grounding line: sedimentological evidence from a former ice-dammed lake, Achnasheen, Scotland. *Boreas* **25**, 23–36.

Benn, D.I., 2002. Clast fabric development in a shearing granular material: implications for subglacial till and fault gouge – Discussion. *Bulletin of the Geological Society of America* **114**, 382–383.

Benn, D.I., 2004a. Macrofabric. In: Evans, D.J.A. and Benn, D.I. (eds), *A Practical Guide to the Study of Glacial Sediments*. Arnold, London, pp. 93–114.

Benn, D.I., 2004b. Clast morphology. In: Evans, D.J.A., Benn, D.I. (Eds.), *A Practical Guide to the Study of Glacial Sediments*. Arnold, London, pp. 77–92.

Benn, D.I., 2006. Interpreting glacial sediments. In: Knight, P.G. (Ed.), *Glacier Science and Environmental Change*. Blackwell, Oxford, pp. 434–439.

Benn, D.I., Ballantyne, C.K., 1993. The description and representation of clast shape. *Earth Surface Processes and Landforms* **18**, 665–672.

Benn, D.I., Ballantyne, C.K., 1994. Reconstructing the transport history of glacigenic sediments: a new approach based on the co-variance of clast form indices. *Sedimentary Geology* **91**, 215–227.

Benn, D.I., Evans, D.J.A., 1996. The interpretation and classification of subglacially-deformed materials. *Quaternary Science Reviews* **15**, 23–52.

Benn, D.I., Evans, D.J.A., 1998. *Glaciers and Glaciation*. Hodder Education.

Benn, D.I., Evans, D.J.A., 2010. *Glaciers and Glaciation*. 2nd Edition. Hodder Education.

Benn, D.I., Lukas, S., 2006. Younger Dryas glacial landsystems in North West Scotland: an assessment of modern analogues and palaeoclimatic implications. *Quaternary Science Reviews* **25**, 2390–2408.

Benn, D.I., Prave, A.R., 2006. Subglacial and proglacial glacitectonic deformation in the Neoproterozoic Port Askaig Formation, Scotland. *Geomorphology* **75**, 266–280.

Benn, D.I., Ringrose, T.J., 2001. Random variation of fabric eigenvalues: implications for the use of A-axis fabric data to differentiate till facies. *Earth Surface Processes and Landforms* **26**, 295–306.

Benn, D.I., Evans, D.J.A., Phillips, E.R., Hiemstra, J.F., Walden, J., Hoey, T.B., 2004. The research project – a case study of Quaternary glacial sediments. In: Evans, D.J.A., Benn, D.I. (Eds.), *A Practical Guide to the Study of Glacial Sediments*. Arnold, London, pp. 209–234.

Benn, D.I., Kirkbride, M.P., Owen, L.A., Brazier, V., 2003. Glaciated valley landsystems. In: Evans, D.J.A. (Ed.), *Glacial Landsystems*. Arnold, London, pp. 372–406.

Bennett, M.R., Doyle, P., 1994. Carruthers and the theory of glacial undermelt: lessons from a pamphleteer? *Geology Today Sept/Oct* **1994**, 191–194.

Bennett, M.R., Huddart, D., Thomas, G.S.P., 2002. Facies architecture within a regional glaciolacustrine basin: Copper River, Alaska. *Quaternary Science Reviews* **21**, 2237–2279.

Bennett, M.R., Waller, R.I., Glasser, N.F., Hambrey, M.J., Huddart, D., 1999. Glacigenic clast fabric: genetic fingerprint or wishful thinking? *Journal of Quaternary Science* **14**, 125–135.

Bennett, M.R., Huddart, D., Waller, R.I., 2006. Diamict fans in subglacial water-filled cavities – a new glacial environment. *Quaternary Science Reviews* **25**, 3050–3069.

Bergersen, O.F., Garnes, K., 1983. Glacial deposits in the culmination zone of the Scandinavian Ice Sheet. In: Ehlers, J. (Ed.), *Glacial Deposits in North West Europe*. Balkema, Rotterdam, pp. 29–40.

Berthelsen, A., 1978. The methodology of kineto-stratigraphy as applied to glacial geology. *Bulletin of the Geological Society of Denmark* 27.

Berthelsen, A., 1974. Nogle forekomster af intrusivt moræneler i NØ-Sjælland. *Dansk Geologisk Foreningens* **1973**, 118–131.

Berthelsen, A., 1979. Recumbent folds and boudinage structures formed by subglacial shear: an example of gravity tectonics. *Geologie en Mijnbouw* **58**, 253–260.

Beskow, G., 1935. Praktiska och kvartargeologiska resultat av grusinventeringen i Norrbottens lan Foredragsreferat. *Geologiska Foreningens i Stockholm, Forhandlingar* **57**, 120–123.

Bindschadler, R.A., Vornberger, P.L., King, M.A., Padman, L., 2003. Tidally driven stick-slip motion in the mouth of Whillans Ice Stream, Antarctica. *Annals of Glaciology* **36**, 263–272.

Bingham, E.C., 1933. The new science of rheology. *Review of Scientific Instruments* **4**, 473–476.

Blake, E.W., Clarke, G.K.C., Gerin, M.C., 1992. Tools for examining subglacial bed deformation. *Journal of Glaciology* **38**, 388–396.

Blankenship, D.D., Bentley, C.R., Rooney, S.T., Alley, R.B., 1986. Seismic measurements reveal a saturated porous layer beneath an active Antarctic ice stream. *Nature* **322**, 54–57.

Blankenship, D.D., Bentley, C.R., Rooney, S.T., Alley, R.B., 1987. Till beneath ice stream B: 1. Properties derived from seismic travel times. *Journal of Geophysical Research* **92**, 8903–8911.

Boone, S.J., Eyles, N., 2001. Geotechnical model for great plains hummocky moraine formed by till deformation below stagnant ice. *Geomorphology* **38**, 109–124.

Boston, C.M., Evans, D.J.A., O' Cofaigh, C., 2010. Styles of till deposition at the margin of the Last Glacial Maximum North Sea lobe of the British Irish Ice Sheet: an assessment based on geochemical properties of glacigenic deposits in eastern England. *Quaternary Science Reviews* **29**, 3184–3211.

Bouchard, M.A., Cadieux, B., Goutier, F., 1984. L'Origine et les caracteristiques des lithofacies du till dans le secteur nord of Lac Albanel, Quebec: Une etude de la dispersion glaciare clastiques. Chibougamou – Stratigraphy and Mineralization. Canadian Institute of Mining, *Special Publication* **34**, pp. 244–260.

Bougamont, M., Tulaczyk, S., Joughin, I., 2003. Numerical investigations of the slow-down of Whillans Ice Stream, West Antarctica: is it shutting down like Ice Stream C? *Annals of Glaciology* **37**, 239–246.

Boulton, G.S., 1967. The development of a complex supraglacial moraine at the margin of Sorbreen, Ny Friesland, Vestspitzbergen. *Journal of Glaciology* **6**, 717–736.

Boulton, G.S., 1968. Flow tills and related deposits on some Vestspitsbergen glaciers. *Journal of Glaciology* **7**, 391–412.

Boulton, G.S., 1970a. On the origin and transport of englacial debris in Svalbard glaciers. *Journal of Glaciology* **9**, 213–229.

Boulton, G.S., 1970b. On the deposition of subglacial and melt-out tills at the margins of certain Svalbard glaciers. *Journal of Glaciology* **9**, 231–245.

Boulton, G.S., 1971. Till genesis and fabric in Svalbard, Spitsbergen. In: Goldthwait, R.P. (Ed.), *Till: A Symposium*. Ohio State University Press, Columbus, pp. 41–72.

Boulton, G.S., 1972a. Modern artic glaciers as a depositional model for former ice-sheets. *Journal of the Geological Society of London* **128**, 361–393.

Boulton, G.S., 1972b. The role of the thermal regime in glacial sedimentation. In: Price, R.J., Sugden, D.E. (Eds.), *Polar Geomorphology. Institute of British Geographers*, Special Publication, vol. 4, pp. 1–19.

Boulton, G.S., 1974. Processes and patterns of subglacial erosion. In: Coates, D.R. (Ed.), *Glacial Geomorphology*. University of New York, Binghamton, pp. 41–87.

Boulton, G.S., 1975. Processes and patterns of subglacial sedimentation: a theoretical approach. In: Wright, A.E., Moseley, F. (Eds.), *Ice Ages: Ancient and Modern*. Seel House Press, Liverpool, pp. 7–42.

Boulton, G.S., 1976. The origin of glacially fluted surfaces: observations and theory. *Journal of Glaciology* **17**, 287–309.

Boulton, G.S., 1977. A multiple till sequence formed by a Late Devensian Welsh icecap: Glanllynnau, Gwynedd. *Cambria* **4**, 10–31.

Boulton, G.S., 1978. Boulder shapes and grain size distributions of debris as indicators of transport paths through a glacier and till genesis. *Sedimentology* **25**, 773–799.

Boulton, G.S., 1979. Processes of glacier erosion on different substrata. *Journal of Glaciology* **23**, 15–38.

Boulton, G.S., 1982. Subglacial processes and the development of glacial bedforms. In: Davidson-Arnott, R., Nickling, W., Fahey, B.D. (Eds.), *Research in Glacial, Glacio-fluvial and Glacio-lacustrine Systems*. Geo Books, Norwich, pp. 1–31.

Boulton, G.S., 1986. A paradigm shift in glaciology? *Nature* **322**, 18.

Boulton, G.S., 1987. A theory of drumlin formation by subglacial deformation. In: Rose, J., Menzies, J. (Eds.), *Drumlin Symposium*. Balkema, Rotterdam, pp. 25–80.

Boulton, G.S., 1996a. Theory of glacial erosion, transport and deposition as a consequence of subglacial sediment deformation. *Journal of Glaciology* **42**, 43–62.

Boulton, G.S., 1996b. The origin of till sequences by subglacial sediment deformation beneath mid-latitude ice sheets. *Annals of Glaciology* **22**, 75–84.

Boulton, G.S., 2006. Glaciers and their coupling with hydraulic and sedimentary processes. In: Knight, P.G. (Ed.), *Glacier Science and Environmental Change*. Blackwell, Oxford, pp. 3–22.

Boulton, G.S., Caban, P.E., 1995. Groundwater flow beneath ice sheets: Part II. Its impact on glacier tectonic structures and moraine formation. *Quaternary Science Reviews* **14**, 563–587.

Boulton, G.S., Dent, D.L., 1974. The nature and rates of post-depositional changes in recently deposited till from south-east Iceland. *Geografiska Annaler* **56A**, 121–134.

Boulton, G.S., Eyles, N., 1979. Sedimentation by valley glaciers: a model and genetic classification. In: Schluchter, C. (Ed.), *Moraines and Varves*, Balkema, Rotterdam, pp. 11–23.

Boulton, G.S., Dobbie, K.E., 1998. Slow flow of granular aggregates: the deformation of sediments beneath glaciers. *Philosophical Transactions of the Royal Society of London. A* **356**, 2713–2745.

Boulton, G.S., Hindmarsh, R.C.A., 1987. Sediment deformation beneath glaciers: rheology and geological consequences. *Journal of Geophysical Research* **92**, 9059–9082.

Boulton, G.S., Jones, A.S., 1979. Stability of temperate ice sheets resting on beds of deformable sediment. *Journal of Glaciology* **24**, 29–43.

Boulton, G.S., Paul, M.A., 1976. The influence of genetic processes on some geotechnical properties of tills. *Journal of Engineering Geology* **9**, 159–194.

Boulton, G.S., Dent, D.L., Morris, E.M., 1974. Subglacial shearing and crushing, and the role of water pressures in tills from southeast Iceland. *Geografiska Annaler* **56A**, 135–145.

Boulton, G.S., Morris, E.M., Armstrong, A.A., Thomas, A., 1979. Direct measurement of stress at the base of a glacier. *Journal of Glaciology* **22**, 3–24.

Boulton, G.S., Dobbie, K.E., Zatsepin, S., 2001. Sediment deformation beneath glaciers and its coupling to the subglacial hydraulic system. *Quaternary International* **86**, 3–28.

Boyce, J.I., Eyles, N., 1991. Drumlins carved by deforming till streams below the Laurentide Ice Sheet. *Geology* **19**, 787–790.

Boyce, J.I., Eyles, N., 2000. Architectural element analysis applied to glacial deposits: internal geometry of a late Pleistocene till sheet, Ontario, Canada. *Bulletin of the Geological Society of America* **112**, 98–118.

Brandes, C., Le Heron, D.P., 2010. The glaciotectonic deformation of Quaternary sediments by fault-propagation folding. *Proceedings of the Geologists' Association* **121**, 270–280.

Brett, C., Baird, G.C., 1986. Symmetrical and upward shallowing cycles in the Middle Devonian of New York State and their implications for the punctuated aggradational cycle hypothesis. *Paleoceanography* **1**, 431–445.

Broster, B.E., Dreimanis, A., 1981. Deposition of multiple lodgement tills by competing glacial flows in a common ice sheet: Cranbrook, British Columbia. *Arctic and Alpine Research* **13**, 197–204.

Broster, B.E., Seaman, A.A., 1991. Glacigenic rafting of weathered granite: Charlie Lake, New Brunswick. *Canadian Journal of Earth Sciences* **28**, 649–654.

Broster, B.E., Dreimanis, A., White, J.C., 1979. A sequence of glacial deformation, erosion and deposition at the ice–rock interface during the last glaciation: Cranbrook, British Columbia, Canada. *Journal of Glaciology* **23**, 283–295.

Brown, N.E., Hallet, B., Booth, D.B., 1987. Rapid soft bed sliding of the Puget glacial lobe. *Journal of Geophysical Research* **92**, 8985–8997.

Brzesowsky, R.H., 1995. Micromechanics of Sand Grain Failure and Sand Compaction. Unpublished PhD thesis, University of Utrecht.

Burke, H., Phillips, E., Lee, J.R., Wilkinson, I.P., 2009. Imbricate thrust stack model for the formation of glaciotectonic rafts: an example from the Middle Pleistocene of north Norfolk, UK. *Boreas* **38**, 620–637.

Busfield, M.E., Le Heron, D.P., 2013. Glacitectonic deformation in the Chuos Formation of northern Namibia: implications for Neoproterozoic ice dynamics. *Proceedings of the Geologists' Association* **124**, 778–789.

Carlson, A.E., 2004. Genesis of dewatering structures and its implications for melt-out till identification. *Journal of Glaciology* **50**, 17–24.

Carr, S.J., 1999. The micromorphology of Last Glacial Maximum sediments in the Southern North Sea. *Catena* **35**, 123–145.

Carr, S.J., 2001. Micromorphological criteria for distinguishing subglacial and glacimarine sediments: evidence from a contemporary tidewater glacier, Spitsbergen. *Quaternary International* **86**, 71–79.

Carr, S.J., 2004. Micro-scale features and structures. In: Evans, D.J.A., Benn, D.I. (Eds.), *A Practical Guide to the Study of Glacial Sediments*. Arnold, London, pp. 115–144.

Carr, S.J., Rose, J., 2003. Till fabric patterns and significance: particle response to subglacial stress. *Quaternary Science Reviews* **22**, 1415–1426.

Carr, S.J., Goddard, M.A., 2007. Role of particle size in till fabric characteristics: systematic variation in till fabric from Vestari Hagafellsjökull, Iceland. *Boreas* **36**, 371–385.

Carr, S.J., Haflidason, H., Sejrup, H.P., 2000. Micromorphological evidence supporting late Weichselian glaciation of the northern North Sea. *Boreas* **29**, 315–328.

Carruthers, R.G., 1939. On northern glacial drifts: some peculiarities and their significance. *Quarterly Journal of the Geological Society of London* **95**, 299–333.

Carruthers, R.G., 1947–48. The secret of the glacial drifts. *Proceedings of the Yorkshire Geological Society* **27**, 43–57 and 129–172.

Carruthers, R.G., 1953. *Glacial Drifts and the Undermelt Theory*. Hill and Son, Newcastle upon Tyne.

Carto, S.L., Eyles, N., 2012a. Sedimentology of the Neoproterozoic (c. 580 Ma) Squantum 'Tillite', Boston Basin, USA: Mass flow deposition in a deep-water arc basin lacking direct glacial influence. *Sedimentary Geology* **269**, 1–14.

Carto, S.L., Eyles, N., 2012b. Identifying glacial influences on sedimentation in tectonically-active, mass flow dominated arc basins with reference to the Neoproterozoic Gaskiers glaciation (c. 580 Ma) of the Avalonian–Cadomian Orogenic Belt. *Sedimentary Geology* **161**, 1–14.

Catto, N.R., 1990. Clast fabric of diamictons associated with some roches moutonnées. *Boreas* **19**, 289–296.

Catto, N.R., 1998. Comparative study of striations and basal till clast fabrics, Malpeque-Bedeque region, Price Edward Island, Canada. *Boreas* **27**, 259–274.

Chamberlin, T.C., 1883. Preliminary paper on the terminal moraine of the second glacial epoch. *USGS 3rd Annual Report*, pp. 291–402.

Chamberlin, T.C., 1894a. Proposed genetic classification of Pleistocene glacial formations. *Journal of Geology* **2**, 517–538.

Chamberlin, T.C., 1894b. Glacial phenomena of North America. In: Geikie, J., *The Great Ice Age*. Stanford. London, pp. 724–775.

Chamberlin, T.C., 1895. Recent glacial studies in Greenland. *Bulletin of the Geological Society of America* **6**, 199–220.

Chandler, B.M.P., Evans, D.J.A., Roberts, D.H., 2016. Characteristics of recessional moraines at a temperate glacier in SE Iceland: Insights into patterns, rates and drivers of glacier retreat. *Quaternary Science Reviews* **135**, 171–205.

Channon, H., 2012. Multiscale analysis of the landforms and sediments of palaeo-ice streams. *Unpublished PhD thesis, Queen Mary*, University of London.

Chapman, L.J., Putnam D.F., 1951. *The Physiography of Southern Ontario*, Toronto: University of Toronto Press.

Charlesworth, J.K., 1957. *The Quaternary Era*. Edward Arnold, London.

Christiansen, E.A., 1968. A thin till in west-central Saskatchewan, Canada. *Canadian Journal of Earth Sciences* **5**, 329–336.

Christiansen, E.A., 1971. Tills in southern Saskatchewan, Canada. In: Goldthwait, R.P. (Ed.), *Till – A Symposium*. Ohio State University Press, Columbus, pp. 167–183.

Christoffersen, P., Tulaczyk, S., 2003. Thermodynamics of basal freeze-on: predicting basal and subglacial signatures of stopped ice streams and interstream ridges. *Annals of Glaciology* **36**, 233–243.

Christoffersen, P., Piotrowski, J.A., Larsen, N.K., 2005. Basal processes beneath an arctic glacier and their geomorphic imprint after a surge, Elisebreen, Svalbard. *Quaternary Research* **64**, 125–137.

Christoffersen, P., Tulaczyk, S., Behar, A., 2010. Basal ice sequences in Antarctic ice stream: Exposure of past hydrologic conditions and a principal mode of sediment transfer. *Journal of Geophysical Research* **115**, F03034, doi:10.1029/2009JF001430, 2010.

Clark, C.D., 2010. Emergent drumlins and their clones: from till dilatancy to flow instabilities. *Journal of Glaciology* **56**, 1011–1025.

Clark, C.D., Tulaczyk, S.M., Stokes, C.R., Canals, M., 2003. A groove-ploughing theory for the production of mega-scale glacial lineations, and implications for ice-stream mechanics. *Journal of Glaciology* **49**, 240–256.

Clark, P.U., 1991. Striated clast pavements, products of deforming subglacial sediment? *Geology* **19**, 530–533.

Clark, P.U., 1992. Comment and reply on 'Striated clast pavements: products of deforming subglacial sediment?' *Geology* **20**, 285–286.

Clark, P.U., Hansel, A.K., 1989. Clast ploughing, lodgement and glacier sliding over a soft glacier bed. *Boreas* **18**, 201–207.

Clark, P.U., Walder, J.S., 1994. Subglacial drainage, eskers, and deforming beds beneath the Laurentide and Eurasian ice sheets. *Bulletin of the Geological Society of America* **106**, 304–314.

Clarke, G.K.C., 1987. Subglacial till: a physical framework for its properties and processes. *Journal of Geophysical Research* **92**, 9023–9036.

Clarke, G.K.C., 2005. Subglacial processes. *Annual Review of Earth and Planetary Sciences* **33**, 247–276.

Clarke, G.K.C., Collins, S.G., Thompson, D.E., 1984. Flow, thermal structure and subglacial conditions of a surge-type glacier. *Canadian Journal of Earth Sciences* **21**, 232–240.

Clayton, L., Mickelson, D.M., Attig, J.W., 1989. Evidence against pervasively deformed bed material beneath rapidly moving lobes of the southern Laurentide ice sheet. *Sedimentary Geology* **62**, 203–208.

Clayton, L., Moran, S.R., 1974. A glacial process–form model. In: Coates, D.R. (Ed.), *Glacial Geomorphology*. State University of New York, Binghamton, pp. 89–120.

Cohen, D., Iverson, N.R., Hooyer, T.S., Fischer, U.H., Jackson, M., Moore, P.L., 2005. *Debris-bed friction of hard-bedded glaciers. Journal of Geophysical Research* 110, doi:10.1029/2004JF000228.

Colgan, P.M., Mickelson, D.M., 1997. Genesis of streamlined landforms and flow history of the Green Bay Lobe, Wisconsin, USA. *Sedimentary Geology* **111**, 7–25.

Colgan, P.M., Mickelson, D.M., Cutler, P.M., 2003. Ice-marginal terrestrial Landsystems: Southern Laurentide Ice Sheet. In: Evans, D.J.A. (Ed.), *Glacial Landsystems*. Edward Arnold, London, pp. 111–142.

Committee on Soil Dynamics, 1978. Definition of terms related to liquefaction. *Journal of Geotechnical Engineering Division ASCE* **104**, 1197–1200.

Cook, S.J., Knight, P.G., Richard, R.I., Robinson, Z.P., Adam, W.G., 2007. The geography of basal ice and its relationship to glaciohydraulic supercooling: Svínafellsjökull, southeast Iceland. *Quaternary Science Reviews* **26**, 2309–2315.

Cook, S.J., Graham, D.J., Swift, D.A., Midgeley, N.G., Adam, W.G., 2011b. Sedimentary signatures of basal ice formation and their preservation in ice-marginal sediments. *Geomorphology* **125**, 122–131.

Cook S.J., Robinson, Z.P., Fairchild, I.J., Knight, P.G., Waller, R.I., Boomer, I., 2010. Role of glaciohydraulic supercooling in the formation of stratified facies basal ice: Svínafellsjökull and Skaftafellsjökull, southeast Iceland. *Boreas* **39**, 24–38.

Cook, S.J., Swift, D.A., Graham, D.J., Midgley, N.G., 2011a. Origin and significance of 'dispersed facies' basal ice: Svínafellsjökull, Iceland. *Journal of Glaciology* **57**, 710–720.

Cowan, D.S., 1985. Structural styles in Mesozoic and Cenozoic mélanges in the western Cordillera of North America. *Geological Society of America Bulletin* **96**, 451–462.

Craddock, C., Anderson, J.J., Webers G.F., 1964. Geologic outline of the Ellsworth Mountains. In: Adie, R.J. (Ed.), *Antarctic Geology*, New York: Wiley, pp. 156–170.

Creyts, T.T., Schoof, C.G., 2009. Drainage through subglacial water sheets, *Journal of Geophysical Research* **114**, F04008, doi:10.1029/2008JF001215.

Croot, D.G., 1988. Morphological, structural and mechanical analysis of neoglacial ice-pushed ridges in Iceland. In: Croot, D.G. (Ed.). *Glaciotectonics Forms and Processes*. Balkema, Rotterdam, pp. 49–61.

Croot, D.G., Sims, P.C., 1996. Early stages of till genesis: an example from Fanore, County Clare, Ireland. *Boreas* **25**, 37–46.

Crosby, W.O., 1890. Composition of the till or bolder clay. *Proceedings of the Boston Society of Natural History* **XXV**, 115–140.

Crosby, W.O., 1896. Englacial drift. *Technology Quarterly* **9**, 116–144.

Crosby, W.O., 1900. *Geology of the Boston Basin, the Blue Hills Complex*. Boston Society for Natural History, Occasional Paper IV, pp. 289–563.

Crowell, J.C., 1957. Origin of pebbly mudstones. *Geological Society of America Bulletin* **68**, 993–1010.

Cuffey, K., Alley, R.B., 1996. Is erosion by deforming subglacial sediments significant? (Toward till continuity). *Annals of Glaciology* **22**, 17–24.

Cuffey, K.M., Paterson, W.S.B., 2010. *The Physics of Glaciers*. 4th Edition. Elsevier, Amsterdam.

Cuffey, K.M., Conway, H., Hallet, B., Gades, A.M., Raymond, C.F., 1999. Interfacial water in polar glaciers, and sliding at −17°C. *Geophysical Research Letters* **26**, 751–754.

Curry, A.M., 1999. Paraglacial modification of slope form. *Earth Surface Processes and Landforms* **24**, 1213–1228.

Curry, A.M., 2000a. Observations on the distribution of paraglacial reworking of glacigenic drift in western Norway. *Norsk Geografisk Tidsskrift* **54**, 139–147.

Curry, A.M., 2000b. Holocene reworking of drift-mantled hillslopes in the Scottish Highlands. *Journal of Quaternary Science* **15**, 529–541.

Curry, A.M., Ballantyne, C.K., 1999. Paraglacial modification of glacigenic sediments. *Geografiska Annaler* **81A**, 409–419.

Dalland, A., 1976. Erratic clasts in the Lower Tertiary deposits in Svalbard: evidence of transport by winter ice. *Norsk Polarinstitutt Arbok* **1976**, 151–165.

Damsgaard, A., Egholm, D.L., Piotrowski, J.A., Tulaczyk, S., Larsen, N.K., Tylmann, K., 2013. Discrete element modeling of subglacial sediment deformation. *Journal of Geophysical Research-Earth* **118**, 2230–2242.

Damsgaard, A., Egholm, D.L., Piotrowski, J.A., Tulaczyk, S., Larsen, N.K., Brædstrup, C.F., 2015. A new methodology to simulate subglacial deformation of water-saturated granular material. *The Cryosphere* **9**, 2183–2200.

Dardis, G.F., McCabe, A.M., Mitchell, W.I., 1984. Characteristics and origins of lee-side stratification sequences in late pleistocene drumlins, Northern Ireland. *Earth Surface Processes and Landforms* **9**, 409–424.

Davies, B.J., Roberts, D.H., O' Cofaigh, C., Bridgland, D.R., 2009. Subglacial and ice marginal controls on sediment deposition during the last glacial maximum: interlobate ice sheet dynamics at Whitburn Bay, County Durham, England. *Boreas* **38**, 555–578.

Davis, W.M., 1926. The value of outrageous geological hypotheses. *Science* **63**, 463–468.

Dean, W., 1953. The drumlinoid land forms of the 'barren grounds', NWT. *Canadian Geographer* **1**, 19–30.

Deane, R.E., 1950. Pleistocene geology of the Lake Simcoe District, *Ontario. Geological Survey of Canada Memoir* **256**.

Derbyshire, E., Owen, L.A., 1990. Quaternary alluvial fans in the Karakoram Mountains. In: Rachocki, A.H., Church, M. (Eds.), *Alluvial Fans – A Field Approach*. Wiley, Chichester, pp. 27–53.

Dionne, J.-C., 1979. Ice action in the lacustrine environment. A review with particular reference to subarctic Quebec, Canada. *Earth Science Reviews* **15**, 185–212.

Dionne J.-C., 1985. Drift-ice abrasion marks along rocky shores. *Journal of Glaciology* **31**, 237–241.

Dionne J.-C., Poitras S., 1998. Geomorphic aspects of mega-boulders at Mitis Bay, Lower St Lawrence Estuary, Quebec, Canada. *Journal of Coastal Research* **14**, 1054–1064.

Dionne, J.-C., Shilts, W.W., 1974. A Pleistocene clastic dike, Upper Chaudiere Valley, Quebec. *Canadian Journal of Earth Sciences* **11**, 1594–1605.

Domack, E.W., Lawson, D.E., 1985. Pebble fabric in an ice rafted diamicton. *Journal of Geology* **93**, 577–591.

Dowdeswell, J.A., 1982. Scanning electron micrographs of quartz sand grains from cold environments examined using Fourier shape analysis. *Journal of Sedimentary Petrology* **52**, 1315–1323.

Dowdeswell, J.A., Sharp, M., 1986. Characterization of pebble fabrics in modern terrestrial glacigenic sediments. *Sedimentology* **33**, 699–710.

Dowdeswell, J.A., Hambrey, M.J., Wu, R., 1985. A comparison of clast fabric and shape in Late Precambrian and modern glacigenic sediments. *Journal of Sedimentary Petrology* **55**, 691–704.

Dowdeswell, J.A., Ó Cofaigh, C., Pudsey, C.J., 2004. Thickness and extent of the subglacial till layer beneath an Antarctic palaeo-ice stream. *Geology* **32**, 13–16.

Dowdeswell, J.A., Ó Cofaigh, C., Noormets, R., Larter, R.D., Hillenbrand, C.-D., Benetti, S., Pudsey, C.J., 2008. A major trough-mouth fan on the continental margin of the Bellingshausen Sea, West Antarctica: The Belgica Fan. *Marine Geology* **252**, 129–140.

Dowdeswell, J.A., Whittington, R.J., Marienfeld, P., 1994. The origin of massive diamicton facies by iceberg rafting and scouring, Scoresby Sund, East Greenland. *Sedimentology* **41**, 21–35.

Dreimanis, A., 1969. Selection of genetically significant parameters for investigation of tills. *Geografia* **8**, 15–29.

Dreimanis, A., 1973. The first report on till wedges in Europe – a reply. *Geologiska Foreningens i Stockholm Forhandlingar* **95**, 156–157.

Dreimanis, A., 1976. Tills: Their origin and properties. In: Legget, R.F. (Ed.), *Glacial till*. Royal Society of Canada Special Publication **12**, pp. 11–49.

Dreimanis, A., 1979. The problem of waterlain tills. In: Schluchter, C. (Ed.), *Moraines and Varves*. Balkema, Rotterdam, pp. 167–177.

Dreimanis, A., 1980. Terminology and development of genetic classifications of materials transported and deposited by glaciers. In: Stankowski, W. (Ed.), *Tills and Glacigene Deposits*. Uniwersytet Imiena Adama Mickiewicza Poznan: Prace. Seria geografia **20**, pp. 5–10.

Dreimanis, A., 1982. Two origins of the stratified Catfish Creek till at Plum Point, Ontario, Canada. *Boreas* **11**, 173–180.

Dreimanis, A., 1983. Penecontemporaneous partial disaggregation and/or resedimentation during the formation and deposition of subglacial till. *Acta Geologia Hispanica* **18**, 153–160.

Dreimanis, A., 1984. Comments on 'Sedimentation in a large lake, an interpretation of the late Pleistocene stratigraphy at Scarborough Bluffs, Ontario, Canada'. *Geology* **12**, 185–186.

Dreimanis, A., 1989. Tills, their genetic terminology and classification. In: Goldthwait, R.P., Matsch, C.L. (Eds.), *Genetic Classification of Glacigenic Deposits*. Balkema, Rotterdam, pp. 17–84.

Dreimanis, A., 1992. Downward injected till wedges and upward injected till dikes. *Sveriges Geologiska Undersøgelse, Serie Ca* **81**, 91–96.

Dreimanis, A., Gibbard, P.L., 2005. Stratigraphy and sedimentation of the stratotype sections of the Catfish Creek Drift Formation between Bradtville and Plum Point, north shore, Lake Erie, southwestern Ontario, Canada. *Boreas* **34**, 101–122.

Dreimanis, A., Lundqvist, J., 1984. What should be called a till? In: Konigsson, L.K. (Ed.), *Ten years of Nordic till research*. Striae **20**, 5–10.

Dreimanis, A., Rappol, M., 1997. Late Wisconsinan sub-glacial clastic intrusive sheets along Lake Erie bluffs, at Bradtville, Ontario, Canada. *Sedimentary Geology* **111**, 225–248.

Dreimanis, A., Reavely, G.H., 1953. Differentiation of the Lower and Upper Till along the north shore of Lake Erie. *Journal of Sedimentary Petrology* **23**, 238–259.

Dreimanis, A., Schluchter, C., 1985. Field criteria for the recognition of till or tillite. *Palaeogeography, Palaeoclimatology, Palaeoecology* **51**, 7–14.

Dreimanis, A., Vagners, U.J., 1971. Bimodal distribution of rock and mineral fragments in basal till. In: Goldthwait, R.P. (Ed.), *Till: a Symposium*. Ohio State University Press, Columbus, pp. 237–250.

Dreimanis, A., Hamilton, J.P., Kelly, P.E., 1987, Complex subglacial sedimentation of Catfish Creek Till at Bradtville, Ontario, Canada. In: van der Meer, J.J.M. (Ed.), *Tills and glaciotectonics*. Balkema, Rotterdam, pp. 73–87.

von Drygalski, E., 1897. Grönland Expedition (1891–1893). *Der Gesellschaft fur Erdkunde zu Berlin*, **2**.

von Drygalski, E., 1898. Die Eisbewegung, *ihre physikalischen Ursachen und ihre geographischen Wirkungen. Petermanns Mitteilungen (Gotha)*, Bd. **44**, pp. 55–64.

Dunlop, P., Clark, C.D., Hindmarsh, R.C.A., 2008. The Bed Ribbing Instability Explanation (BRIE): Testing a numerical model of ribbed moraine formation arising from coupled flow of ice and subglacial sediment. *Journal of Geophysical Research* **113**: F03005.

Dyke, A.S., Evans, D.J.A., 2003. Ice-marginal terrestrial landsystems: northern Laurentide and Innuitian ice sheet margins. In: Evans, D.J.A. (Ed.), *Glacial Landsystems*. Arnold, London, pp. 143–165.

Echelmeyer, K., Wang, Z., 1987. Direct observation of basal sliding and deformation of basal drift at sub-freezing temperatures. *Journal of Glaciology* **33**, 83–98.

Eden, D.J., Eyles, N., 2001. Description and numerical model of Pleistocene iceberg scours and ice-keel turbated facies at Toronto, Canada. *Sedimentology* **48**, 1079–1102.

Ehlers, J., Stephan, H.-J., 1979. Forms at the base of till strata as indicators of ice movement. *Journal of Glaciology* **22**, 345–355.

Eklund, A., Hart, J.K., 1996. Glaciotectonic deformation within a flute from the Isfallsglaciaren, Sweden. *Journal of Quaternary Science* **11**, 299–310.

Ekstrom, G., Nettles, M., Abers, G.A., 2003. Glacial earthquakes. *Science* **302**, 622–624.

Ekstrom, G., Nettles, M., Tsai, V.C., 2006. Seasonality and increasing frequency of Greenland glacial earthquakes. *Science* **311**, 1756–1758.

Elson, J.A., 1961. The geology of tills. In: Penner, E., Butler, J. (Eds.), *Proceedings of the 14th Canadian soil mechanics conference*. Commission for Soil and Snow Mechanics, Technical Memoir, vol. 69, pp. 5–36.

Elson, J.A., 1975. Origin of a clastic dike at St Ludger, Quebec: an alternative hypothesis. *Canadian Journal of Earth Sciences* **12**, 1048–1053.

Elson, J.A., 1989. Comment on glacitectonite, deformation till and comminution till. In: Goldthwait, R.P., Matsch, C.L. (Eds.), *Genetic Classification of Glacigenic Deposits*. Balkema, Rotterdam, pp. 85–88.

Engelhardt, H.F., Kamb, B., 1997. Basal hydraulic system of a West Antarctic ice stream: constraints from borehole observations. *Journal of Glaciology* **43**, 207–230.

Engelhardt, H.F., Kamb, B., 1998. Sliding velocity of Ice Stream B. *Journal of Glaciology* **44**, 223–230.

Engelhardt, H.F., Harrison, W.D., Kamb, B., 1978. Basal sliding and conditions at the glacier bed as revealed by bore-hole photography. *Journal of Glaciology* **20**, 469–508.

Engelhardt, H.F., Humphrey, N., Kamb, B., Fahnestock, M., 1990. Physical conditions at the base of a fast moving Antarctic ice stream. *Science* **248**, 57–59.

Evans, D.J.A., 1989. Apron entrainment at the margins of sub-polar glaciers, northwest Ellesmere Island, Canadian high arctic. *Journal of Glaciology* **35**, 317–324.

Evans, D.J.A., 1994. The stratigraphy and sedimentary structures associated with complex subglacial thermal regimes at the southwestern margin of the Laurentide Ice Sheet, southern Alberta, Canada. In: Warren, W.P., Croot, D.G. (Eds.), *Formation and Deformation of Glacial Deposits*. Balkema, Rotterdam, pp. 203–220.

Evans, D.J.A., 2000a. A gravel outwash/deformation till continuum, Skalafellsjokull, Iceland. *Geografiska Annaler* **82A**, 499–512.

Evans, D.J.A., 2000b. Quaternary geology and geomorphology of the Dinosaur Provincial Park area and surrounding plains, Alberta, Canada: the identification of former glacial lobes, drainage diversions and meltwater flood tracks. *Quaternary Science Reviews* **19**, 931–958.

Evans, D.J.A., 2009. Controlled moraine: origins, characteristics and palaeoglaciological implications. *Quaternary Science Reviews* **28**, 183–308.

Evans, D.J.A., Benn, D.I., 2001. Earth's giant bulldozers. *Geography Review* **14**, 29–33.

Evans, D.J.A., Benn, D.I., 2004. Facies description and the logging of sedimentary exposures. In: Evans, D.J.A., Benn, D.I. (Eds.), *A Practical Guide to the Study of Glacial Sediments*. Arnold, London, pp. 11–51.

Evans, D.J.A., England, J.H., 1991. Canadian landform examples 19: high arctic thrust block moraines. *Canadian Geographer* **35**, 93–97.

Evans, D.J.A., Hiemstra, J.F., 2005. Till deposition by glacier submarginal, incremental thickening. *Earth Surface Processes and Landforms* **30**, 1633–1662.

Evans, D.J.A., Ó Cofaigh, C., 2003. Depositional evidence for marginal oscillations of the Irish Sea ice stream in southeast Ireland during the last glaciation. *Boreas* **32**, 76–101.

Evans, D.J.A., O' Cofaigh, C., 2008. The sedimentology of the Late Pleistocene Bannow Till stratotype, County Wexford, southeast Ireland. *Proceedings of the Geologists' Association* **119**, 329–338.

Evans, D.J.A., Rea, B.R., 1999a. The geomorphology and sedimentology of surging glaciers: A land-systems approach. *Annals of Glaciology* **28**, 75–82.

Evans, D.J.A., Rea, B.R., 2003. Surging glacier landsystem. In: Evans, D.J.A. (Ed.), *Glacial Landsystems*. London, Arnold, pp. 259–288.

Evans, D.J.A., Thomson, S.A., 2010. Glacial sediments and landforms of Holderness, eastern England: A glacial depositional model for the North Sea Lobe of the British–Irish Ice Sheet. *Earth Science Reviews* **101**, 147–189.

Evans, D.J.A., Twigg, D.R., 2002. The active temperate glacial landsystem: a model based on Breiðamerkurjökull and Fjallsjokull, Iceland. *Quaternary Science Reviews* **21**, 2143–2177.

Evans, D.J.A., Clark, C.D., Rea, B.R., 2008. Landform and sediment imprints of fast glacier flow in the southwest Laurentide Ice Sheet. *Journal of Quaternary Science* **23**, 249–272.

Evans, D.J.A., Ewertowski, M., Orton, C., 2015a. Fláajökull (north lobe), *Iceland: active temperate piedmont lobe glacial landsystem, Journal of Maps, DOI*: 10.1080/17445647.2015.1073185.

Evans, D.J.A., Hiemstra, J.F., Boston, C.M., Leighton, I., Cofaigh, C., Rea, B.R., 2012a. Till stratigraphy and sedimentology at the margins of terrestrially terminating ice streams: case study of the western Canadian prairies and high plains. *Quaternary Science Reviews* **46**, 80–125.

Evans, D.J.A., Hiemstra, J.F., Ó Cofaigh, C., 2007. An assessment of clast macrofabrics in glaciogenic sediments based on A/B plane data. *Geografiska Annaler* **A89**, 103–120.

Evans D.J.A., Hiemstra J.F., Ó Cofaigh C., 2012b. Stratigraphic architecture and sedimentology of a Late Pleistocene subaqueous moraine complex, southwest Ireland. *Journal of Quaternary Science* **27**, 51–63.

Evans, D.J.A., Nelson, C.D., Webb, C., 2010a. An assessment of fluting and till esker formation on the foreland of Sandfellsjökull, Iceland. *Geomorphology* **114**, 453–465.

Evans, D.J.A., Owen, L.A., Roberts, D., 1995. Stratigraphy and sedimentology of Devensian (Dimlington Stadial) glacial deposits, east Yorkshire, England. *Journal of Quaternary Science* **10**, 241–265.

Evans, D.J.A., Phillips, E.R., Hiemstra, J.F., Auton, C.A., 2006b. Subglacial till: formation, sedimentary characteristics and classification. *Earth Science Reviews* **78**, 115–176.

Evans, D.J.A., Rea, B.R., Benn, D.I., 1998. Subglacial deformation and bedrock plucking in areas of hard bedrock. *Glacial Geology and Geomorphology* (rp04/1998 – http://ggg.qub.ac.uk/ggg/papers/full/1998/rp041998/rp04.html).

Evans, D.J.A., Rea, B.R., Hiemstra, J.F., Ó Cofaigh, C., 2006a. A critical assessment of subglacial megafloods: a case study of glacial sediments and landforms in south-central Alberta, Canada. *Quaternary Science Reviews* **25**, 1638–1667.

Evans, D.J.A., Roberts, D.H., Evans, S.C., 2016. *Multiple subglacial till deposition: a modern exemplar for Quaternary palaeoglaciology.* Quaternary Science Reviews **145**, 183–203.

Evans, D.J.A., Roberts, D.H., Ó Cofaigh C., 2015b. Drumlin sedimentology in a hard-bed, lowland setting, Connemara, western Ireland: implications for subglacial bedform generation in areas of sparse till cover. *Journal of Quaternary Science* **30**, 537–557.

Evans, D.J.A., Rother, H., Hyatt, O.M., Shulmeister, J., 2013. The glacial sedimentology and geomorphological evolution of an outwash head/morainedammed lake, South Island, New Zealand. *Sedimentary Geology* **284–285**, 45–75.

Evans, D.J.A., Salt, K., Allen, C.S., 1999. Glacitectonized lake sediments, Barrier Lake, Kananaskis Country, Canadian Rocky Mountains. *Canadian Journal of Earth Sciences* **36**, 395–407.

Evans, D.J.A., Shulmeister, J., Hyatt, O.M., 2010b. Sedimentology of latero-frontal moraines and fans on the west coast of South Island, New Zealand. *Quaternary Science Reviews* **29**, 3790–3811.

Evans, D.J.A., Young, N.J., Cofaigh, C., 2014. Glacial geomorphology of terrestrial terminating fast flow lobes/ice stream margins in the southwest Laurentide ice sheet. *Geomorphology* **204**, 86–113.

Evans, H.M., Lee, J.R., Riding, J.B., 2011. A thrust-stacked origin for interstratified till sequences: an example from Weybourne Town Pit, north Norfolk, UK. *Bulletin of the Geological Society of Norfolk* **61**, 23–49.

Evenson, E.B., 1970. A method for 3-dimensional microfabrics analysis of tills obtained from exposures or cores. *Journal of Sedimentary Petrology* **40**, 762–764.

Evenson, E.B., 1971. The relationship of macro- to microfabrics of tills and the genesis of glacial landforms in Jefferson County, Wisconsin. In: Goldthwait, R.P. (Ed.), *Till, a Symposium.* Ohio State University Press, pp. 345–364.

Evenson, E.B., Dreimanis, A., Newsome, J.W., 1977. Subaquatic flow tills: a new interpretation for the genesis of some laminated till deposits. *Boreas* **6**, 115–133.

Evenson, E.B., Lawson, D.E., Strasser, J.C., Larson, G.J., Alley, R.B., Ensminger, S.L., Stevenson, W.E., 1999. Field evidence for the recognition of glaciohydrologic supercooling. In: Mickelson, D.M., Attig, J.W. (Eds.), *Glacial Processes Past and Present.* Geological Society of America Special Paper, vol. 337, pp. 23–35.

Evenson, E.B., Schluchter, C., Rabassa, J. (Eds.), 1983. *Tills and Related Deposits.* Balkema, Rotterdam.

Eyles, C.H., 1988. A model for striated boulder pavement formation on glaciated, shallow-marine shelves: An example from the Yakataga Formation, Alaska. *Journal of Sedimentary Petrology* **58**, 62–71.

Eyles, C.H., Eyles, N., 1983a. Glaciomarine model upper Precambrian diamictites of the Port Askaig Formation, Scotland. *Geology* **11**, 692–696.

Eyles, C.H., Eyles, N., 1983b. Sedimentation in a large lake: a reinterpretation of the late Pleistocene stratigraphy at Scarborough Bluffs, Ontario, Canada. *Geology* **11**, 146–152.

Eyles, C.H., Eyles, N., 2000. Subaqueous mass flow origin for Lower Permian diamictites and associated facies of the Grant Group, Barbwire Terrace, Canning Basin, Western Australia. *Sedimentology* **47**, 343–356.

Eyles, C.H., Eyles, N., 1984a. Glaciomarine sediments of the Isle of Man as a key to late Pleistocene stratigraphic investigations in the Irish Sea basin. *Geology* **12**, 359–364.

Eyles, C.H., Eyles, N., 1984b. Sedimentation in a large lake – reply. *Geology* **12**, 188–190.

Eyles, C.H., Eyles, N., Miall, A.D., 1985. Models of glaciomarine sedimentation and their application to the interpretation of ancient glacial sequences. *Palaeogeography, Palaeoclimatology, Palaeoecology* **51**, 15–84.

Eyles C.H., Lagoe M.B., 1990. Sedimentation patterns and facies geometries on a temperate glacially-influenced continental shelf: the Yakataga Formation, Middleton Island, Alaska. In,

Dowdeswell J.A. and Scourse J.D. (eds.), *Glacimarine Environments: Processes and Sediments.* Geological Society Special Publication **53**, London, 363–386.

Eyles, N., 1978. Scanning electron microscopy and particle size analysis of debris from a British Columbian glacier: a comparative report. In: Whalley, W.B. (Ed.), *Scanning Electron Microscopy in the Study of Sediments.* Geo Abstracts, Norwich, pp. 227–242.

Eyles, N., 1979. Facies of supraglacial sedimentation on Icelandic and Alpine temperate glaciers. *Canadian Journal of Earth Sciences* **16**, 1341–1361.

Eyles, N., 1987. Late Pleistocene debris-flow deposits in large glacial lakes in British Columbia and Alaska. *Sedimentary Geology* **53**, 33–71.

Eyles, N., Boyce, J.I., 1998. Kinematic indicators in fault gouge: tectonic analog for soft-bedded ice sheets. *Sedimentary Geology* **116**, 1–12.

Eyles, N., Doughty, M., 2016. *Glacially-streamlined hard and soft beds of the Ontario Ice Stream in Southern Ontario and New York State.* Sedimentary Geology (in press).

Eyles, N., Kocsis, S., 1988. Sedimentology and clast fabric of subaerial debris flow facies in a glacially-influenced alluvial fan. *Sedimentary Geology* **59**, 15–28.

Eyles, N., McCabe, A., 1989a. The Late Devensian (<22,000 BP) Irish Sea Basin: the sedimentary record of a collapsed ice sheet margin. *Quaternary Science Reviews* **8**, 307–351.

Eyles, N., McCabe, A., 1989b. Glaciomarine facies within subglacial tunnel valleys – the sedimentary record of glacio-isostatic downwarping in the Irish Sea Basin. *Sedimentology* **36**, 431–448.

Eyles, N., McCabe, A.M., 1991. Glaciomarine deposits of the Irish Sea Basin: the role of glacio-isostatic disequilibrium. In: Ehlers, J., Gibbard, P.L., Rose, J. (Eds.), *Glacial Deposits in Great Britain and Ireland.* Balkema, Rotterdam, pp. 311–331.

Eyles, N., Boyce, J.I., Barendregt, R.W., 1999. Hummocky moraine: sedimentary record of stagnant Laurentide Ice Sheet lobes resting on soft beds. *Sedimentary Geology* **123**, 163–174.

Eyles, N., Boyce, J., Putkinen, N., 2015. Neoglacial (<3000 years) till and flutes at Saskatchewan Glacier, Canadian Rocky Mountains, formed by subglacial deformation of a soft bed. *Sedimentology* **62**, 182–203.

Eyles, N., Clark, B.N., Clague, J.J., 1987. Coarse-grained sediment gravity flow facies in a large supraglacial lake. *Sedimentology* **34**, 193–216.

Eyles, N., Eyles, C.H., McCabe, A.M., 1989. Sedimentation in an ice-contact subaqueous setting: the mid Pleistocene 'North Sea Drifts' of Norfolk, UK. *Quaternary Science Reviews* **8**, 57–74.

Eyles, N., Eyles, C.H., Menzies, J., Boyce, J., 2011. End moraine construction by incremental till deposition below the Laurentide Ice Sheet: Southern Ontario, Canada. *Boreas* **40**, 92–104.

Eyles, N., Eyles, C.H., Miall, A.D., 1983a. Lithofacies types and vertical profile models; an alternative approach to the description and environmental interpretation of glacial diamict and diamictite sequences. *Sedimentology* **30**, 393–410.

Eyles, N., Eyles, C.H., Woodworth-Lynas, C.M.T., Randall, T.A., 2005. The sedimentary record of drifting ice (early Wisconsin Sunnybrook deposit) in an ancestral ice-dammed Lake Ontario, Canada. *Quaternary Research* **63**, 171–181.

Eyles, N., McCabe, A.M., Bowen, D.Q., 1994. The stratigraphic and sedimentological significance of Late Devensian ice sheet surging in Holderness, Yorkshire, UK. *Quaternary Science Reviews* **13**, 727–759.

Eyles, N., Putkinen, N., Sookhan, S., Arbelaez-Moreno, L., 2016. Erosional origin of drumlins and megaridges. *Sedimentary Geology* **338**, 2–23.

Eyles, N., Sladen, J.A., Gilroy, S., 1982. A depositional model for stratigraphic complexes and facies superimposition in lodgement tills. *Boreas* **11**, 317–333.

Eyles, N., Eyles, C.H., Day, T.E., 1983b. Sedimentologic and palaeomagnetic characteristics of glaciolacustrine diamict assemblages at Scarborough Bluffs, Ontario, Canada. In: Evenson, E.B., Schluchter, C., Rabassa, J. (Eds.), *Tills and Related Deposits*. Balkema, Rotterdam, pp. 23–45.

Eyles, N., Clark, B.M., Clague, J.J., 1988a. Coarse-grained sediment gravity flow facies in a large supraglacial lake – reply. *Sedimentology* **35**, 529–530.

Eyles, N., Eyles, C.H., McCabe, A.M. 1988b. Late Pleistocene subaerial debris flow facies of the Bow Valley near Banff, Canadian Rocky Mountains. *Sedimentology* **35**, 465–480.

Eyles, N., Eyles, C.H., McCabe, A.M. 1990. Late Pleistocene subaerial debris flow sediments near Banff, Canada – reply. *Sedimentology* **37**, 544–547.

Eyles, N., McCabe, A.M., Bowen, D.Q., 1994. The stratigraphic and sedimentological significance of late Devensian ice sheet surging in Holderness, Yorkshire, UK. *Quaternary Science Reviews* **8**, 727–759.

Ferrians, O.N., 1963. Glaciolacustrine diamicton deposits in the Copper River Basin, Alaska. *USGS Professional Paper* **475-C**, pp. C129–C135.

Fischer, U., Clarke, G.K.C., 1994. Ploughing of subglacial sediment. *Journal of Glaciology* **40**, 97–106.

Fischer, U., Clarke, G.K.C., 1997. Stick-slip sliding behaviour at the base of a glacier. *Annals of Glaciology* **24**, 390–396.

Fischer, U., Clarke, G.K.C., 2001. Review of subglacial hydromechanical coupling: Trapridge Glacier, Yukon Territory, Canada. *Quaternary International* **86**, 29–43.

Fischer, U., Clarke, G.K.C., Blatter, H., 1999. Evidence for temporally varying 'sticky spots' at the base of Trapridge Glacier, Yukon Territory, Canada. *Journal of Glaciology* **45**, 352–360.

Fischer, U., Porter, P.R., Schuler, T., Evans, A.J., Gudmundsson, G.H., 2001. Hydraulic and mechanical properties of glacial sediments beneath Unteraargletscher, Switzerland: implications for glacier basal motion. *Hydrological Processes* **15**, 3525–3540.

Fitzsimons, S.J., 1990. Ice-marginal depositional processes in a polar maritime environment, Vestfold Hills, Antarctica. *Journal of Glaciology* **36**, 279–286.

Fleming, E.J., Stevenson, C.T.E., Petronic, M.S., 2013. New insights into the deformation of a Middle Pleistocene glaciotectonised sequence in Norfolk, England through magnetic and structural analysis. *Proceedings of the Geologists' Association* **124**, 834–854.

Flint, R.F., 1947. *Glacial Geology and the Pleistocene Epoch*. Wiley, New York.

Flint, R.F., 1955. Pleistocene geology of eastern South Dakota: U.S. Geological Survey Professional Paper **262**, 173 p.

Flint, R.F., 1957. *Glacial and Pleistocene Geology*. John Wiley and Sons, Inc., New York.

Flint, R.F., 1971. *Glacial and Quaternary Geology*. John Wiley and Sons, Inc., New York.

Flint, R.F., Sanders, J.E., Rodgers, J., 1960. Diamictite: a substitute term for symmictite. *Bulletin of the Geological Society of America* **71**, 1809.

Fountain, A.G., Walder, J.S., 1998. Water flow through temperate glaciers. *Reviews of Geophysics* **36**, 299–328.

Fowler, A.C., 2000. An instability mechanism for drumlin formation. In: Maltman, A.J., Hubbard, B., Hambrey, M.J. (Eds.), *Deformation of Glacial Materials*. Geological Society, London, Special Publications, vol. 176, pp. 307–319.

Fowler, A.C., 2002. Rheology of subglacial till. *Journal of Glaciology* **48**, 631–632.

Fowler, A.C., 2003. On the rheology of till. *Annals of Glaciology* **37**, 55–59.

Fowler, A.C., 2009. Instability modelling of drumlin formation incorporating leeside cavity growth. *Proceedings of the Royal Society of London, Series A* **465**, 2681–2702.

Fowler, A.C., 2010. The instability theory of drumlin formation applied to Newtonian viscous ice of finite depth. *Proceedings of the Royal Society of London, Series A* **466**, 2673–2694.

Francis, E.A., 1975. Glacial sediments: A selective review. In: Wright, A.E., Moseley, F. (Eds.), *Ice Ages: Ancient and Modern*. Seel House Press, Liverpool, pp. 43–68.

French, H.M., Bennett, L., Hayley, D.W., 1986. Ground ice conditions near Rea Point and on Sabine Peninsula, eastern Melville Island. *Canadian Journal of Earth Sciences* **23**, 1389–1400.

Frye, J.C., Glass, H.D., Kempton, J.P., Willman, H.B., 1969. Glacial Tills of Northwestern Illinois. *Illinois State Geological Survey, Circular* **437**.

Fuller, M.L., 1914. *The geology of Long Island*, New York. USGS Professional Paper **82**.

Fuller, S., Murray, T., 2000. Evidence against pervasive bed deformation during the surge of an Icelandic glacier. In: Maltman, A.J., Hubbard, B., Hambrey, M.J. (Eds.), *Deformation of Glacial Materials. Geological Society*, London, Special Publications, vol. 176, pp. 203–216.

Fullerton, D.S., Bush, C.A., Pennell, J.N., 2003. Map of surficial deposits and materials in the Eastern and Central United States (East of 102° West Longitude). USGS Geological Investigations Series I-2789, 1:2,500,000 scale.

Garwood, E.J., Gregory, J.W., 1898. Contributions to the glacial geology of Spitsbergen. *Quarterly Journal of the Geological Society, London* **54**, 197–227.

Gee, M., Gawthorpe, R., Friedmann, J., 2005. Giant striations at the base of a submarine landslide. *Marine Geology* **214**, 287–294.

Geikie, A., 1863. On the glacial drift of Scotland. *Transactions of the Geological Society of Glasgow* **1**, 1–190.

Geikie, A., 1903. *Text Book of Geology*. Fourth Edition. MacMillan, London.

Geikie, J., 1894. *The Great Ice Age*. Edward Stanford, London.

Geinitz, E., 1903. *In Das Quartar*. Stuttgart.

Gentoso, M.J., Evenson, E.B., Kodama, K.P., Iverson, N.R., Alley, R.B., Berti, C., Kozlowski, A., 2012. Exploring till bed kinematics using AMS magnetic fabrics and pebble fabrics: the Weedsport drumlin field, New York State, USA. *Boreas* **41**, 31–41.

Gibbard, P.L., 1980. The origin of stratified Catfish Creek till by basal melting. *Boreas* **9**, 71–85.

Gilbert, G.K., 1898. Boulder pavement at Wilson, NY. *Journal of Geology* **6**, 771–775.

Gilbert, R., 1990. Rafting in glacimarine environments. In: Dowdeswell, J.A., Scourse, J.D. (Eds.), Glacimarine Environments: Processes and Sediments: Journal of the Geological Society, London, Special Publications, **53**, pp. 105–120.

Goldthwait, R.P., 1951. Development of end moraines in east-central Baffin Island. *Journal of Geology* **59**, 567–577.

Goldthwait, R.P. (Ed.), 1971a. *Till: a Symposium*. Ohio State University Press, Columbus.

Goldthwait, R.P., 1971b. Introduction to till today. In: Goldthwait, R.P. (Ed.), *Till: a Symposium*. Ohio State University Press, Columbus, pp. 3–26.

Goldthwait, R.P., Matsch, C.L. (Eds.), 1989. *Genetic Classification of Glacigenic Deposits*. Balkema, Rotterdam.

Goldthwaite, L., 1948. Glacial till in New Hampshire. Mineral Resources Survey, New Hampshire State Planning and Development Committee, Part 10.

Goodchild, J.G., 1875. Glacial phenomena of the Eden Valley and the western part of the Yorkshire Dale district. *Quarterly Journal of the Geological Society of London* **31**, 55–99.

Gordon, J.E., Whalley, W.B., Gellatly, A.F., Vere, D.M., 1992. The formation of glacial flutes: assessment of models with evidence from Lyngsdalen, north Norway. *Quaternary Science Reviews* **11**, 709–731.

Goughnour, R.R., Andersland, O.B., 1968. Mechanical properties of a sand–ice system. *Journal of the Soil Mechanics and Foundations Division, American Society of Civil Engineers*, vol. 94, No. SM4, pp. 923–950.

Gow, A., Ueda, H., Garfield, D., 1968. Antarctic Ice Sheet: preliminary results of first core hole to bedrock. *Science* **161**, 1011–1013.

Gravenor, C.P., 1953. The origin of drumlins. *American Journal of Science* **251**, 674–681.

Gravenor, C.P., 1957. Surficial geology of the Lindsay–Peterborough area. *Geological Survey of Canada Memoir* **288**, 121 pp.

Gravenor, C.P., von Brunn, V., Dreimanis, A., 1984. Nature and classification of waterlain glaciogenic sediments, exemplified by Pleistocene, Late Palaeozoic and Late Precambrian deposits. *Earth Science Reviews* **20**, 105–166.

Green, S.M., Obermeier, S.F., Olson, S.M., 2005. Engineering geologic and geotechnical analysis of paleoseismic shaking using liquefaction effects: field examples. *Engineering Geology* **76**, 263–293.

Greenly, E., 1919. Geology of Anglesey, Memoirs of the Geological Survey of Britain 2.

Griggs, G.B., Kulm, L.D., 1969. Glacial marine sediments from the northern Pacific. *Journal of Sedimentary Petrology* **39**, 1142–1148.

Haldorsen, S., 1981. Grain-size distribution of subglacial till and its relation to glacial crushing and abrasion. *Boreas* **10**, 91–105.

Haldorsen, S., 1982. The genesis of tills from Astadalen, south-eastern Norway. *Norsk Geologisk Tidsskrift* **62**, 17–38.

Haldorsen, S., Shaw, J., 1982. The problem of recognizing melt out till. *Boreas* **11**, 261–277.

Hallet, B., 1976. Deposits formed by subglacial precipitation of $CaCo_3$. *Geological Society of America Bulletin* **87**, 1003–1015.

Hallet, B., 1979. Subglacial regelation water film. *Journal of Glaciology* **23**, 321–334.

Hallet, B., 1981. Glacial abrasion and sliding: their dependence on the debris concentration in basal ice. *Annals of Glaciology* **2**, 23–28.

Ham, N.R., Mickelson, D.M., 1994. Basal till fabric and deposition at Burroughs Glacier, Glacier Bay, Alaska. *Bulletin of the Geological Society of America* **106**, 1552–1559.

Hambrey, M.J., Ehrmann, W., 2004. Modification of sediment characteristics during glacial transport in high-alpine catchments: Mount Cook area, New Zealand. *Boreas* **33**, 300–318.

Hambrey, M.J., Harland, W.B. (eds.), 1981. *Earth's Pre-Pleistocene Glacial Record*. Cambridge University Press, Cambridge.

Hampton, M.A., 1972. The role of subaqueous debris flow in generating turbidity currents. *Journal of Sedimentary Petrology* **42**, 277–292.

Hampton, M.A., 1975. Competence of fine-grained debris flows. *Journal of Sedimentary Petrology* **49**, 753–793.

Hampton, M.A., 1979. Buoyancy in debris flows. *Journal of Sedimentary Petrology* **49**, 753–758.

Hansel, A.K., Johnson, W.H., 1987. Ice marginal sedimentation in a late Wisconsinan end moraine complex, northeastern Illinois, USA. In: van der Meer, J.J.M. (Ed.), *Tills and Glaciotectonics*. Balkema, Rotterdam, 97–104.

Hansen, S., 1930. Om forekomster af glacialflager af paleocaen mergel paa Sjaelland. *Meddelelser fra Dansk Geologisk Forening* **7**, 391–410.

Hansom J.D., 1983. Ice-formed intertidal boulder pavements in the sub-Antarctic. *Journal of Sedimentary Petrology* **53**, 135–145.

Harland, W.B., Herod, K.N., Krinsley D.H., 1966. The definition and identification of tills and tillites. *Earth Science Reviews* **2**, 225–256.

Harris, C., 1991. Glacially deformed bedrock at Wylfa Head, Anglesey, North Wales. In: Forster, A., Culshaw, M.G., Cripps, J.C., Little, J.A., Moon, C.F. (Eds.), *Quaternary Engineering Geology*. Geological Society Engineering Geology Special Publication 7, London, pp. 135–142.

Harris, C., Bothamley, K., 1984. Englacial deltaic sediments as evidence for basal freezing and marginal shearing, Leirbreen, Norway. *Journal of Glaciology* **30**, 30–34.

Harrison, P.W., 1957. A clay-till fabric: its character and origin. *Journal of Geology* **65**, 275–307.

Harrison, S., 1996. Paraglacial or periglacial? The sedimentology of slope deposits in upland Northumberland. In: Anderson, M.G., Brooks, S.M. (Eds.), *Advances in Hillslope Processes*. Wiley, Chichester, 1197–1218.

Harrison, S., Winchester, V., 1997. Age and nature of paraglacial debris cones along the margins of the San Rafael Glacier, Patagonian Chile. *The Holocene* **7**, 481–487.

Harrison, S., Whalley, B., Anderson, E., 2008. Relict rock glaciers and protalus lobes in the British Isles: implications for Late Pleistocene mountain geomorphology and palaeoclimate. *Journal of Quaternary Science* **23**, 287–304.

Hart, J.K., 1990. Proglacial glaciotectonic deformation and the origin of the Cromer Ridge push moraine complex, North Norfolk, England. *Boreas* **19**, 165–180.

Hart, J.K., 1994. Till fabric associated with deformable beds. *Earth Surface Processes and Landforms* **19**, 15–32.

Hart, J.K., 1995. Subglacial erosion, deposition and deformation associated with deformable beds. *Progress in Physical Geography* **19**, 173–191.

Hart, J.K., 1997. The relationship between drumlins and other forms of subglacial glaciotectonic deformation. *Quaternary Science Reviews* **16**, 93–107.

Hart, J.K., 2006. Deforming layer erosion over rigid bed rock at Athabasca Glacier, Canada. *Earth Surface Processes and Landforms* **31**, 65–80.

Hart, J.K., 2007. An investigation of subglacial shear zone processes from Weybourne, Norfolk, UK. *Quaternary Science Reviews* **26**, 2354–2374.

Hart, J.K., Boulton, G.S., 1991. The inter-relation of glaciotectonic and glaciodepositional processes within the glacial environment. *Quaternary Science Reviews* **10**, 335–350.

Hart, J.K., Roberts, D.H., 1994. Criteria to distinguish between subglacial glaciotectonic and glaciomarine sedimentation: I. Deformation styles and sedimentology. *Sedimentary Geology* **91**, 191–213.

Hart, J.K., Roberts, D.H., 2005. The deforming bed characteristics of a stratified till assemblage in north East Anglia, UK: investigation controls on sediment rheology and strain signatures. *Quaternary Science Reviews* **25**, 123–140.

Hart, J.K., Rose, J. (Eds.), 2001. Glacier deforming-bed processes. *Quaternary International* **86**, 1–150.

Hart, J.K., Hindmarsh, R.C.A., Boulton, G.S., 1990. Styles of subglacial glaciotectonic deformation within the context of the Anglian ice sheet. *Earth Surface Processes and Landforms* **15**, 227–241.

Hart, J.K., Rose, K.C., Martinez, K., Ong, R., 2009. Subglacial clast behaviour and its implications for till fabric development: new results derived from wireless subglacial probe experiments. *Quaternary Science Reviews* **28**, 597–607.

Hart, J.K., Martinez, K., Ong, R., Riddoch, A., Rose, K.C., Padhy, P., 2006. An autonomous multi-sensor subglacial probe: design and preliminary results from Briksdalsbreen, Norway. *Journal of Glaciology* **51**, 389–397.

Hart, J.K., Rose, K.C., Martinez, K., 2011. Subglacial till behaviour derived from in situ wireless multi-sensor subglacial probes: Rheology, hydro-mechanical interactions and till formation. *Quaternary Science Reviews* **30**, 234–247.

Hartshorn, J.H., 1958. Flow-till in southeastern Massachusetts. *Bulletin of the Geological Society of America* **69**, 477–482.

Hayashi, T., 1966. Clastic dykes in Japan. *Transactions of the Japanese Journal of Geology and Geography* **37**, 1–20.

Heim, A., 1885. *Handbuch der Gletscherkunde*. Stuttgart.

Hein, F.J., 1982. Depositional mechanisms of deep-sea coarse clastic sediments, Cap Enrage Formation, Quebec. *Canadian Journal of Earth Sciences* **19**, 267–287.

Hemingway, J.E., Riddler, G.P., 1980. Glacially transported Liassic rafts at Upgang, Near Whitby. *Proceedings of the Yorkshire Geological Society* **43**, 183–189.

Heron, R., Woo, M.-K., 1994. Decay of a High Arctic lake-ice cover: observations and modelling. *Journal of Glaciology* **40**, 283–292.

Hershey, O.H., 1897. Mode of formation of till, as illustrated by the Kansas drift of northern Illinois. *Journal of Geology* **5**, 50–62.

Hicock, S.R., 1990. Genetic till prism. *Geology* **18**, 517–519.

Hicock, S.R., 1991. On subglacial stone pavements in till. *Journal of Geology* **99**, 607–619.

Hicock, S.R., 1992. Lobal interactions and rheologic superposition in subglacial till near Bradtville, Ontario, Canada. *Boreas* **21**, 73–88.

Hicock, S.R., 1993. Glacial octahedron. *Geografiska Annaler* **75A**, 35–39.

Hicock, S.R., Dreimanis, A., 1989. Sunnybrook drift indicates a grounded early Wisconsin glacier in the Lake Ontario basin. *Geology* **17**, 169–172.

Hicock, S.R., Dreimanis, A., 1992a. Sunnybrook drift in the Toronto area, Canada: reinvestigation and reinterpretation. In: Clark, P.U., Lea, P.D. (Eds.), *The Last Interglacial–Glacial Transition in North America*. Geological Society of America Special Paper, vol. 270, pp. 139–161.

Hicock, S.R., Dreimanis, A., 1992b. Deformation till in the Great Lakes region: implications for rapid flow along the south-central margin of the Laurentide Ice Sheet. *Canadian Journal of Earth Sciences* **29**, 1565–1579.

Hicock, S.R., Fuller, E.A., 1995. Lobal interactions, rheologic superposition, and implications for a Pleistocene ice stream on the continental shelf of British Columbia. *Geomorphology* **14**, 167–184.

Hicock, S.R., Lian, O.B., 1999. Cordilleran Ice Sheet lobal interactions and glaciotectonic superposition through stadial maxima along a mountain front in southwestern British Columbia, Canada. *Boreas* **28**, 531–542.

Hicock, S.R., Goff, J.R., Lian, O.B., Little, E.C., 1996. On the interpretation of subglacial till fabric. *Journal of Sedimentary Research* **66**, 928–934.

Hiemstra, J.F., Rijsdijk, K.F., 2003. Observing artificially induced strain: implications for subglacial deformation. *Journal of Quaternary Science* **18**, 373–383.

Hiemstra, J.F., van der Meer, J.J.M., 1997. Porewater controlled grain fracturing as an indicator for subglacial shearing in tills. *Journal of Glaciology* **43**, 446–454.

Hiemstra, J.F., Rijsdijk, K.F., Evans, D.J.A., van der Meer, J.J.M., 2005. Integrated micro- and macro-scale analyses of last glacial maximum Irish Sea diamicts from Abermaw and Treath y Mwnt, Wales, UK. *Boreas* **34**, 61–74.

Hiemstra, J.F., Evans, D.J.A., Ó Cofaigh, C., 2007. The role of glaciotectonic rafting and comminution in the production of subglacial tills: examples from SW Ireland and Antarctica. *Boreas* **36**, 386–399.

Hiemstra, J.F., Zaniewski, K., Powell, R.D., Cowan, E.A., 2004. Strain signatures of fjord sediment sliding: micro-scale examples from Yakutat Bay and Glacier Bay, Alaska, USA. *Journal of Sedimentary Research* **74**, 760–769.

Hiemstra, J.F., Rijsdijk, K.F., Shakesby, R., McCarroll, D., 2009. Reinterpreting Rotherslade, Gower Peninsula: implications for Last Glacial ice limits and Quaternary stratigraphy of the British Isles. *Journal of Quaternary Science* **24**, 399–410.

Hillefors, A., 1973. The stratigraphy and genesis of stoss- and lee-side moraines. *Bulletin of the Geological Institute of the University of Uppsala* **5**, 139–154.

Hind, H.Y., 1859. A preliminary and general report on the Assiniboine and Saskatchewan exploring expedition. *Canada Legislative Assembly Journal 19*, Appendix 36.

Hindmarsh, R.C.A., 1996. Sliding of till over bedrock: scratching, polishing, comminution and kinematic wave theory. *Annals of Glaciology* **22**, 41–48.

Hindmarsh, R.C.A., 1997. Deforming beds: viscous and plastic scales of deformation. *Quaternary Science Reviews* **16**, 1039–1056.

Hindmarsh, R.C.A., 1998a. Drumlinization and drumlin forming instabilities: viscous till mechanisms. *Journal of Glaciology* **44**, 293–314.

Hindmarsh, R.C.A., 1998b. The stability of a viscous till sheet coupled with ice flow, considered at wavelengths less than the ice thickness. *Journal of Glaciology* **44**, 285–292.

Hitchcock, C.H., 1879. *The Geology of New Hampshire*. New Hampshire Geological Survey.

Hobbs W.H., 1931. Loess, pebble bands and boulders from the glacial outwash of the Greenland continental glacier. *Journal of Geology* **39**, 381–385.

Hoffmann, K., Piotrowski, J.A., 2001. Till melange at Amsdorf, central Germany: sediment erosion, transport and deposition in a complex, soft-bedded subglacial system. *Sedimentary Geology* **140**, 215–234.

Hoffman, P.F., Schrag, D.P., 2000. Snowball Earth. *Scientific American* **282**, 62–75.

Hoffman, P.F., Schrag, D.P., 2002. The Snowball Earth hypothesis: testing the limits of global change. *Terra Nova* **14**, 129–155.

Hoffman, P.F., Kaufman, A.J., Halverson, G.P., Schrag, D.P., 1998. A Neoproterozoic Snowball Earth. *Science* **281**, 1342–1346.

Hollingworth, S.E., 1931. The glaciation of western Edenside and the Solway Basin. *Quarterly Journal of the Geological Society, London* **87**, 281–359.

Holmes, C.D., 1941. Till fabric. *Geological Society of America Bulletin* **52**, 1301–1352.

Holmes, C.D., 1944. 'Pavement-boulders' as interglacial evidence. *American Journal of Science* **242**, 431–435.

Holmes, C.D., 1960. Evolution of till stone shapes, New York. *Geological Society of America Bulletin* **71**, 1645–1660.

Holzer, T.L., Hanks, T.C., Youd, T.L., 1989. Dynamics of Liquefaction during the 1987 Superstition Hills, California, Earthquake. *Science* **244**, 56–59.

Hooke, R. LeB., Iverson, N.R., 1995. Grain-size distribution in deforming subglacial tills: role of grain fracture. *Geology* **23**, 57–60.

Hooke, R. LeB., Hanson, B., Iverson, N.R., Jansson, P., Fischer, U.H., 1997. Rheology of till beneath Storglaciaren, Sweden. *Journal of Glaciology* **43**, 172–179.

Hooyer, T.S., Iverson, N.R., 2000. Diffusive mixing between shearing granular layers: constraints on bed deformation from till contacts. *Journal of Glaciology* **46**, 641–651.

Hooyer, T.S., Iverson, N.R., Lagroix, F., Thomason, J.F., 2008. Magnetic fabric of sheared tills: a strain indicator for evaluating the bed deformation model of glacier flow. *Journal of Geophysical Research* **113**, F02002, doi:10.1029/2007JF000757.

Hopson, M., 1995. Chalk rafts in Anglian till in Hertfordshire. *Proceedings of the Geologists' Association* **106**, 151–215.

Hoppe, G., 1959. Glacial morphology and inland ice recession in northern Sweden. *Geografiska Annaler* **41**, 193–212.

Hsu, K.I., 1974. Melanges and their distinction from olistostromes. In: Dott, R.H., Shaver, R.H. (Eds.), *Modern and Ancient Geosynclinal Sedimentation*. SEPM Special Publication 19, pp. 321–333.

Hubbard, B., Glasser, N.F., 2005. *Field Techniques in Glaciology and Glacial Geomorphology*. Wiley.

Hubbard, B., Nienow, P., 1997. Alpine subglacial hydrology. *Quaternary Science Reviews* **16**, 939–955.

Hubbard, B., Sharp, M., 1989. Basal ice formation and deformation: a review. *Progress in Physical Geography* **13**, 529–558.

Humlum, O., 1978. Genesis of layered lateral moraines – implications for palaeoclimatology and lichenometry. *Geografisk Tidsskrift* **77**, 65–72.

Humphrey, N., Kamb, B., Fahnestock, M., Engelhardt, H., 1993. Characteristics of the bed of the lower Columbia Glacier, Alaska. *Journal of Geophysical Research* **98**, 837–846.

Ildefonse, B., Mancktelow, N.S., 1993. Deformation around rigid particles: the influence of slip at the particle/matrix interface. *Tectonophysics* **221**, 345–359.

Ildefonse, B., Launeau, P., Bouchez, J.L., Fernandez, A., 1992. Effects of mechanical interactions on the development of preferred orientations: a two-dimensional experimental approach. *Journal of Structural Geology* **14**, 73–83.

Ingólfsson, Ó., Lokrantz, H., 2003. Massive ground ice body of glacial origin at Yugorski Peninsula, arctic Russia. *Permafrost and Periglacial Processes* **14**, 199–215.

Iverson, N.R., 1993. Regelation of ice through debris at glacier beds: implications for sediment transport. *Geology* **21**, 559–562.

Iverson, N.R., 1999. Coupling between a glacier and a soft bed: II. *Model results. Journal of Glaciology* **45**, 41–53.

Iverson, N.R., 2000. Sediment entrainment by a soft-bedded glacier: a model based on regelation into the bed. *Earth Surface Processes and Landforms* **25**, 881–893.

Iverson, N.R., 2010. Shear resistance and continuity of subglacial till: hydrology rules. *Journal of Glaciology* **56**, 1104–1114.

Iverson, N.R., Hooyer, T.S., 2002. Clast fabric development in a shearing granular material: implications for subglacial till and fault gouge – reply. *Bulletin of the Geological Society of America* **114**, 383–384.

Iverson, N.R., Iverson, R.M., 2001. Distributed shear of subglacial till due to Coulomb slip. *Journal of Glaciology* **47**, 481–488.

Iverson, N.R., Semmens, D., 1995. Intrusion of ice into porous media by regelation: a mechanism of sediment entrainment by glaciers. *Journal of Geophysical Research* **100**, 10219–10230.

Iverson, N.R., Hooyer, T.S., Thomason, J.F., Graesch, M., Shumway, J.R., 2008. The experimental basis for interpreting particle and magnetic fabrics of sheared till. *Earth Surface Processes and Landforms* **33**, 627–645.

Iverson, N.R., Jansson, P., Hooke, R.LeB., 1994. In situ measurements of the strength of deforming subglacial till. *Journal of Glaciology* **40**, 497–503.

Iverson, N.R., Hanson, B., Hooke, R.LeB., Jansson, P., 1995. Flow mechanism of glaciers on soft beds. *Science* **267**, 80–81.

Iverson, N.R., Baker, R.W., Hooyer, T.S., 1997. A ring-shear device for the study of till deformation tests on tills with contrasting clay contents. *Quaternary Science Reviews* **16**, 1057–1066.

Iverson, N.R., Hooyer, T.S., Baker, R.W., 1998. Ring-shear studies of till deformation: Coulomb-plastic behaviour and distributed strain in glacier beds. *Journal of Glaciology* **44**, 634–642.

Iverson, N.R., Baker, R.W., Hooke, R.LeB., Hanson, B., Jansson, P., 1999. Coupling between a glacier and a soft bed: I. A relation between effective pressure and local shear stress determined from till elasticity. *Journal of Glaciology* **45**, 31–40.

Iverson, N.R., Hooyer, T.S., Hooke, R.L., 1996. A laboratory study of sediment deformation: stress heterogeneity and grain-size evolution. *Annals of Glaciology* **22**, 167–175.

Iverson, N.R., Cohen, D., Hooyer, T.S., Fischer, U.H., Jackson, M., Moore, P.L., Lappegard, G., Kohler, J., 2003. Effects of basal debris on glacier flow. *Science* **301**, 81–84.

Iverson, N.R., Hooyer, T.S., Fischer, U.H., Cohen, D., Moore, P.L., Jackson, M., Lappegard, G., Kohler, J., 2007. Soft-bed experiments beneath Engabreen, Norway: regelation infiltration, basal slip and bed deformation. *Journal of Glaciology* **53**, 323–340.

Iverson, R.M., 1997. The physics of debris flows. *Reviews of Geophysics* **35**, 245–296.

Jackson, M., Kamb, B., 1997. The marginal shear stress of Ice Stream B, West Antarctica. *Journal of Glaciology* **43**, 415–426.

Jaeger, H.M., Nagel, S.R., 1992. Physics of the granular state. *Science* **255**, 1523–1531.

Jeffery, G.B., 1922. The motion of ellipsoidal particles immersed in a viscous fluid. *Proceedings of the Royal Society of London*, Series A, **102**, 161–179.

Jensen, J.W., Clark, P.U., MacAyeal, D.R., Ho, C., Vela, J.C., 1995. Numerical modelling of advective transport of saturated deforming sediment beneath the Lake Michigan Lobe, Laurentide Ice Sheet. *Geomorphology* **14**, 157–166.

Jensen, J.W., MacAyeal, D.R., Clark, P.U., Ho, C., Vela, J.C., 1996. Numerical modelling of subglacial sediment deformation: implications for the behaviour of the Lake Michigan Lobe, Laurentide Ice Sheet. *Journal of Geophysical Research* **101**, 8717–8728.

Jewtuchowicz, S., 1956. Wokolicach Zbójna. Panstwowe Wydawn, Naukowe 73 pp.

Johnson, A.M., 1965. A model for debris flow. Unpublished Ph.D. thesis, Pennsylvania State University.

Johnson, A.M., 1970. *Physical Processes in Geology*. Freeman, New York.

Johnson, A.M., Rodine, J.R., 1984. Debris flow. In: Brunsden, D., Prior, D.B. (Eds.), *Slope Instability*. Wiley, Chichester, pp. 257–361.

Johnson, M.D., Benediktsson, Í.Ö., Björklund, L., 2013. The Ledsjö end moraine – a subaquatic push moraine composed of glaciomarine clay in central Sweden. *Proceedings of the Geologists' Association* **124**, 738–752.

Johnson, M.D., 1983. The origin and microfabric of Lake Superior Clay. *Sedimentary Petrology* **53**, 859–873.

Johnson, M.D., Gillam, M.L., 1995. Composition and construction of late Pleistocene end moraines, Durango, Colorado. *Geological Society of America Bulletin* **107**, 1241–1253.

Johnson, M.D., Mickelson, D.M., Clayton, L., Attig, J.W., 1995. Composition and genesis of glacial hummocks, western Wisconsin, USA. *Boreas* **24**, 97–116.

Johnson, M.D., Schomacker, A., Benediktsson, Í.Ö., Geiger, A.J., Ferguson, A., Ingólfsson, Ó., 2010. Active drumlin field revealed at the margin of Múlajökull, Iceland: a surge-type glacier. *Geology* **38**, 943–946.

Johnson, W.H., Hansel, A.K., 1990. Multiple Wisconsinan glacigenic sequences at Wedron, Illinois. *Journal of Sedimentary Petrology* **60**, 26–41.

Johnson, W.H., Hansel, A.K., 1999. Wisconsin episode glacial landscape of central Illinois: a product of subglacial deformation processes? In: Mickelson, D.M., Attig, J.W. (Eds.), *Glacial Processes: Past and Present*. Geological Society of America, Special Paper, vol. 337, pp. 121–135.

Johnstrup, F., 1874. Uber die Lagerungsverhaltnisse und die Hebungsphanomene in den Kreidefelsen auf Mon und Rugen. *Zeitschrift der Deutschen Geoligischen Gesellschaft* **1874**, 533–585.

Jonk, R., Durant, D., Parnell, J., Hurst, A., Fallick, A.E., 2003. The structural and diagenetic evolution of injected sandstones: examples from the Kimmeridgian of NE Scotland. *Journal of the Geological Society London* **160** 881–894.

Jørgensen, F., Piotrowski, J.A., 2003. Signature of the Baltic Ice Stream on Funen Island, Denmark during the Weichselian glaciation. *Boreas* **32**, 242–255.

Kamb, B., 1991. Rheological nonlinearity and flow instability in the deforming bed mechanism of ice stream motion. *Journal of Geophysical Research* **96**, 16585–16595.

Kamb, B., 2001. Basal zone of the West Antarctic Ice Streams and its role in lubrication of their rapid motion. In: Alley, R.B., Bindschadler, R.A. (Eds.), *The West Antarctic Ice Sheet: Behaviour and Environment*. American Geophysical Union, Antarctic Research Series, 77, pp. 157–199.

Kaplyanskaya, F.A., Tarnogradskiy, V.D., 1986. Remnants of the Pleistocene ice sheets in the permafrost zone as an object for paleoglaciological research. *Polar Geography and Geology* **10**, 257–266.

Kavanaugh, J.L., Clarke, G.K.C., 2006. Discrimination of the flow law for subglacial sediment using in situ measurements and an interpretation model. *Journal of Geophysical Research-Earth* **111**, F01002, doi:10.1029/2005JF000346.

Kay G.F., 1931. Classification and duration of the Pleistocene period. *Geological Society of America Bulletin* **48**, 425–466.

Kelly, R.I., Martini, I.P., 1986. Pleistocene glacio-lacustrine deltaic deposits of the Scarborough Formation, Ontario, Canada. *Sedimentary Geology* **47**, 27–52.

Kemmis, T.J., 1981. Importance of the regelation process to certain properties of basal tills deposited by the Laurentide ice sheet in Iowa and Illinois. *Annals of Glaciology* **2**, 147–152.

Kerr, M., Eyles, N., 2007. Origin of drumlins on the floor of Lake Ontario and in upper New York State. *Sedimentary Geology* **193**, 7–20.

Kerr, R.J., 1978. The nature and derivation of glacial till in the upper part of the Tweed Basin. Unpublished PhD thesis, University of Edinburgh.

Kilfeather, A.A., van der Meer, J.J.M., 2008. Pore size, shape and connectivity in tills and their relationship to deformation processes. *Quaternary Science Reviews* **27**, 250–266.

Kilfeather, A.A., Ó Cofaigh, C., Dowdeswell, J.A., van der Meer, J.J.M., Evans, D.J.A., 2010. Micromorphological characteristics of glacimarine sediments: implications for distinguishing genetic processes of massive diamicts. *Geo-Marine Letters* **30**, 77–97.

Kindle, E.M., 1924. Observations on ice-borne sediments by the Canadian and other arctic expeditions. *American Journal of Science* **7**, 251–286.

King, E.C., Hindmarsh, R.C., Stokes, C., 2009. Formation of mega-scale glacial lineations observed beneath a west Antarctic ice stream. *Nature Geoscience* **2**, 585–588.

King, E.C., Woodward, J., Smith, A.M., 2004. Seismic evidence for a water-filled canal in deforming till beneath Rutford Ice Stream, West Antarctica. *Geophysical Research Letters* **31**, L20401, doi:10.1029/2004GL020379.

King, E.C., Woodward, J., Smith, A.M., 2007. Seismic and radar observations of subglacial bedforms beneath the onset zone of Rutford Ice Stream, Antarctica. *Journal of Glaciology* **53**, 665–672.

Kjær, K.H., Krüger, J., 1998. Does clast size influence fabric strength? *Journal of Sedimentary Research* **68**, 746–749.

Kjær, K.H., Krüger, J., 2001. The final phase of dead ice moraine development: processes and sediment architecture, Kotlujökull, Iceland. *Sedimentology* **48**, 935–952.

Kjær, K.H., Larsen, E., van der Meer, J.J.M., Ingólfsson, Ó., Krüger, J., Benediktsson, Í.Ö., Knudsen, C.G., Schomacker, A., 2006. Subglacial decoupling at the sediment/bedrock interface: a new mechanism for rapid flowing ice. *Quaternary Science Reviews* **25**, 2704–2712.

Kjær, K.H., Korsgaard, N.J., Schomacker, A., 2008. Impact of multiple glacier surges – a geomorphological map from Bruarjökull, east Iceland. *Journal of Maps* **2008**, 5–20.

Klassen, R.W., 1989. Quaternary geology of the southern Canadian interior plains. In: Fulton, R.J. (Ed.), *Quaternary Geology of Canada and Greenland*. Geological Survey of Canada, Geology of Canada 1, pp. 138–174.

Knight, J., 2014. Subglacial hydrology and drumlin sediments in Connemara, western Ireland. *Geografiska Annaler* **96A**, 403–415.

Knill, J.L., 1968. Geotechnical significance of certain glacially induced discontinuities in rock. *Bulletin of the International Association of Engineering Geology* **5**, 49–62.

Kozarski, S., 1959. O genezie chodzieskiej moreny czolowej. *Badania Fizjograficzne nad Polska Zachodnia* **5**, 45–69. (English translation: 'On the origin of the Chodziez end moraine', in Evans, D.J.A. (Ed.), Cold Climate Landforms, Wiley, Chichester, pp. 293–312.)

Kozarski, S., Kasprzak, L., 1994. Dynamics of the last Scandinavian ice sheet and glacio-dislocation metamorphism of unconsolidated deposits in west central Poland: a terminological approach. *Zeitschrift fur Geomorphologie* **95**, 49–58.

Krinsley, D.H., Doornkamp, J.C., 1973. *Atlas of Quartz Sand Surface Textures.* Cambridge University Press.

Krinsley, D.H., Smalley, I.J., 1972. *Sand. American Science* **60**, 286–291.

Kristensen, L., Benn, D.I., Hormes, A., Ottesen, D., 2009. Mud aprons in front of Svalbard surge moraines: evidence of subglacial deforming layers or proglacial glaciotectonics? *Geomorphology* **111**, 206–221.

Krüger, J., 1979. Structures and textures in till indicating subglacial deposition. *Boreas* **8**, 323–340.

Krüger, J., 1984. Clasts with stoss-lee form in lodgement tills: a discussion. *Journal of Glaciology* **30**, 241–243.

Krüger, J., 1993. Moraine ridge formation along a stationary ice front in Iceland. *Boreas* **22**, 101–109.

Krüger, J., 1994. Glacial processes, sediments, landforms and stratigraphy in the terminus region of Myrdalsjokull, Iceland. *Folia Geographica Danica* **21**, 1–233.

Krüger, J., 1995. Origin, chronology and climatological significance of annual moraine ridges at Myrdalsjökull, Iceland. *The Holocene* **5**, 420–427.

Krüger, J., 1996. Moraine ridges formed from subglacial frozen-on sediment slabs and their differentiation from push moraines. *Boreas* **25**, 57–63.

Krüger, J., Thomsen, H.J., 1984. Morphology, stratigraphy, and genesis of small drumlins in front of the glacier Myrdalsjokull, south Iceland. *Journal of Glaciology* **30**, 94–105.

Krumbein, W.C., 1933. Textural and lithological variations in glacial till. *Journal of Geology* **41**, 382–408.

Laberg, J.S., Vorren, T.O., 2000. Flow behaviour of the submarine glacigenic debris flows on the Bear Island trough-mouth fan, western Barents Sea. *Sedimentology* **47**, 1105–1117.

Lacelle, D., Lauriol, B., Clark, I.D., Cardyn, R., Zdanowicz, C., 2007. Nature and origin of a Pleistocene-age massive ground ice body exposed in the Chapman Lake moraine complex, central Yukon Territory, Canada. *Quaternary Research* **68**, 249–260.

Lachniet, M.S., Larson, G.J., Lawson, D.E., Evenson, E.B., Alley, R.B., 2001. Microstructures of sediment flow deposits and subglacial sediments: a comparison. *Boreas* **30**, 254–262.

Lachniet, M.S., Larson, G.J., Strasser, J.C., Lawson, D.E., Evenson, E.B., 1999. Microstructures of glacigenic sediment flow deposits, Matanuska Glacier, Alaska. In: Mickleson, D.M., Attig, J.W. (Eds.), *Glacial Processes Past and Present.* Geological Society of America, Boulder, CO Special Paper, 337 pp.

Lamplugh, G.W., 1881a. On glacial sections near Bridlington. *Proceedings of the Yorkshire Geological Society* **7**, 383–397.

Lamplugh, G.W., 1881b. On the Bridlington and Dimlington glacial shell beds. *Geological Magazine* **8**, 535–546.

Lamplugh, G.W., 1881c. On a shell bed at the base of the drift at Speeton, near Filey, on the Yorkshire coast. *Geological Magazine* **8**, 174–180.

Lamplugh, G.W., 1882. Glacial sections near Bridlington. Part II. Cliff section extending 900 yards south of the harbour. *Proceedings of the Yorkshire Geological Society* **8**, 27–38.

Lamplugh, G.W., 1884a. Glacial sections near Bridlington. Part III. The drainage sections. *Proceedings of the Yorkshire Geological Society* **8**, 240–254.

Lamplugh, G.W., 1884b. On a recent exposure of the shelly patches in the boulder clay at Bridlington Quay. *Quarterly Journal of the Geological Society, London* **40**, 312–328.

Lamplugh, G.W., 1890. Glacial sections near Bridlington. Part IV. *Proceedings of the Yorkshire Geological Society* **11**, 275–307.

Lamplugh, G.W., 1911. On the shelly moraine of the Sefstromglacier and other Spitsbergen phenomena illustrative of the British glacial conditions. *Proceedings of the Yorkshire Geological Society* **17**, 216–241.

Lamplugh, G.W., 1919. On a boring at Kilnsea, Holderness. *Geological Survey Summary of Progress* (**1918**), 63–64.

Larsen, N.K., Piotrowski, J.A., 2003. Fabric pattern in a basal till succession and its significance for reconstructing subglacial processes. *Journal of Sedimentary Research* **73**, 725–734.

Larsen, N.K., Piotrowski, J.A., Christiansen, F., 2006. Microstructures and microshears as proxy for strain in subglacial diamicts: implications for basal till formation. *Geology* **34**, 889–892.

Larsen, N.K., Piotrowski, J.A., Kronborg, C., 2004. A multiproxy study of a basal till: a time-transgressive accretion and deformation hypothesis. *Journal of Quaternary Science* **19**, 9–21.

Larson, G.J., Lawson, D.E., Evenson, E.B., Alley, R.B., Knudsen, O., Lachniet, M.S., Goetz, S.L., 2006. Glaciohydraulic supercooling in former ice sheets? *Geomorphology* **75**, 20–32.

Larson, G.J., Menzies, J., Lawson, D.E., Evenson, E.B. & Hopkins, N.R., 2016. Macro- and micro-sedimentology of a modern melt-out till – Matanuska Glacier, Alaska, USA. *Boreas* **45**, 235–251.

Lavrushin, J.A., 1968. Features of deposition and structures of the glacial–marine deposits under conditions of a fjord coast. *Litologiya: Poloznyye Iskopayemyye* **3**, 63–79.

Lawson, D.E., 1979a. Sedimentological analysis of the western terminus region of the Matanuska Glacier, Alaska. *CRREL Report 79-9*, Hanover, NH.

Lawson, D.E., 1979b. A comparison of the pebble orientations in ice and deposits of the Matanuska Glacier, Alaska. *Journal of Geology* **87**, 629–645.

Lawson, D.E., 1981a. Distinguishing characteristics of diamictons at the margin of the Matanuska Glacier, Alaska. *Annals of Glaciology* **2**, 78–84.

Lawson, D.E., 1981b. *Sedimentological characteristics and classification of depositional processes and deposits in the glacial environment CRREL Report 81-27*, Hanover, NH.

Lawson, D.E., 1982. Mobilization, movement and deposition of active subaerial sediment flows. Matanuska Glacier, Alaska. *Journal of Geology* **90**, 279–300.

Lawson, D.E., 1989. Glacigenic resedimentation: classification concepts and application to mass movement processes and deposits. In: Goldthwait, R.P., Matsch, C.L. (Eds.), *Genetic Classification of Glacigenic Deposits*. Balkema, Rotterdam, pp. 147–169.

Lawson, D.E., Strasser, J.C., Evenson, E.B., Alley, R.B., Larson, G.J., Arcone, S.A., 1998. Glaciohydraulic supercooling: a freeze-on mechanism to create stratified, debris-rich basal ice: I. Field evidence. *Journal of Glaciology* **44**, 547–561.

Lee, J.R., 2001. Genesis and palaeogeographical significance of the Corton Diamicton (basal member of the North Sea Drift Formation), East Anglia, UK. *Proceedings of the Geologists' Association* **112**, 29–43.

Lee, J.R., Phillips, E.R., 2008. Progressive soft sediment deformation within a subglacial shear zone – a hybrid mosaic-pervasive deformation model for Middle Pleistocene glaciotectonised sediments from eastern England. *Quaternary Science Reviews* **27**, 1350–1362.

Lee, J.R., Phillips, E.R., Rose, J., Vaughan-Hirsch, D., 2017. The Middle Pleistocene glacial evolution of northern East Anglia, UK: a dynamic tectonostratigraphic–parasequence approach. *Journal of Quaternary Science* **32**, 231–260.

Legget, R.F., 1942. An engineering study of glacial drift for an earth dam near Fergus, Ontario. *Economic Geology* **37**, 531–556.

Legget, R.F. (Ed.), 1976. Glacial till. Royal Society of Canada Special Publication 12.

Le Heron, D.P., Etienne, J.L., 2005. A complex subglacial clastic dyke swarm, Solheimajökull, southern Iceland. *Sedimentary Geology* **181**, 25–37.

Levson, V.M., Rutter, N.W., 1989a. A lithofacies analysis and interpretation of depositonal environments of montane glacial diamictons, Jasper, Alberta, Canada. In: Goldthwait, R.P., Matsch, C.L. (Eds.), *Genetic Classification of Glacigenic Deposits*. Balkema, Rotterdam, pp. 117–140.

Levson, V.M., Rutter, N.W., 1989b. Late Quaternary stratigraphy, sedimentology and history of the Jasper townsite area, Alberta, Canada. *Canadian Journal of Earth Sciences* **26**, 1325–1342.

Leysinger-Vieli, G.J.M.C., Gudmundsson, G.H., 2010. A numerical study of glacier advance over deforming till. *The Cryosphere* **4**, 359–372.

Li, D., Yi, C., Ma, B., Wang, P., Ma, C. and Cheng, G., 2006. Fabric analysis of till clasts in the upper Urumqi River, Tian Shan, China. *Quaternary International* **154–155**, 19–25.

Lian, O.B., Hicock, S.R., 2000. Thermal conditions beneath parts of the last Cordilleran Ice Sheet near its centre as inferred from subglacial till, associated sediments and bedrock. *Quaternary International* **68–71**, 147–162.

Licht, K.J., Dunbar, N.W., Andrews, J.T., Jennings, A.E., 1999. Distinguishing subglacial till and glacial marine diamictons in the western Ross Sea, Antarctica: implications for Last Glacial Maximum grounding line. *Geological Society of America Bulletin* **111**, 91–103.

Lin, S., Williams, P.F., 1992. Ridge-in-groove slickenside striae in S-C mylonite. *Journal of Structural Geology* **14**, 315–321.

Lovell, H., Fleming, E.J., Benn, D.I., Hubbard, B., Lukas, S., Naegeli, K., 2015. Evidence of former dynamic flow of a currently cold-based valley glacier on Svalbard revealed by basal ice and structural glaciology investigations. *Journal of Glaciology* **61**, 309–328.

Lowe, D.R., 1975. Water escape structures in coarse-grained sediments. *Sedimentology* **22**, 157–204.

Lowe, D.R., 1976a. Grain flow and grain flow deposits. *Journal of Sedimentary Petrology* **46**, 188–199.

Lowe, D.R., 1976b. Subaqueous liquefied and fluidized sediment flows and their deposits. *Sedimentology* **23**, 285–308.

Lowe, D.R., 1979. Sediment gravity flows: Their classification and some problems of application to natural flows and deposits. In; Doyle, L.J., Pilkey, O.H. (Eds.), *Geology of continental slopes: Society of Economic Paleontologists and Mineralogists Special Publication 27*, pp. 75–84.

Lowe, D.R., 1982. Sediment gravity flows: II. Depositional models with special reference to the deposits of high-density turbidity currents. *Journal of Sedimentary Petrology 52*, 279–297.

Lukas, S., 2005. A test of the englacial thrusting hypothesis of 'hummocky' moraine formation – case studies from the north-west Highlands, Scotland. *Boreas* **34**, 287–307.

Lukas, S., 2012. Processes of annual moraine formation at a temperate alpine valley glacier: insights into glacier dynamics and climatic controls. *Boreas* **41**, 463–480.

Lukas, S., Sass, O., 2011. The formation of Alpine lateral moraines inferred from sedimentology and radar reflection patterns – a case study from Gornergletscher, Switzerland. *Geological Society of London Special Publications*, **354**, 77–92.

Lukas, S., Benn, D.I., Boston, C.M., Brook, M., Coray, S., Evans, D.J.A., Graf, A., Kellerer-Pirklbauer, A., Kirkbride, M.P., Krabbendam, M., Lovell, H., Machiedo, M., Mills, S.C., Nye, K., Reinardy, B.T.I., Ross, F.H., Signer, M., 2013. Clast shape analysis and clast transport paths in glacial environments: a critical review of methods and the role of lithology. *Earth-Science Reviews* **121**, 96–116.

Lukas, S., Graf, A., Coray, S., Schlüchter, C., 2012. Genesis, stability and preservation potential of large lateral moraines of Alpine valley glaciers – towards a unifying theory based on Findelengletscher, Switzerland. *Quaternary Science Reviews* **38**, 27–48.

Lundqvist, J., 1967. Submoräna sediment i Jämtlands län. *Sveriges Geologiska Undersökning* **C618**, 1–267.

Lundqvist, J., 1969a. Beskrivning till jordartskarta över Jämtlands Län. *Sveriges Geologiske Undersoekelse* **C45**, 418 pp.

Lundqvist, J., 1969b. Problems of the so-called Rogen moraine. *Sveriges Geologiske Undersoekelse C 648*, 32 pp.

Lundqvist, J., 1983. The glacial history of Sweden. In: Ehlers, J. (Ed.), *Glacial Deposits in North West Europe*. Balkema, Rotterdam, pp. 77–82.

Lundqvist, J., 1989. Till and glacial landforms in a dry, polar region. *Zeitschrift fur Geomorphologie* **33**, 27–41.

Lunkka, J.P., 1994. Sedimentation and lithostratigraphy of the North Sea Drift and Lowestoft Till Formations in the coastal cliffs of northeast Norfolk, England. *Journal of Quaternary Science* **9**, 209–233.

MacAyeal, D.R., 1992. The basal stress distribution of Ice Stream E, Antarctica, inferred by control methods. *Journal of Geophysical Research* **97**, 595–603.

MacAyeal, D.R., Bindschadler, R.A., Scambos, T.A., 1995. Basal friction of Ice Stream E, Antarctica. *Journal of Glaciology* **41**, 247–262.

MacClintock, P., Dreimanis, A., 1964. Reorientation of till fabric by overriding glacier in the St. Lawrence Valley. *American Journal of Science* **262**, 133–142.

Mackay, J.R., 1956. Deformation by glacier-ice at Nicholson Peninsula, N.W.T., Canada. *Arctic* **9**, 218–228.

Mackay, J.R., 1959. Glacier ice-thrust features of the Yukon coast. *Geographical Bulletin* **13**, 5–21.

Mackay, J.R., Rampton, V.N., Fyles, J.G., 1972. Relic Pleistocene permafrost, western Arctic, Canada. *Science* **176**, 1321–1323.

Mager, S., Fitzsimons, S., 2007. Formation of glaciolacustrine Late Pleistocene end moraines in the Tasman Valley, New Zealand. *Quaternary Science Reviews* **26**, 743–758.

Mahaney, W.C., 1995. Glacial crushing, weathering and diagenetic histories of quartz grains inferred from scanning electron microscopy. In: Menzies, J. (Ed.), *Modern Glacial Environments: Processes, Dynamics and Sediments*. Butterworth-Heinemann, Oxford, pp. 487–506.

Mahaney, W.C., Kalm, V., 2000. Comparative scanning electron microscopy study of oriented till blocks, glacial grains and Devonian sands in Estonia and Latvia. *Boreas* **29**, 35–51.

Mahaney, W.C., Vortisch, W., Julig, P., 1988. Relative differences between glacially crushed quartz transported by mountain and continental ice – some examples from North America and East Africa. *American Journal of Science* **288**, 810–826.

Mahaney, W.C., Stewart, A., Kalm, V., 2001. Quantification of SEM microtextures useful in sedimentary environmental discrimination. *Boreas* **30**, 165–171.

Major, J.J., 1997. Depositional processes in large scale debris flow experiments. *Journal of Geology* **105**, 345–366.

Major, J.J., 2000. Gravity-driven consolidation of granular slurries – implications for debris flow deposition and deposit characteristics. *Journal of Sedimentary Research* **70**, 64–83.

Mandl, G., Harkness, R.M., 1987. Hydrocarbon migration by hydraulic fracturing. In: Jones, M.E., Preston, R.M.F. (Eds.), *Deformation of Sediments and Sedimentary Rocks*. Geological Society of London, Special Publication, vol. 29, pp. 39–53.

Mandryk, G.B., Rutter, N.W., 1990. Discussion: Late Pleistocene subaerial debris flow sediments near Banff, Canada. *Sedimentology* **37**, 541–544.

March, A.,1932. Mathematische Theorie der Regelung nach der Korngestalt bei affiner Deformation. *Zeitschrift fur Kristallographie* **81**, 285–297.

Martin, H., Porada, H., Walliser, O.H., 1985. Mixtite deposits of the Demara sequence, Namibia, problems of interpretation. *Palaeogeography, Palaeoclimatology, Palaeoecology* **51**, 159–196.

Mark, D.M., 1973. Analysis of axial orientation data, including till fabrics. *Geological Society of America Bulletin* **84**, 1369–1374.

Mark, D.M., 1974. On the interpretation of till fabrics. *Geology* **2**, 101–104.

Mathews, W.H., Mackay, J.R., 1960. Deformation of soils by glacier ice and the influence of pore pressures and permafrost. *Transactions of the Royal Society of Canada, Series* **3**, 54, 27–36.

Matthews, J.A., McCarroll, D., Shakesby, R.A., 1995. Contemporary terminal moraine ridge formation at a temperate glacier: Styggedalsbreen, Jotunheimen, southern Norway. *Boreas* **24**, 129–139.

May, R.W., Dreimanis, A., Stankowski, W., 1980. Quantitative evaluation of clast fabrics within the Catfish Creek Till, Bradtville, Ontario. *Canadian Journal of Earth Sciences* **17**, 1064–1074.

McCabe, A.M., 1986. Glaciomarine facies deposited by retreating tidewater glaciers: an example from the late Pleistocene of Northern Ireland. *Journal of Sedimentary Petrology* **56**, 880–894.

McCabe, A.M., Dardis, G.F., 1989. Sedimentology and depositional setting of Late Pleistocene drumlins, Galway Bay, western Ireland. *Journal of Sedimentary Petrology* **59**, 944–959.

McCabe, A.M., Dardis, G.F., 1994. Glaciotectonically induced water-throughflow structures in a Late Pleistocene drumlin, Kanrawer, County Galway, western Ireland. *Sedimentary Geology* **91**, 173–190.

McCabe, A.M., Dardis, G.F., Hanvey, P.M., 1987. Sedimentation at the margins of a Late Pleistocene ice-lobe terminating in shallow marine environments, Dundalk Bay, eastern Ireland. *Sedimentology* **34**, 473–493.

McCarroll, D., 2001. Deglaciation of the Irish Sea Basin: a critique of the glacimarine model. *Journal of Quaternary Science* **16**, 393–404.

McCarroll, D., Harris, C., 1992. The glacigenic deposits of western Lleyn, north Wales: terrestrial or marine? *Journal of Quaternary Science* **7**, 19–29.

McCarroll, D., Rijsdijk, K.F., 2003. Deformation styles as a key for interpreting glacial depositional environments. *Journal of Quaternary Science* **18**, 473–489.

McGee, W.J., 1894. Glacial canons. *Journal of Geology* **2**, 350–364.

Mead, W.J., 1925. The geologic role of dilatancy. *Journal of Geology* **33**, 685–698.

Means, W.D., 1987. A newly recognized type of slickenside striation. *Journal of Structural Geology* **9**, 585–590.

Meehan, R.T., Warren, W.P., Gallagher, C.J.D., 1997. The sedimentology of a late Pleistocene drumlin near Kingscourt, Ireland. *Sedimentary Geology* **111**, 91–105.

Meneley, W.A., 1964. *Geology of the Melfort area (73-A)*, Saskatchewan. Unpublished PhD thesis, University of Illinois.

Menzies, J., 1987. Towards a general hypothesis on the formation of drumlins. In: Menzies, J., Rose, J. (Eds.), *Drumlin Symposium*. A.A. Balkema, Rotterdam, pp. 9–24.

Menzies, J., 1989. Subglacial hydraulic conditions and their possible impact upon subglacial bed formation. *Sedimentary Geology* **62**, 125–150.

Menzies, J., 1990a. Sand intraclasts within a diamicton melange, southern Niagara Peninsula, Ontario, Canada. *Journal of Quaternary Science* **5**, 189–206.

Menzies, J., 1990b. Brecciated diamictons from Mohawk Bay, S. Ontario, Canada. *Sedimentology* **37**, 481–493.

Menzies, J., 2000. Micromorphological analyses of microfabrics and microstructures indicative of deformation processes in glacial sediments. In: Maltman, A.J., Hubbard, B., Hambrey, M.J. (Eds.), *Deformation of Glacial Materials*. Geological Society, Special Publication, vol. 176, pp. 245–257.

Menzies, J., Maltman, A.J., 1992. Microstructures in diamictons – evidence of subglacial bed conditions. *Geomorphology* **6**, 27–40.

Menzies, J., Shilts, W.W., 1996. Subglacial environments. In, Menzies, J. (Ed.), *Past Glacial Environments: Sediments, Forms and Techniques*. Butterworth-Heinemann, Oxford, 15–136.

Menzies, J., Taylor, J.M., 2003. Seismically induced soft-sediment microstructures (seismites) from Meileour, western Strathmore, Scotland. *Boreas* **32**, 314–327.

Menzies, J., van der Meer, J.J.M., Rose, J., 2006. Till – as a glacial 'tectomict', its internal architecture, and the development of a 'typing' method for till differentiation. *Geomorphology* **75**, 172–200.

Menzies, J., Zaniewski, K., 2003. Microstructures within a modern debris flow deposit derived from Quaternary glacial diamicton – a comparative micromorphological study. *Sedimentary Geology* **157**, 31–48.

Meriano, M., Eyles, N., 2009. Quantitative assessment of the hydraulic role of subglaciofluvial interbeds in promoting deposition of deformation till (Northern Till, Ontario). *Quaternary Science Reviews* **28**, 608–620.

Merrill, F.J.H., 1886. On the geology of Long Island. *Annals of the New York Academy of Sciences* **3**, 341–364.

Mickelson, D.M., 1971. *Glacial geology of the Burroughs Glacier area, southeastern Alaska*. Ohio State University, Polar Studies Report 40.

Mickelson, D.M., 1973. Nature and rate of basal till deposition in a stagnating ice mass, Burroughs Glacier, Alaska. *Arctic and Alpine Research* **5**, 17–27.

Mickelson, D.M., Ham, N.R., Ronnert, L., 1992. Comment and reply on 'Striated clast pavements: products of deforming subglacial sediment?' *Geology* **20**, 285.

Middleton, G.V., Hampton, M.A., 1973. Sediment gravity flows: Mechanics of flow and deposition. In: Middleton, G.V., Bouma, A.H. (Eds.), *Turbidites and deep water sedimentation*. Society of Economic Paleontologists and Mineralogists Pacific Section Short Course Lecture Notes, pp. 1–38.

Miller, D.J., 1953. Late Cenozoic marine glacial sediments and marine terraces of Middleton Island, Alaska. *Journal of Geology* **61**, 17–40.

Miller, H., 1884. On boulder glaciation. *Royal Physical Society of Edinburgh Proceedings* **8**, 156–189.

Miwa, S., Ikedaa, T., Satob, T., 2006. Damage process of pile foundation in liquefied ground during strong ground motion. *Soil Dynamics and Earthquake Engineering* **26**, 325–336.

Möller, P., 2006. Rogen moraine: an example of glacial reshaping of pre-existing landforms. *Quaternary Science Reviews* **25**, 362–389.

Möller, P., 2010. Melt-out till and ribbed moraine formation, a case study from south Sweden. *Sedimentary Geology* **232**, 161–180.

Möller, P., Dowling, T.P.F., 2015. The importance of thermal boundary transitions on glacial geomorphology; mapping of ribbed/hummocky moraine and streamlined terrain from LiDAR, over Småland, South Sweden. *Geologiska Foreningens i Stockholm Forhandlingar* **137**, 252–283.

Moncrieff, A.C.M., 1989. Classification of poorly sorted sedimentary rocks. *Sedimentary Geology* **65**, 191–194.

Money, M.S., 1983. Dam and reservoir construction in glaciated valleys. In: Eyles, N. (Ed.), *Glacial Geology*. Pergamon, Oxford, pp. 313–348.

Mooers, H.D., 1989. On the formation of the tunnel valleys of the Superior Lobe, central Minnesota. *Quaternary Research* **32**, 24–35.

Moore, P.L., Iverson, N.R., 2002. Slow episodic shear of granular materials regulated by dilatant strengthening. *Geology* **30**, 843–846.

Moran, S.R., 1971. Glacitectonic structures in drift. In: Goldthwaite, R.P. (Ed.), *Till: A Symposium*. Ohio State University Press, Ohio, pp. 127–148.

Mörner, N.A., 1972. The first report on till wedges in Europe and Late Weichselian ice flows over southern Sweden. *Geologiska Foreningens i Stockholm Forhandlingar* **94**, 581–587.

Mörner, N.A., 1973a. A new find of till wedges in Nova Scotia, Canada. *Geologiska Foreningens i Stockholm Forhandlingar* **95**, 272–273.

Mörner, N.A., 1973b. The first report on 'till wedges in Europe': a reply. *Geologiska Foreningens i Stockholm Forhandlingar* **95**, 273–276.

Mörner, N.A., 1974. Facts and fiction in the till wedge discussion. *Geologiska Foreningens i Stockholm Forhandlingar* **96**, 282–283.

Mulder, T., Alexander, J., 2001. The physical character of subaqueous sedimentary density flows and their deposits. *Sedimentology* **48**, 269–299.

Mulugeta, G., Koyi, H., 1987. Three-dimensional geometry and kinematics of experimental piggyback thrusting. *Geology* **15**, 1052–1056.

Munro-Stasiuk, M.J., 1999. Evidence for water storage and drainage at the base of the Laurentide Ice Sheet. *Annals of Glaciology* **28**, 175–180.

Munro-Stasiuk, M.J., 2000. Rhythmic till sedimentation: evidence for repeated hydraulic lifting of a stagnant ice mass. *Journal of Sedimentary Research* **70**, 94–106.

Munro-Stasiuk, M.J., 2003. Subglacial Lake McGregor, south-central Alberta, Canada. *Sedimentary Geology* **160**, 325–350.

Murray, T., Corr, H., Forieri, A., Smith, A.M., 2008. Contrasts in hydrology between regions of basal deformation and sliding beneath Rutford Ice Stream, West Antarctica, mapped using radar and seismic data. *Geophyscial Research Letters 35*, doi:10.1029/2008GL033681.

Murton, J.B., 1996. Near-surface brecciation of chalk, Isle of Thanet, southeast England: a comparison with ice-rich brecciated bedrock in Canada and Spitsbergen. *Permafrost and Periglacial Processes 7*, 153–164.

Murton, J.N., Coutard, J.P., Lautidou, J.P., Ozouf, J.C., Robinson, D.A., Williams, R.G.B., 2001. Physical modelling of bedrock brecciation by ice segregation in permafrost. *Permafrost and Periglacial Processes* **12**, 255–266.

Murton, J.B., Waller, R.I., Hart, J.K., Whiteman, C.A., Pollard, W.H., Clark, I.D., 2004. Stratigraphy and glaciotectonic structures of a relict deformable bed of permafrost at the northwestern margin of the Laurentide ice sheet, Tuktoyaktuk Coastlands, Canada. *Journal of Glaciology* **50**, 399–412.

Murton, J.B., Whiteman, C.A., Waller, R.I., Pollard, W.H., Clark, I.D., Dallimore, S.R., 2005. Basal ice facies and supraglacial melt-out till of the Laurentide Ice Sheet, Tuktoyaktuk Coastlands, western Arctic Canada. *Quaternary Science Reviews* **24**, 681–708.

Nelson, A.E., Willis, I.C., Ó Cofaigh, C., 2005. Till genesis and glacier motion inferred from sedimentological evidence associated with the surge-type glacier, Bruarjökull, Iceland. *Annals of Glaciology* **42**, 14–22.

Nemec, W., 1990. Aspects of sediment movement on steep delta slopes. In: Colella, A., Prior, D. (Eds.), *Coarse-Grained Deltas*. International Association of Sedimentologists, Special Publication 10, pp. 29–73.

Neudorf, C.M., Brennand, T.A., Lian, O.B., 2013. Till-forming processes beneath parts of the Cordilleran Ice Sheet, British Columbia, Canada: macroscale and microscale evidence and a new statistical technique for analysing microstructure data. *Boreas* **42**, 848–875.

Newman, W.A., Mickelson, D.M., 1994. Genesis of Boston harbor drumlins, Massachusetts. *Sedimentary Geology* **91**, 333–343.

Ng, F.S.L., 2000. Canals under sediment-based ice sheets. *Annals of Glaciology* **30**, 146–152.

Nichols, R.J., Sparks, R.S.J., Wilson, C.J.N., 1994. Experimental studies of the fluidization of layered sediments and the formation of fluid escape structures. *Sedimentology* **41**, 233–253.

Nickling, W.G., Bennett, L., 1984. The shear strength characteristics of frozen coarse granular debris. *Journal of Glaciology* **30**, 348–357.

Nobles, L.H., Weertman, J., 1971. Influence of irregularities of the bed of an ice sheet on deposition rate of till. In: Goldthwait, R.P. (Ed.), *Till: A Symposium*. Ohio State University Press, Columbus, pp. 117–126.

Obermeier, S.F., 1998. Liquefaction evidence for strong earthquakes of Holocene and latest Pleistocene ages in the states of Indiana and Illinois, USA. *Engineering Geology* **50**, 227–254.

Obermeier, S.F., Olson, S.M., Green, R.A., 2005. Field occurrences of liquefaction-induced features: a primer for engineering geologic analysis of paleoseismic shaking. *Engineering Geology* **76**, 209–234.

Ó Cofaigh, C., Dowdeswell, J.A., 2001. Laminated sediments in glacimarine environments: diagnostic criteria for their interpretation. *Quaternary Science Reviews* **20**, 1411–1436.

Ó Cofaigh, C., Evans, D.J.A., 2001a. Deforming bed conditions associated with a major ice stream of the last British ice sheet. *Geology* **29**, 795–798.

Ó Cofaigh, C., Evans, D.J.A., 2001b. Sedimentary evidence for deforming bed conditions associated with a grounded Irish Sea glacier, southern Ireland. *Journal of Quaternary Science* **16**, 435–454.

Ó Cofaigh, C., Dowdeswell, J.A., Allen, C.S., Hiemstra, J.F., Pudsey, C.J., Evans, J., Evans, D.J.A., 2005. Flow dynamics and till genesis associated with a marine-based Antarctic palaeo-ice stream. *Quaternary Science Reviews* **24**, 709–740.

Ó Cofaigh, C., Evans, D.J.A., Hiemstra, J.F., 2011. Formation of a stratified subglacial 'till' assemblage by ice-marginal thrusting and glacier overriding. *Boreas* **40**, 1–14.

Ó Cofaigh, C., Stokes, C., Lian, O., Clark, C., Tulacyzk, S., 2013. Formation of mega-scale glacial lineations on the Dubawnt Lake Ice Stream bed: 2. Sedimentology and stratigraphy. *Quaternary Science Reviews* **77**, 210–227.

Ó Cofaigh, C., Telfer, M.W., Bailey, R.M., Evans, D.J.A., 2012. Late Pleistocene chronostratigraphy and ice sheet limits, southern Ireland. *Quaternary Science Reviews* **44**, 160–179.

Ottesen, D., Dowdeswell, J.A., Benn, D.I., Kristensen, L., Christiansen, H.H., Christensen, O., Hansen, L., Lebesbye, E., Forwick, M., Vorren, T.O., 2008. Submarine landforms characteristic of glacier surges in two Spitsbergen fjords. *Quaternary Science Reviews* **27**, 1583–1599.

Owen, L.A., 1989. Terraces, uplift and climate in the Karakoram Mountains, northern Pakistan: Karakoram intermontane basin evolution. *Zeitschrift fur Geomorphologie, Supp.* **76**, 117–146.

Owen, L.A., 1991. Mass movement deposits in the Karakoram Mountains. *Zeitschrift für Geomorphologie* **35**, 401–424.

Owen, L.A., 1994. Glacial and non-glacial diamictons in the Karakoram Mountains and Western Himalayas. In: Warren, W.P., Croot, D.G. (Eds.), *The Formation and Deformation of Glacial Deposits*. Balkema, Rotterdam, pp. 9–28.

Owen, L.A., Derbyshire, E., 1988. Glacially-deformed diamictons in the Karakoram Mountains, northern Pakistan. In: Croot, D.G. (Ed.), *Glacitectonics: Forms and Processes*. Balkema, Rotterdam, pp. 149–176.

Owen, L.A., Derbyshire, E., 1989. The Karakoram glacial depositional system. *Zeitschrift für Geomorphologie Supp.* **76**, 33–73.

Owen, L.A., Derbyshire, E., 1993. Quaternary and Holocene intermontane basin sedimentation in the Karakoram Mountains. In: Shroder, J.F. (Ed.), *Himalaya to the Sea*. Routledge, London, pp. 108–131.

Owen, L.A., Sharma, M.C., 1998. Rates and magnitudesof paraglacial fan formation in the Garhwal Himalaya: implications for landscape evolution. *Geomorphology* **26**, 171–184.

Passchier, C.W., Trouw, R.A.J., 1996. *Microtectonics*. Springer, Berlin. p. 289.

Patterson, C.J., Hooke, R. LeB., 1996. Physical environment of drumlin formation. *Journal of Glaciology* **41**, 30–38.

Patterson, C.J., Hansel, A.K., Mickelson, D.M., Quade, D.J., Bettis, E.A., Colgan, P.M., McKay, E.D., Stumpf, A.J., 2003. Contrasting glacial landscapes created by ice lobes of the Southern Laurentide Ice Sheet. In: Easterbrook, D.J. (Ed.), *Quaternary Geology of the US*. INQUA Field Guide, Desert Research Institute, Reno, NV, pp. 135–153.

Paul, M.A., Eyles, N., 1990. Constraints on the preservation of diamict facies (melt-out tills) at the margins of stagnant glaciers. *Quaternary Science Reviews* **9**, 51–69.

Pedersen, S.A.S., 1989. Glaciotectonite: brecciated sediments and cataclastic sedimentary rocks formed subglacially. In: Goldthwait, R.P., Matsch, C.L. (Eds.), *Genetic Classification of Glacigenic Deposits.* Balkema, Rotterdam, pp. 89–91.

Pedersen, S.A.S., 1996. Progressive glaciotectonic deformation in Weichselian and Palaeogene deposits at Feggeklit, northern Denmark. *Bulletin of the Geological Society of Denmark* **42**, 153–174.

Penck, A., 1882. *Die grosse Eiszeit.* Himmel u. Erde IV, Berlin.

Perrin, R.M.S., Rose, J., Davies, H., 1979. The Distribution, variation and origins of pre-Devensian tills in eastern England. *Philosophical Transactions of the Royal Society of London B287*, 535–570.

Peterson, D.N., 1970. *Glaciological investigations on the Casement Glacier south-east Alaska.* Ohio State University, Institute of Polar Studies, p. 36.

Pettijohn, F.J., 1949. *Sedimentary Rocks.* Harper, New York, pp. 526.

Pettijohn, F.J., 1957. *Sedimentary Rocks.* 2nd Edition. Harper, New York.

Phillips, E.R., 2006. Micromorphology of a debris flow deposit: evidence of basal shearing, hydrofracturing, liquefaction and rotational deformation during emplacement. *Quaternary Science Reviews* **25**, 720–738.

Phillips, E.R., Auton, C.A., 2000. Micromorphological evidence for polyphase deformation of glaciolacustrine sediments from Strathspey, Scotland. In: Maltman, A.J., Hubbard, B., Hambrey, M.J. (Eds.), *Deformation of Glacial Materials, 176.* The Geological Society of London, Special Publication, pp. 279–291.

Phillips, E.R., Auton, C.A., 2008. Microtextural analysis of glacially 'deformed' bedrock: implications for inheritance of preferred clast orientations in diamictons. *Journal of Quaternary Science* **23**, 229–240.

Phillips, E.R., Merritt, J., 2008. Evidence for multiphase water-escape during rafting of shelly marine sediments at Clava, Inverness-shire, NE Scotland. *Quaternary Science Reviews* **27**, 988–1011.

Phillips, E.R., Evans, D.J.A., Auton, C.A., 2002. Polyphase deformation at an oscillating ice margin following the Loch Lomond Readvance, central Scotland, UK. *Sedimentary Geology* **149**, 157–182.

Phillips, E.R., Evans, D.J.A., van der Meer, J.J.M., Lee, J.R., in press. Microscale evidence of liquefaction and its potential triggers during soft-bed deformation within subglacial traction tills. Quaternary Science Reviews.

Phillips, E., Lee, J.R., Burke, H., 2008. Progressive proglacial to subglacial deformation and syntectonic sedimentation at the margins of the Mid-Pleistocene British Ice Sheet: evidence from north Norfolk, UK. *Quaternary Science Reviews* **27**, 1848–1871.

Phillips, E., Lee, J.R., Riding, J.B., Kendall, R. & Hughes, L., 2013a. Periglacial disruption and subsequent glacitectonic deformation of bedrock: an example from Anglesey, North Wales, UK. Proceedings of the Geologists' Association. Proceedings of the Geologists Association, 124. 802–817.

Phillips, E.R., Lipka, E., van der Meer, J.J.M., 2013b. Micromorphological evidence of liquefaction and sediment deposition during basal sliding of glaciers. *Quaternary Science Reviews* **81**, 114–137.

Phillips, E.R., Merritt, J.W., Auton, C.A., Golledge, N.R., 2007. Microstructures developed in subglacially and proglacially deformed sediments: faults, folds and fabrics, and the influence of water on the style of deformation. *Quaternary Science Reviews* **26**, 1499–1528.

Phillips, E. van der Meer, J.J.M., Ferguson, A., 2011. A new 'microstructural mapping' methodology for the identification, analysis and interpretation of polyphase deformation within subglacial sediments. *Quaternary Science Reviews* **30**, 2570–2596.

Piotrowski, J.A., 1994. Waterlain and lodgement till facies of the lower sedimentary complex from the Danischer-Wohld cliff, Schleswig-Holstein, North Germany. In: Warren, W.P., Croot, D.G. (Eds.), *Formation and Deformation of Glacial Deposits.* Balkema, Rotterdam, pp. 3–8.

Piotrowski, J.A., Kraus, A.M., 1997. Response of sediment to ice sheet loading in northwestern Germany: effective stresses and glacier bed stability. *Journal of Glaciology* **43**, 495–502.

Piotrowski, J.A., Tulaczyk, S., 1999. Subglacial conditions under the last ice sheet in northwest Germany: ice–bed separation and enhanced basal sliding? *Quaternary Science Reviews* **18**, 737–751.

Piotrowski, J.A., Larsen, N.K., Menzies, J., Wysota, W., 2006. Formation of subglacial till under transient bed conditions: deposition, deformation and basal decoupling under a Weichselian ice sheet lobe, central Poland. *Sedimentology* **53**, 83–106.

Piotrowski, J.A., Mickelson, D.M., Tulaczyk, S., Krzyszowski, D., Junge, F.W., 2001. Were deforming subglacial beds beneath past ice sheets really widespread? *Quaternary International* **86**, 139–150.

Piotrowski, J.A., Mickelson, D.M., Tulaczyk, S., Krzyszowski, D., Junge, F.W., 2002. Reply to comments by G.S. Boulton, K.E. Dobbie, S. Zatsepin on: deforming soft beds under ice sheets: how extensive were they? *Quaternary International* **97–98**, 173–177.

Piotrowski, J.A., Larsen, N.J., Junge, F.W., 2004. Reflections on soft subglacial beds as a mosaic of deforming and stable spots. *Quaternary Science Reviews* **23**, 993–1000.

Porter, P.R., Murray, T., 2001. Mechanical and hydrological properties of till beneath Bakaninbreen, Svalbard. *Journal of Glaciology* **47**, 167–175.

Postma, G., 1983. Water escape structures in the context of a depositional model of a mass flow dominated conglomerate fan-delta (Abrioja Formation, Pliocene, Almeria Basin, SE Spain). *Sedimentology* **30**, 91–103.

Postma, G., 1986. Classification for sediment gravity-flow deposits based on flow conditions during sedimentation. *Geology* **14**, 291–294.

Postma, G., Nemec, W., Kleinspehn, K., 1988. Large floating clasts in turbidites: a mechanism for their emplacement. *Sedimentary Geology* **58**, 47–61.

Powell, R.D., 1984. Glacimarine processes and inductive lithofacies modelling of ice shelf and tidewater glacier sediments based on Quaternary examples. *Marine Geology* **57**, 1–52.

Price, R.J., 1970. Moraines at Fjallsjokull, Iceland. *Arctic and Alpine Research* **2**, 27–42.

Rains, R.B., Kvill, D., Shaw, J., 1999. Evidence and some implications of coalescent Cordilleran and Laurentide glacier systems in western Alberta. In: Smith, P.J. (Ed.), *A World of Real Places: Essays in Honour of William C. Wonders*. University of Alberta, Edmonton, pp. 147–161.

Ramsden, J., Westgate, J.A., 1971. Evidence for reorientation of a till fabric in the Edmonton area, Alberta. In: *Till-a symposium*. R.P. Goldthwait (Ed.). Ohio State University Press, Columbus, OH, pp. 335–344.

Rappol, M., 1985. Clast-fabric strength in tills and debris flows compared for different environments. *Geologie en Mijnbouw* **64**, 327–332.

Rappol, M., Stoltenberg, H.M.P., 1985. Compositional variability of Saalian till in the Netherlands. *Boreas* **14**, 33–50.

Rathbun, A.P., Marone, C., Alley, R.B., Anandakrishnan, S., 2008. Laboratory study of the frictional rheology of sheared till. *Journal of Geophysical Research* **113**, F02020, doi:10.1029/2007JF000815.

Rea, B.R., Evans, D.J.A., 2011. An assessment of surge – induced crevassing and the formation of crevasse squeeze ridges. *Journal of Geophysical Research* **116**, F04005, doi:10.1029/2011JF001970.

Rea, B.R., Whalley, W.B., 1994. Subglacial observations from Øksfjordjøkelen, north Norway. *Earth Surface Processes and Landforms* **19**, 659–673.

Reid, C., 1885. *Geology of Holderness*. Geological Survey Memoir. HMSO.

Reimnitz, E., Kempema, E.W., 1988. Ice rafting: an indication of glaciation? *Journal of Glaciology* **34**, 254–255.

Reimnitz, E., Kempema, E.W., Barnes, P.W., 1987. Anchor ice, seabed freezing, and sediment dynamics in shallow arctic seas. *Journal of Geophysical Research* **92**, 14, 671–614, 678.

Reinardy, B.T.I., Lukas, S., 2009. A comparison of the sedimentary signature of ice-contact sedimentation and deformation at macro- and micro-scale: a case study from NW Scotland. *Sedimentary Geology* **221**, 87–98.

Rijsdijk, K.F., 2001. Density-driven deformation structures in glacigenic consolidated diamicts: examples from Traeth y Mwnt, Cardiganshire, Wales, UK. *Journal of Sedimentary Research* **71**, 122–135.

Rijsdijk, K.F., Owen, G., Warren, W.P., McCarroll, D., van der Meer, J.J.M., 1999. Clastic dykes in over-consolidated tills: evidence for subglacial hydrofracturing at Killiney Bay, eastern Ireland. *Sedimentary Geology* **129**, 111–126.

Rijsdijk, K.F., Warren, W.P, van der Meer, J.J.M., 2010. The deglacial sequence at Killiney, SE Ireland: terrestrial sedimentation and glaciotectonics. *Quaternary Science Reviews* **29**, 696–719.

Ringberg, B., Holland, B., Miller, U., 1984. Till stratigraphy and provenance of the glacial chalk rafts at Kvarnby and Angdala, southern Sweden. *Striae* **20**, 79–90.

Roberts, D.H., Hart, J.K., 2005. The deforming bed characteristics of a stratified till assemblage in north East Anglia, UK: investigating controls on sediment rheology and strain signatures. *Quaternary Science Reviews* **24**, 123–140.

Roberts, D.H., Evans, D.J.A., Lodwick, J., Cox, N.J., 2013. The subglacial and ice-marginal signature of the North Sea Lobe of the British–Irish Ice Sheet during the Last Glacial Maximum at Upgang, North Yorkshire, UK. *Proceedings of the Geologists' Association* **124**, 503–519.

Roberts, M.J., Tweed, F.S., Russell, A.J., Knudsen, O., Lawson, D.E., Larson, G.J., Evenson, E.B., Björnsson, H., 2002. Glaciohydraulic supercooling in Iceland. *Geology* **30**, 439–442.

Robin, G. de Q., 1955. Ice movement and temperature distribution in glaciers and ice sheets. *Journal of Glaciology* **2**, 523–532.

Rodine, J.D., Johnson, A.M., 1976. The ability of debris heavily freighted with coarse clastic materials to flow on gentle slopes, *Sedimentology* **23**, 213–224.

Ronnert, L., Mickelson, D.M., 1992. High porosity of basal till at Burroughs Glacier, southeastern Alaska. *Geology* **20**, 849–852.

Rooney, S.T., Blankenship, D.D., Bentley, C.R., Alley, R.B., 1987. Till beneath Ice Stream B: 2. Structure and continuity. *Journal of Geophysical Research* **92** (B9), 8913–8920.

Rose, J., 1989. Glacier stress patterns and sediment transfer associated with the formation of superimposed flutes. *Sedimentary Geology* **62**, 151–176.

Rose, J., 1992. Boulder clusters in glacial flutes. *Geomorphology* **6**, 51–58.

Rotnicki, K., 1976. The theoretical basis for and a model of glaciotectonic deformations. *Quaestiones Geographicae* **3**, 103–139.

Rousselot, M., Fischer, U.H., 2007. A laboratory study of ploughing. *Journal of Glaciology* **53**, 225–231.

Russell, I.C., 1895. The influence of debris on the flow of glaciers. *Journal of Geology* **3**, 823–332.

Ruszczynska-Szenajch, H., 1976. Glacitektoniczne depresje i kry lodowcowe na tle budowy geologicznej poludniowo-wschodniego Mazowsza i poludniowego Podlasia. *Studia Geologica Polonica* **50**, 1–106.

Ruszczynska-Szenajch, H., 1987. The origin of glacial rafts: detachment, transport, deposition. *Boreas* **16**, 101–112.

Ruszczynska-Szenajch, H., 2001. 'Lodgement till' and 'deformation till'. *Quaternary Science Reviews* **20**, 579–581.

Salisbury, R.D., 1896. Stratified drift. *Journal of Geology* **4**, 948–970.

Salisbury, R.D., 1900. The local origin of glacial drift. *Journal of Geology* **8**, 426–432.

Salisbury, R.D., 1902. *The Glacial Geology of New Jersey*. Trenton.

Sane, S.M., Desai, C.S., Jenson, J.W., Contractor, D.M., Carslon, A.E., Clark, P.U., 2008. Disturbed state constitutive modelling of two Pleistocene tills. *Quaternary Science Reviews* **27**, 267–283.

Sardeson, F.W., 1906. The folding of subjacent strata by glacial action. *Journal of Geology* **14**, 226–232.

Sauer, E.K., 1974. Geotechnical implications of Pleistocene deposits in southern Saskatchewan. *Canadian Geotechnical Journal* **11**, 359–373.

Sauer, E.K., 1978. The engineering significance of glacier ice thrusting. *Canadian Geotechnical Journal* **15**, 457–472.

Scherer, R.P., Aldaham, A., Tulaczyk, S., Kamb, B., Engelhardt, H., Possnert, G., 1998. Pleistocene collapse of the West Antarctic ice sheet. *Science* **281**, 82–85.

Schermerhorn, L.J.G., 1966. Terminology of mixed coarse-fine sediments. *Journal of Sedimentary Petrology* **36**, 831–835.

Schermerhorn, L.J.G., 1974. Late Precambrian mixtites: glacial and/or non-glacial? *American Journal of Science* **274**, 673–824.

Schomacker, A., Kruger, J., Kurth, K., 2006. Ice-cored drumlins at the surge-type glacier Bruarjökull, Iceland: a transitional-state landform. *Journal of Quaternary Science* **21**, 85–93.

Schoof, C., 2007. Pressure-dependent viscosity and interfacial instability in coupled ice–sediment flow. *Journal of Fluid Mechanics* **570**, 227–252.

Schweizer, J., Iken, A., 1992. The role of bed separation and friction in sliding over an undeformable bed. *Journal of Glaciology* **38**, 77–92.

Scourse, J.D., Furze, M.F.A., 2001. A critical review of the glaciomarine model for Irish Sea deglaciation: evidence from southern Britain, the Celtic shelf and adjacent continental slope. *Journal of Quaternary Science* **16**, 419–434.

Sedgewick, A., 1825. Diluvial Formations. *Annals of Philosophy* **10**, 18–37.

Seret, G., 1993. Microstructures in thin sections of several kinds of till. *Quaternary International* **18**, 97–101.

Sergienko, O.V., Hindmarsh, R.C.A., 2013. Regular patterns in frictional resistance of ice stream beds seen by surface data inversion. *Science* **342**, 1086–1089.

Shaler, N.S., 1870. On the parallel ridges of glacial drift in eastern Massachusetts, with some remarks on the glacial period. *Proceedings of the Boston Society of Natural History* **13**, 196–204.

Shaler, N.S., 1889. The geology of Cape Ann, Massachusetts. *Ninth Annual Report United States Geological Survey*, **1887–1888**, pp. 529–611.

Sharp, M.J., 1982. Modification of clasts in lodgement tills by glacial erosion. *Journal of Glaciology* **28**, 475–481.

Sharp, M.J., 1984. Annual moraine ridges at Skalafellsjökull, south-east Iceland. *Journal of Glaciology* **30**, 82–93.

Sharp, M.J., 1985. 'Crevasse-fill' ridges – a landform type characteristic of surging glaciers? *Geografiska Annaler* **67A**, 213–220.

Sharp, M.J., Gomez, B., 1986. Processes of debris comminution in the glacial environment and implications for quartz sand-grain micromorphology. *Sedimentary Geology* **46**, 33–47.

Sharp, R.P., 1949. Studies of supraglacial debris on valley glaciers. *American Journal of Science* **247**, 289–315.

Shaw, E.W., 1912. Description of the Murphysboro and Herrin (Illinois) quadrangles. US Geological Survey Atlas of the United States, Folio 185.

Shaw, J., 1972. Sedimentation in the ice-contact environment, with examples from Shropshire, England. *Sedimentology* **18**, 23–62.

Shaw, J., 1977. Tills deposited in arid polar environments. *Canadian Journal of Earth Sciences* **14**, 1239–1245.

Shaw, J., 1979. Genesis of the Sveg tills and Rogen moraines of central Sweden: a model of basal melt-out. *Boreas* **8**, 409–426.

Shaw, J., 1982. Melt out till in the Edmonton area, Alberta, Canada. *Canadian Journal of Earth Sciences* **19**, 1548–1569.

Shaw, J., 1983. Forms associated with boulders in melt-out till. In: Evenson, E.B., Schluchter, C., Rabassa, J. (Eds.), *Tills and Related Deposits*. Balkema, Rotterdam, pp. 3–12.

Shaw, J., 1987. Glacial sedimentary processes and environmental reconstruction based on lithofacies. *Sedimentology* **34**, 103–116.

Shaw, J., 1988. Discussion – Coarse-grained sediment gravity flow facies in a large supraglacial lake. *Sedimentology* **35**, 527–530.

Shaw, J., 1989. Sublimation till. In: Goldthwait, R.P., Matsch, C.L. (Eds.), *Genetic Classification of Glacigenic Deposits*. Balkema, Rotterdam, pp. 141–142.

Shepps, V.C., 1953. Correlation of the tills of northeastern Ohio by size analysis: *Journal of Sedimentary Petrology* **23**, 34–48.

Shepps, V.C., 1958. 'Size factors' – a means of analysis of data from textural studies of till. *Journal of Sedimentary Petrology* **28**, 482–485.

Shumway, J.R., Iverson, N.R., 2009. Magnetic fabrics of the Douglas Till of the Superior Lobe: exploring bed deformation kinematics. *Quaternary Science Reviews* **28**, 107–119.

Slater, G., 1926. Glacial tectonics as reflected in disturbed drift deposits. *Proceedings of the Geologists' Association* **37**, 392–400.

Slater, G., 1927a. The structure of the disturbed deposits in the lower part of the Gipping Valley near Ipswich. *Proceedings of the Geologists' Association* **38**, 157–182.

Slater, G., 1927b. The structure of the disturbed deposits of the Hadleigh Road area, Ipswich. *Proceedings of the Geologists' Association* **38**, 183–261.

Slater, G., 1927c. The disturbed glacial deposits in the neighbourhood of Lonstrup, near Hjorring, north Denmark. *Transactions of the Royal Society of Edinburgh* **55**, 303–315.

Slater, G., 1927d. Structure of the Mud Buttes and Tit Hills in Alberta. *Bulletin of the Geological Society of America* **38**, 721–730.

Slater, G., 1927e. The structure and disturbed deposits of Moens Klint, Denmark. *Transaction of the Royal Society of Edinburgh* **55**, 289–302.

Small, R.J., 1983. Lateral moraines of Glacier de Tsidjiore Nouve: form, development and implications. *Journal of Glaciology* **29**, 250–259.

Smalley, I.J., 1966. Drumlin formation: a rheological model. *Science* **151**, 1379–1380.

Smalley, I.J., Unwin, D.J., 1968. The formation and shape of drumlins and their distribution and orientation in drumlin fields. *Journal of Glaciology* **7**, 377–390.

Smith, A.M., Murray, T., 2009. Bedform topography and basal conditions beneath a fast-flowing West Antarctic ice stream. *Quaternary Science Reviews* **28**, 584–596.

Smith, A.M., Murray, T., Nicholls, K.W., Makinson, K., Adalgeirsdóttir, G., Behar, A., Vaughan, D., 2007. Rapid erosion, drumlin formation, and changing hydrology beneath an Antarctic ice stream. *Geology* **35**, 127–130.

Smith, I.R., 2000. Diamictic sediments within high Arctic lake sediment cores: evidence for lake ice rafting along the lateral glacial margin. *Sedimentology* **47**, 1157–1179.

Sohn, Y.K., Kim, S.B., Hwang, I.G., Bahk, J.J., Choe, M.Y., Chough, S.K., 1997. Characteristics and depositional processes of large-scale gravelly Gilbert-type foresets in the Miocene Doumsan fan delta, Pohang Basin, SE Korea. *Journal of Sedimentary Research* **67**, 130–141.

Sohn, Y.K., Chose, M.Y., Jo, H.R., 2002. Transition from debris flow to hyperconcentrated flow in a submarine channel (the Cretaceous Cerro Toro Formation, southern Chile). *Terra Nova* **14**, 405–415.

Sollas, W.J., Praeger, R. Ll., 1894. A walk along the glacial cliffs of Killiney Bay. *Irish Naturalist* **3**, 13–18.

Spagnolo, M., Phillips, E.R, Piotrowski, J.A., Rea, B.R., Clark, C.D., Stokes, C.R., Carr, S.J., Ely, J.C., Ribolini, A., Wysota, W., Szuman, I., 2016. Ice stream motion facilitated by a shallow-deforming and accreting bed. *Nature Communications DOI*: 10.1038/ncomms10723.

Spedding, N., Evans, D.J.A., 2002. Sediments and landforms at Kviarjokull, south-east Iceland: a reappraisal of the glaciated valley landsystem. *Sedimentary Geology* **149**, 21–42.

Spencer, A.M., 1971. Late Precambrian glaciation in Scotland. *Geological Society of London*, Memoir No. **6**.

Spencer, A.M., 1985. Mechanisms and environments of deposition of late Precambrian geosynclinal tillites. *Palaeogeography, Palaeoclimatology, Palaeoecology* **51**, 143–157.

Squyres, S.W., Andersen, D.W., Nedell, S.S., Wharton R.A., 1991. Lake Hoare, Antarctica: sedimentation through thick perennial ice cover. *Sedimentology* **38**, 363–379.

Stalker, A. MacS., 1960. Ice-pressed drift forms and associated deposits in Alberta. *Geological Survey of Canada Bulletin* 57.

Stalker, A. MacS., 1963. Quaternary stratigraphy in southern Alberta. *Geological Survey of Canada*, Paper **62–34**, 52.

Stalker A. MacS., 1969. Quaternary Stratigraphy in Southern Albertad Report II: Sections near Medicine Hat. *Geological Survey of Canada* Paper 69–26.

Stalker, A. MacS., 1973. The large interdrift bedrock blocks of the Canadian Prairies. *Geological Survey of Canada* Paper 75-1A, 421–422.

Stalker, A. MacS., 1976. Megablocks, or the enormous erratics of the Albertan prairies. Geological Survey of Canada Paper 76-1C, 185–188.

Stalker, A. MacS., 1983. Quaternary stratigraphy in southern Alberta report 3: The Cameron Ranch section. *Geological Survey of Canada, Paper* **83–10**, 20.

Stalker, A. MacS., Wyder, J.E., 1983. Borehole and outcrop stratigraphy compared with illustrations from the Medicine Hat area of Alberta. *Geological Survey of Canada, Bulletin* **296**, 28.

Stankowski, W. (Ed.), 1976. *Till – its Genesis and Diagenesis*. Uniwersytet Imiena Adama Mickiewicza Poznan: Prace. Seria geografia **12**.

Stankowski, W. (Ed.), 1980. *Tills and Glacigene Deposits*. Uniwersytet Imiena Adama Mickiewicza Poznan: Prace. Seria geografia **20**.

Stea, R.R., Brown, Y., 1989. Variation in drumlin orientation, form and stratigraphy relating to successive ice flows in Southern and Central Nova Scotia. *Sedimentary Geology* **62**, 223–240.

Stephan, H.-J., 1987. Form, composition and origin of drumlins in Schleswig-Holstein. In: Menzies, J., Rose, J. (Eds.), *Drumlin Symposium*. A.A. Balkema, Rotterdam, pp. 335–344.

Stephan, H.-J., 1989. Origin of a till-like diamicton by shearing. In: Goldthwait, R.P., Matsch, C.L. (Eds.), *Genetic Classification of Glacigenic Deposits*. Balkema, Rotterdam, pp. 93–96.

Stephan, H.-J., Ehlers, J., 1983. North German till types. In: Ehlers, J. (Ed.), *Glacial Deposits in North-West Europe*. Balkema, Rotterdam, pp. 239–248.

Stoddart, O.N., 1859. Diluvial striae on fragments in situ. *American Journal of Science* **28**, 227–228.

Stokes, C.R., Clark, C.D., Lian, O.B., Tulaczyk, S., 2007. Ice stream sticky spots: a review of their identification and influence beneath contemporary and palaeo-ice streams. *Earth Science Reviews* **81**, 217–249.

Stokes, C.R., Fowler, A.C., Clark, C.D., Hindmarsh, R.C.A., Spagnolo, M., 2013. The instability theory of drumlin formation and its explanation of their varied composition and internal structure. *Quaternary Science Reviews* **62**, 77–96.

Stokes, C.R., Spagnolo, M., Clark, C.D., 2011. The composition and internal structure of drumlins: complexity, commonality, and implications for a unifying theory of their formation. *Earth Science Reviews* **107**, 398–422.

Stone, D.B., Clarke, G.K.C., 1993. Estimation of subglacial hydraulic properties from induced changes in basal water pressure: a theoretical framework for borehole-response tests. *Journal of Glaciology* **39**, 327–340.

Talling P.J., Amy L.A., Wynn R.B., Peakall J., Robinson M., 2004. Beds comprising debrite sandwiched within co-genetic turbidite: origin and widespread occurrence in distal depositional environments. *Sedimentology* **51**, 163–194.

Talling, P.J., 2014. On the triggers, resulting flow types and frequencies of subaqueous sediment density flows in different settings. *Marine Geology* **352**, 155–182.

Talling, P.J., Masson, D.G., Sumner, E.J., Malgesini, G., 2012. Subaqueous sediment density flows: depositional processes and deposit types. *Sedimentology* **59**, 1937–2003.

Tarling, D.H., Hrouda, F., 1993. *The Magnetic Anisotropy of Rocks*. Chapman and Hall, London.

Tarplee, M.F.V., van der Meer, J.J.M, Davis, G.R., 2010. The 3D microscopic 'signature' of strain within glacial sediments revealed using X-ray computed microtomography. *Quaternary Science Reviews* **30**, 3501–3532.

Tarr, R.S., 1897. The margin of the Cornell glacier. *American Geologist* **20**, 139–156.

Tarr, R.S., 1909. Some phenomena of the glacier margins in the Yakutat Bay region, Alaska. *Zeitschrift fur Gletscherkunde* **3**, 81–110.

Thomas, G.S.P., Chiverrell, R.C., 2006. A model of subaqueous sedimentation at the margin of the late Midlandian Irish Ice Sheet, Connemara, Ireland, and its implications for regionally high isostatic sea levels. *Quaternary Science Reviews* **25**, 2868–2893.

Thomas, G.S.P., Chiverrell, R.C., 2007. Structural and depositional evidence for repeated ice-marginal oscillation along the eastern margin of the Late Devensian Irish Sea Ice Stream. *Quaternary Science Reviews* **26**, 2375–2405.

Thomas, G.S.P., Kerr, A., 1987. The stratigraphy, sedimentology and palaeontology of the Pleistocene Kocknasilloge Member, Co. Wexford, Ireland. *Geological Journal* **22**, 67–82.

Thomas, G.S.P., Summers, A.J., 1982. Drop-stone and allied structures from Pleistocene waterlain till at Ely House, County Wexford. *Journal of Earth Sciences Royal Dublin Society* **4**, 109–119.

Thomas, G.S.P., Summers A.J., 1983. The Quaternary stratigraphy between Blackwater Harbour and Tinnaberna, Co. Wexford. Journal of Earth Sciences, *Royal Dublin Society* **5**, 121–143.

Thomas, G.S.P., Chester, D.K., Crimes, P., 1998. The late Devensian glaciation of the eastern Lleyn Peninsula, North Wales: evidence for terrestrial depositional environments. *Journal of Quaternary Science* **13**, 255–270.

Thomason, J.F., Iverson, N.R., 2006. Microfabric and microshear evolution in deformed till. *Quaternary Science Reviews* **25**, 1027–1038.

Thomason, J.F., Iverson, N.R., 2008. A laboratory study of particle ploughing and pore-pressure feedback: a velocity-weakening mechanism for soft glacier beds. *Journal of Glaciology* **54**, 169–181.

Thorsteinsson, T., Raymond, C.F., 2000. Sliding versus till deformation in the fast motion of an ice stream over a viscous till. *Journal of Glaciology* **46**, 633–640.

Tjia, H.D., 1967. Sense of fault displacement. *Geologie en Mijnbouw* **46**, 392–396.

Torell, O., 1872. *Undersokningar ofver istiden del I. Aftryck ur Ofversigt af Kungliga Vetenskapsakademiens Fordhandlingar 1872*. P.A. Nordstedt och Soner, Stockholm.

Torell, O., 1873. Undersokningar ofver istiden del II. Skandinaviska landisens utsrackning under isperioden. *Ofversigt af Kungliga Vetenskapsakademiens Fordhandlingar* **1873** 1, 47–64.

Torell, O., 1877. On the glacial phenomena of North America. *American Journal of Science* **13**, 76–79.

Tornquist, O., 1910. *Geologie von Ostpreussen*. Berlin.

Truffer, M., Harrison, W.D., 2006. In situ measurements of till deformation and water pressure. *Journal of Glaciology* **52**, 175–182.

Truffer, M., Harrison, W.D., Echelmeyer, K.A., 2000. Glacier motion dominated by processes deep in underlying till. *Journal of Glaciology* **46**, 213–221.

Truffer, M., Motyka, R.J., Harrison, W.D., Echelmeyer, K.A., Fisk, B., Tulaczyk, S., 1999. Subglacial drilling at Black Rapids Glacier, Alaska, USA: drilling method and sample descriptions. *Journal of Glaciology* **45**, 495–505.

Truffer M., Motyka R.J., Hekkers M., Howat I.M., King M.A., 2009. Terminus dynamics at an advancing glacier: Taku Glacier, Alaska. *Journal of Glaciology* **55**, 1052–1060.

Tsai, V.C., Ekstrom, G., 2007. Analysis of glacial earthquakes. *Journal of Geophysical Research* **112**, doi:10.1029/2006JF000596.

Tsui, P.C., Cruden, D.M., Thomson, S., 1989. Ice thrust terrains and glaciotectonic settings in central Alberta. *Canadian Journal of Earth Sciences* **26**, 1308–1318.

Tulaczyk, S., 1999. Ice sliding over weak, fine-grained tills: dependence of ice–till interactions on till granulometry. In: Mickelson, D.M., Attig, J.W. (Eds.), *Glacial Processes: Past and Present*. Geological Society of America, Special Paper, vol. 337, pp. 159–177.

Tulaczyk, S., Kamb, B., Scherer, R.P., Engelhardt, H.F., 1998. Sedimentary processes at the base of a West Antarctic ice stream: constraints from textural and compositional properties of subglacial debris. *Journal of Sedimentary Research* **68**, 487–496.

Tulaczyk, S., Kamb, B., Engelhardt, H.F., 2000a. Basal mechanics of Ice Stream, B.I. Till mechanics. *Journal of Geophysical Research* **105**, 463–481.

Tulaczyk, S., Kamb, B., Engelhardt, H.F., 2000b. Basal mechanics of Ice Stream, B: II. *Plastic undrained bed model*. *Journal of Geophysical Research* **105**, 483–494.

Tulaczyk, S., Scherer, R.P., Clark, C.D., 2001. A ploughing model for the origin of weak tills beneath ice streams: a qualitative treatment. *Quaternary International* **86**, 59–70.

Tylmann, K., Piotrowski, J.A., Wysota, W., 2013. The ice/bed interface mosaic: deforming spots intervening with stable areas under the fringe of the Scandinavian Ice Sheet at Samplawa, Poland. *Boreas* **42**, 428–441.

Tsytovich, N.A., 1975. *The Mechanics of Frozen Ground*. McGraw-Hill, New York.

Upham W., 1889. The structure of drumlins. *Proceedings of the Boston Society of Natural History* **24**, 228–242.

Upham, W., 1891a. Criteria of englacial and subgiacial drift. *American Geologist* **8**, 376–385.

Upham, W., 1891b. Inequality of distribution of the englacial drift. *Bulletin of the Geological Society of America* **3**, 134–148.

Upham, W., 1892. Conditions of accumulation of drumlins. *American Geologist* **10**, 339–362.

Upham, W., 1894a. Evidences of the derivation of the kames, eskers, and moraines of the North American ice sheet chiefly from its englacial drift. *Bulletin of the Geological Society of America* **5**, 71–86.

Upham, W., 1894b. The Madison type of drumlins. *American Geologist* **14**, 69–83.

Upham, W., 1895. Discrimination of glacial accumulation and invasion. *Bulletin of the Geological Society of America* **6**, 343–352.

van der Meer, J.J.M., 1980. Different types of wedges in deposits of Wurm age from the Murten area (western Swiss Plain). *Eclogae geologicae Helvetiae* **73**, 839–854.

van der Meer, J.J.M. (Ed.), 1987. *Tills and Glaciotectonics*. Balkema, Rotterdam.

van der Meer, J.J.M., 1993. Microscopic evidence of subglacial deformation. *Quaternary Science Reviews* **12**, 553–587.

van der Meer, J.J.M., 1996. Micromorphology. In: Menzies, J. (Ed.), *Glacial Environments, Past Glacial Environments – Processes, Sediments and Landforms*, vol. 2. Butterworth and Heinemann, Oxford, pp. 335–355.

van der Meer, J.J.M., 1997. Particle and aggregate mobility in till: microscopic evidence of subglacial processes. *Quaternary Science Reviews* **16**, 827–831.

van der Meer, J.J.M., Hiemstra, J.F., 1998. Micromorphology of Miocene diamicts, indications of grounded ice. *Terra Antarctica* **5**, 363–366.

van der Meer, J.J.M., Kjaer, K.H., Krüger, J., 1999. Subglacial water escape structures and till structure, Slettjokull, Iceland. *Journal of Quaternary Science* **14**, 191–205.

van der Meer, J.J.M., Kjær, K.H., Krüger, J., Rabassa, J., Kilfeather, A.A., 2009. Under pressure: clastic dykes in glacial settings. *Quaternary Science Reviews* **28**, 708–720.

van der Meer, J.J.M., Menzies, J., Rose, J., 2003. Subglacial till: the deforming glacier bed. *Quaternary Science Reviews* **22**, 1659–1685.

van der Meer, J.J.M., Rabassa, J.O., Evenson, E.B., 1992. Micromorphological aspects of glaciolacustrine sediments in northern Patagonia, Argentina. *Journal of Quaternary Science* **7**, 31–44.

van der Meer, J.J.M., Rappol, M., Semeyn, J.N., 1985. Sedimentology and genesis of glacial deposits in the Goudsberg, central Netherlands. *Mededelingen van de Rijks Geologische Dienst* **39–2**, 1–29.

van der Veen, C.J., 1999. *Fundamentals of Glacier Dynamics*. Balkema, Rotterdam.

van der Wateren, F.M., 1985. A model of glacial tectonics applied to the ice-pushed ridges in the central Netherlands. *Bulletin of the Geological Society of Denmark* **34**, 55–74.

van der Wateren, F.M., 1994. Proglacial subaquatic outwash fan and delta sediments in push moraines – indicators of subglacial meltwater activity. *Sedimentary Geology* **91**, 145–172.

van der Wateren, F.M., 1995a. Processes of glaciotectonism. In: Menzies, J. (Ed.), *Modern Glacial Environments: Processes*, Dynamics and Sediments. Butterworth-Heinemann, Oxford, pp. 309–335.

van der Wateren, F.M., 1995b. Structural geology and sedimentology of push moraines. *Mededelingen Rijks Geologische Dienst* **54**, 168.

van der Wateren, F.M., 2003. Ice-marginal terrestrial landsystems: southern Scandinavian ice sheet margin. In: Evans, D.J.A. (Ed.), *Glacial Landsystems*. Arnold, London, pp. 166–203.

van der Wateren, F.M., Kluiving, S.J., Bartek, L.R., 2000. Kinematic indicators of subglacial shearing. In: Maltman, A.J., Hubbard, B., Hambrey, M.J. (Eds.), *Deformation of Glacial Materials*. Geological Society of London, Special Publication, vol. 176, pp. 259–278.

van Loon, A.J., 2008. Could 'Snowball Earth' have left thick glaciomarine deposits? *Gondwana Research* **14**, 73–81.

Vaughan, D.G., Smith, A.M., Nath, P.C., Le Meur, E., 2003. Acoustic impedance and basal shear stress beneath four Antarctic ice streams. *Annals of Glaciology* **36**, 225–232.

Vaughan-Hirsch, D.P., Phillips, E.R., Lee, J.R., Hart, J.K., 2013. Micromorphological analysis of poly-phase deformation associated with the transport and emplacement of glaciotectonic rafts at West Runton, North Norfolk, UK. *Boreas* **42**, 376–394.

Virkkala, K., 1952. On the bed structure of till in eastern Finland. *Bulletin de la Commission Geologique de Finlande* **157**, 97–109.

Virkkala, K., 1969. On the lithology and provenance of the till of a gabbro area in Finland. In, Ters, M. (Ed.), *Etudes sur le Quaternaire dans le monde*. INQUA, VIII Congress, Paris, pp. 711–714.

Visser, J.N.J., Colliston, W.P., Terblanche, J.C., 1984. The origin of soft-sediment deformation structures in Permo-Carboniferous glacial and proglacial beds, South Africa. *Journal of Sedimentary Petrology* **54**, 1183–1196.

Vivian, R., 1976. *Les Glaciers des Alpes Occidentales*. Allier, Grenoble.

Vogel, S.W., Tulaczyk, S., Joughin, I.R., 2003. Distribution of basal melting and freezing beneath tributaries of Ice Stream C: implication for the Holocene decay of the West Antarctic ice sheet. *Annals of Glaciology* **36**, 273–282.

Wahnschaffe, F., 1901. *Die Ursachen der Oberflachengestaltung des norddeutschen Flachlandes*. Engelhorn, Stuttgart.

Walder, J.S., Fowler, A., 1994. Channelized subglacial drainage over a deformable bed. *Journal of Glaciology* **40**, 3–15.

Waller, R.I., 2001. The influence of basal processes on the dynamic behaviour of cold-based glaciers. *Quaternary International* **86**, 117–128.

Waller, R.I., Murton, J.B., Whiteman, C.A., 2009. Geological evidence for subglacial deformation of Pleistocene permafrost. *Proceedings of the Geologists' Association* **120**, 155–162.

Waller, R.I., Phillips, E.R., Murton, J.B., Lee, J.R., Whiteman, C., 2011. Sand intraclasts as evidence of subglacial deformation of Middle Pleistocene permafrost, North Norfolk, UK. *Quaternary Science Reviews* **30**, 3481–3500.

Warren, W.P., Croot, D.G. (Eds.), 1994. *Formation and Deformation of Glacial Deposits.* Balkema, Rotterdam.

Watts, R.J., Carr, S.J., 2002. A laboratory simulation of the subglacial deforming bed (the deformation tank). *Quaternary Newsletter* **98**, 1–9.

Wayne, W.J., 1963. Pleistocene formations in Indiana. *Indiana Geological Survey Bulletin* **25**.

Weertman, J., 1961. Mechanism for the formation of inner moraines found near the edge of cold ice caps and ice sheets. *Journal of Glaciology* **3**, 965–978.

Weertman, J., 1964. The theory of glacier sliding. *Journal of Glaciology* **5**, 287–303.

Westgate, J.A., 1968. Linear sole markings in Pleistocene till. *Geological Magazine* **105**, 501–505.

Whalley, W.B., 1978. An SEM examination of quartz grains from sub-glacial and associated environments and some methods for their characterization. *Scanning Electron Microscopy* **1**, 355–358.

Whalley, W.B., 2009. On the interpretation of discrete debris accumulations associated with glaciers with special reference to the British Isles. In: Knight. J., Harrison, S. (Eds.), *Periglacial and Paraglacial Processes and Environments.* Geological Society of London Special Publication 320, London, pp. 85–102.

Whalley, W.B., Krinsley, D.H., 1974. A scanning electron microscope study of surface textures of quartz grains from glacial environments. *Sedimentology* **21**, 87–105.

Whalley, W.B., Langway, C.C., 1980. A scanning electron microscope examination of subglacial quartz grains from Camp Century core, Greenland – a preliminary study. *Journal of Glaciology* **25**, 125–131.

Whittecar, G.R., Mickelson, D.M., 1977. Sequence of till deposition and erosion in drumlins. *Boreas* **6**, 213–217.

Whittecar, G.R., Mickelson, D.M., 1979. Composition, internal structures, and a hypothesis of formation for drumlins, Waukesha County, Wisconsin, USA. *Journal of Glaciology* **22**, 357–371.

Wiens, D.A., Anandakrishnan, S., Winberry, J.P., King, M.A., 2008. Simultaneous teleseismic and geodetic observations of the stick–slip motion of an Antarctic ice stream. *Nature* **453**, doi:10.1038/nature06990.

Wingfield, R.T.R., 1992. The Late Devensian (<22,000 BP) Irish Sea Basin: The sedimentary record of a collapsed ice sheet margin – Discussion. *Quaternary Science Reviews* **11**, 377–378.

Woodcock, N.H., 1977. Specification of fabric shapes using an eigenvalue method. *Geological Society of America Bulletin* **88**, 1231–1236.

Woodcock, N.H., Naylor, M.A., 1983. Randomness testing in three-dimensional orientation data. *Journal of Structural Geology* **5**, 539–548.

Woodworth-Lynas, C.M.T., Guigne, J.Y., 1990. Iceberg scours in the geological record: examples from glacial Lake Agassiz. In: Dowdeswell, J.A., Scourse, J.D. (Eds.), *Glacimarine Environments: Processes and Sediments.* Geological Society Special Publication, vol. 53, pp. 217–223.

Wordie, J.M. (Chairman), 1950. Discussion on the origin of glacial drifts. *Journal of Glaciology* **1**, 430–436.

Worsley, P., 1973. The first report on 'till wedges in Europe' – a discussion. *Geologiska Foreningens i Stockholm Forhandlingar* **95**, 152–155.

Yi Chaolu, 1997. Subglacial comminution in till – evidence from microfabric studies and grain-size distributions. *Journal of Glaciology* **43**, 473–479.

Youd, T.L., 1978. Major cause of earthquake damage is ground failure. *Civil Engineering* **48**, 47–51.

Youd, T.L., 2003. Liquefaction mechanisms and induced ground failure. *International Handbook of earthquake and engineering seismology*, volume **81B**, 1159–1173.

Index

a

Ablation (till)   15, 20, 23, 24, 27, 209

Abrasion   12, 15, 16, 54–56, 58, 74, 135, 136, 302, 304, 336

Advection (till)   94, 122, 133, 134, 145, 302, 304, 305, 310–312, 320, 323, 325, 327

A horizon   25, 69–71, 74, 104–106, 108, 118, 146, 150, 301, 302, 304–306, 315, 318, 327, 337

Amalgamation zone   105, 106, 124, 126–128, 146, 148, 152, 153, 156, 160, 231, 233, 238, 320, 331

Anchor ice   213

Anisotropy of magnetic susceptibility (AMS)   52, 106, 108, 156, 158, 247

Apron overriding   86

Architecture   93, 124, 133, 194, 216, 220, 262, 269, 284, 285, 288, 291, 293, 296, 298ff, 333, 339

b

Banks Island   34, 297

Bed deformation model   132–134

B horizon   25, 69–71, 74, 100, 105, 106, 108, 114, 118, 123, 141, 145–147, 157, 165, 176, 184, 300–306, 318, 337, 339, 341

Boudin (boudinage)   25, 38, 44–47, 91, 93, 94, 150, 152, 154, 161, 163, 222, 224, 226, 227, 231, 238, 247, 253, 254, 262, 273, 278, 320, 334

Boulder clay   15, 16, 18, 20

Braided canals/canal fill   81, 82, 152, 179–182, 184, 185, 187, 238, 241, 284, 311, 312, 315, 327, 337, 339

Breccia/brecciation/brecciated   25–27, 45, 91, 97, 150, 163, 167, 196, 197, 208, 224, 228, 230, 231, 233, 269, 272–274, 334

Breidamerkurjökull   25, 40, 52, 62, 69–78, 98–100, 108, 109, 116, 120, 124, 126, 127, 145–147, 223, 298, 300, 305, 337

Brittle (deformation/failure/behaviour)   25, 40, 41, 44, 45, 48, 74, 77, 86, 99, 131, 150, 163, 165, 168, 171, 175, 224, 238, 250, 272, 274, 282, 315, 333, 334, 336

Burst-out structure   188, 189, 192, 194, 196, 253, 254

c

Cannibalisation   18, 26, 38, 58, 81, 86ff, 150–152, 154, 171, 174, 181, 187, 222, 227, 252, 253, 322, 323, 325, 332, 337, 339, 341

Catfish Creek   28, 33, 192

Cavity (subglacial)   28, 59ff, 84, 128, 142ff, 147, 180, 182, 213, 309, 313, 335, 336, 339, 340

Channelised (meltwater) system   81, 82, 142, 157, 179, 180, 187, 213, 315, 325, 336

Clast collision/cluster   12, 45, 59, 60, 110, 114, 119, 122, 130, 136, 139, 149, 192, 253, 302, 304

Clast form/modification   1, 3, 12, 54, 143, 144, 146

Clastic dyke   45, 126, 127, 188ff, 207, 231, 262, 263

Clinoforms   36, 288, 290, 293

Cohesion   33, 40, 79, 80, 269, 275, 277

Comminution   11, 25, 54, 55, 58, 150, 231, 335

Till: A Glacial Process Sedimentology, First Edition. David J A Evans.
© 2018 John Wiley & Sons Ltd. Published 2018 by John Wiley & Sons Ltd.

Printed and bound by CPI Group (UK) Ltd, Croydon, CR0 4YY

16/04/2025

14658553-0005